PHYSICS OF NON-CONVENTIONAL ENERGY SOURCES AND MATERIAL SCIENCE FOR ENERGY

1985 Workshop on the

PHYSICS OF NON-CONVENTIONAL ENERGY SOURCES AND MATERIAL SCIENCE FOR ENERGY

I.C.T.P., Trieste, 2nd-20th, Sept., 1985

Editors:

G Furlan *(University of Trieste/I.C.T.P., Italy)*

N A Mancini *(University of Catania, Italy)*

A A M Sayigh *(O.A.P.E.C., Kuwait)*

B O Seraphin *(University of Arizona, USA)*

World Scientific

Published by

World Scientific Publishing Co. Pte. Ltd.
J Toh Tuck Link, Singapore 596224
USA office: 27 Warren Street, Suite 401-402, Hackensack, NJ 07601
UK office: 57 Shelton Street, Covent Garden, London WC2H 9HE

British Library Cataloguing-in-Publication Data
A catalogue record for this book is available from the British Library.

**1985 WORKSHOP ON THE PHYSICS OF NON-CONVENTIONAL
ENERGY SOURCES AND MATERIAL SCIENCE FOR ENERGY**

ISBN-13 978-9971-5-0252-2
ISBN-10 9971-5-0252-6
ISBN-13 978-9971-5-0346-8 (pbk)
ISBN-10 9971-5-0346-8 (pbk)

PREFACE

It is inevitable that conventional sources of energy will be almost totally depleted after one hundred years and renewable energy will have been developed to replace them. The sun supplies half of the exposed earth with energy, at such a quantity that six hours of sunshine only per year is sufficient to meet the present world energy demand. During the last ten years a scientific breakthrough in utilizing solar energy has been achieved which has led to the reduction in costs as well as materials. In photovoltaic conversion, the cost of a peak watt of single crystal fell from \$25 to \$5 within eight years. In the laboratory scale amorphous silicon cells have increased in efficiency from 6% to 13% within five years. While solar water heaters are now cost effective everywhere in the world.

Since its inception the International Centre for Theoretical Physics (I.C.T.P.) has promoted a well defined programme for the development of science in the Third World countries via courses, workshops and permanent research activities in many branches of Physics and Mathematical Physics. As energy is a principal success of Physics the I.C.T.P. is involved with the energy problem per se. Since 1977, increasing attention has been devoted to the analysis of non-conventional energy sources and the I.C.T.P. has provided the forum for scientists from all parts of the world to discuss, cooperate and exchange ideas and experiences in the field. A regular programme of courses and symposia started in 1977, alternately in English and in French, attracting many leading scholars and is now becoming a world-wide recognized event.

The lectures contained in this volume were delivered in plenary sessions by invited speakers during the workshop on the physics of non-conventional energy sources and material science for energy, held at the I.C.T.P., Miramare-Trieste, Italy from 2-20 September, 1985. Selected aspects of energy management and planning, heat mirrors and solar absorbers, review of solar thermal power systems, recent progress in a-Si solar cell technologies, Lattice defects and lifetime of silicon photovoltaic devices, recent aspects in the technology of amorphous silicon solar cells, development of polycrystalline solar cells for terrestrial use, spectrally selective coatings for energy conserving windows, integrated rural energy systems and agricultural applications of renewable energies, prospects of energy demand and supply in developing countries, and many other topics related to various applications of solar energy are included.

The workshop was sponsored by C.N.R. (National Research Council), Italy, the Dipartimento per la cooperazione allo Sviluppo, Italy, the Kuwait Foundation for the Advancement of Science, Kuwait, the Regione Autonoma Friuli-Venezia Giulia, Italy and the Arab and Italian sections of I.S.E.S.

The moral support and advice given by the I.C.T.P. Director, Nobel laureate, Professor Abdus Salam, represented an invaluable encouragement for all the people who contributed towards the realization of the successful workshop. We should like to thank the members of the International Advisory Committee, all the lecturers and participants for their passionate interest and activity which helped to create a stimulating atmosphere. The tireless efforts of the I.C.T.P. staff is gratefully acknowledged.

The workshop was directed by Professors G. Furlan (University of Trieste and I.C.T.P., Italy), N.A. Mancini (Department of Physics, University of Catania, Italy), A.A.M. Sayigh (O.A.P.E.C., Kuwait) and B.O. Seraphin (University of Arizona, U.S.A.).

CONTENTS

WORKSHOP ON THE PHYSICS OF NON CONVENTIONAL ENERGY SOURCES
AND MATERIAL SCIENCE FOR ENERGY

2 - 20 September 1985

P R O G R A M M E

Week 2 -6 September

Monday, 2 September

11.30 -   O P E N I N G

15.00 - 16.00   Measurement and analysis of energy use
                  M. ROSS (University of Michigan, U.S.A.)

16.30 - 17.30   Acquisition and management of field data for photovoltaic
                  systems : an application to the characterization of the
                  experimental plant of Passo Mandrioli
                  G.C. CARDINALI (LAMBL, Bologna)

17.30 -   Scientific documentary (film)
                  Energy from the sun and the wind: the plant at Passo dei
                  Mandrioli

---

* Some last minute changes may be made.
Additional seminars can be arranged in the framework of the working groups.

Tuesday, 3 September

9.15 - 10.15   Thermal storage for housing and green-houses by phase change materials
                      M. SCHNEIDER (CNRS, Valbonne, France)

10.45 - 11.45   Energy measurement in manufacturing
                      M. ROSS

11.45 - 12.45   Radiative cooling
                      V. SILVESTRINI (University of Naples, Italy)

Afternoon :   (a)   Discussion groups

                      (b)   Seminar: Dust effects on solar devices
                               A.A. SAYIGH (OAPEC, Kuwait)

                      (c)   Working group on analysis of field data

18.00 -   R e c e p t i o n

Wednesday, 4 September

9.15 - 10.15   M. SCHNEIDER - II

10.45 - 11.45   Fundamentals and properties of solar heat mirrors and solar absorbers
                      C. LAMPERT (L.B.L., Berkeley, USA)

11.45 - 12.45   Cooling and heating in Japan
                      K. KIMURA (Waseda University, Japan)

Afternoon :   (a)   Discussion groups

                      (b)   Seminar:

                      (c)   Working group on analysis of field data

Thursday, 5 September

9.15 - 10.15   C. LAMPERT - II

10.45 - 11.45   Review of solar thermal power systems
                      H.P. GARG (I.I.T., New Delhi, India)

11.45 - 12.45   Economic aspects of using non conventional energy sources in
                      developing countries
                      R. BHATHIA (University of Delhi, India)

Afternoon :   (a)   Discussion groups

                      (b)   Seminar: Cooling and heating in Kuwait
                               A.A. SAYIGH (OAPEC, Kuwait)

                      (c)   Working group on analysis of field data

Friday, 6 September

9.15 - 10.15   ENEL's activity in the field of renewable energy sources
                      P. PEISER (ENEL, Rome, Italy)

10.45 - 11.45   A review of storage
                      H.P. GARG

11.45 - 12.45   R. BHATHIA - II

Afternoon :   (a)   Discussion groups

                      (b)   Seminar:

                      (c)   Working group on analysis of field data

A note on the working group on analysis and management of field data

A reliable evaluation of the performance of an operating photovoltaic system needs a careful collection and analysis of field data.

The main aspects of these activities, related both to the elaboration techniques and to the data management, will be discussed in the framework of the working group. To this end the operating data of the plant of Passo Mandrioli (approx. 28 variables monitored since the beginning of 1982 with an acquisition period of 10 minutes) will be directly available on a computer terminal and it will be possible to examine and elaborate them by exploiting the facilities of a purposely designed data base system. For practical reasons the working group will be limited to 10-12 individuals and the activity will take place in the Department of Physics, Central Building of the University of Trieste. Transportation will be provided.

WORKSHOP ON THE PHYSICS OF NON CONVENTIONAL ENERGY SOURCES
AND MATERIAL SCIENCE FOR ENERGY

2 - 20 September 1985

P R O G R A M M E

Week 9 - 14 September

Monday, 9 September

9.15 - 10.15	Recent progress in a-Si solar cell technologies (General Review) Y. HAMAKAWA (Osaka University, Japan)
10.45 - 11.45	Lattice defects and lifetime of silicon photovoltaic devices D. NOBILI (LAMEL, Bologna)
11.45 - 12.45	Optimization of the physical parameters for silicon solar cells F. DEMICHELIS (Politecnico of Turin)
Afternoon	Discussion groups

*Some last minute changes may be made.
Seminars will be arranged in the framework of the working groups.

Tuesday, 10 September

9.15 - 10.15	Device physics and optimum design of the a-Si solar cells Y. HAMAKAWA
10.45 - 11.45	Recent aspects in the technology of amorphous silicon solar cells S. OVSHINSKY Energy Conversion Devices, Inc. Troy, U.S.A.
11.45 - 12.45	Heterojunctions among amorphous semiconductors and relevance for photovoltaic devices F. EVANGELISTI (University of Rome)
Afternoon:	Discussion groups
17.30	Scientific documentary: Solar cells

Wednesday, 11 September

9.15 - 10.15	Technology of crystalline P.V. cells. Theory and technology of photovoltaic concentration A. LUQUE (University of Madrid, Spain)
10.45 - 11.45	Development of polycrystalline solar cells for terrestrial use J. MEAKIN (University of Delaware, USA)
11.45 - 12.45	ENEA activities on amorphous silicon solar cells C. MESSANA (ENEA, Italy)
Afternoon: 15.00	Seminar: Some topics on R&D efforts for efficiency improvement: Y. HAMAKAWA
	Discussion Groups
17.30	Scientific documentary: Solar energy applications at low temperatures.

Thursday, 12 September

9.15 – 10.15	A. LUQUE – II
10.45 – 11.45	J. MEAKIN – II
11.45 – 12.45	CISE experiences on Ga As solar cells for terrestrial and space uses F. PALETTA (CISE, Italy)
Afternoon:	Discussion groups

Friday, 13 September

Visit to outside installations

a) Photovoltaic pumping station at Zambelli (Verona)
   (Rated peak power 70 kWp)

b) "Helios Technology", a firm manufacturing solar cells and panels,
   Galliera Veneta (Padova).

WORKSHOP ON THE PHYSICS OF NON CONVENTIONAL ENERGY SOURCES
AND MATERIAL SCIENCE FOR ENERGY

2 - 20 September 1985

P R O G R A M M E

Week 16 - 20 September

Monday, 16 September

9.15 - 10.15	Spectrally selective coatings for energy conserving windows Ch.GRANQVIST (Chalmers University, Sweden)
10.15 - 10.45	COFFEE BREAK
10.45 - 11.45	Material aspects of photo—electrochemical conversion M. GRÄTZEL (Ecole Polytechnique Fédérale, Switzerland)
11.45 - 12.45	An example of research on solar refrigeration H. FLECHON (University of Nancy, France)

Afternoon

15.00 - 16.00	Discussion session(s)
16.30	Working group on solar refrigeration Contributed papers

Tuesday, 17 September

9.15 - 10.15	Spectrally selective coatings for energy conserving windows II Ch. GRANQVIST (Chalmers University, Sweden)
10.15 - 10.45	COFFEE BREAK
10.45 - 11.45	Material aspects of photo—electrochemical conversion B. SCROSATI (University of Rome)
11.45 -12.45	The R&D programme of the European Economic Community on Solar Energy G. GRASSI (C.E.E., Bruxelles)

Afternoon

15.00	Discussion session(s) Working group on solar refrigeration Contributed papers
16.30	Energy/Agriculture integrated projects G. GRASSI (C.E.E.—Bruxelles)

Wednesday, 18 September

9.15 - 10.15   Integrated rural energy systems and agricultural
                      applications of renewable energies
                      Th. LAWAND (Brace Institute, Canada)

10.15 - 10.45   COFFEE BREAK

10.45 - 11.45   Prospects of energy demand and supply in developing
                      countries
                      A. KHAN (P.A.E.C. Pakistan)

11.45 - 12.45   Qualification and durability tests for solar thermal
                      collectors
                      G. RIESCH (J.R.C. Ispra, Italy)

Afternoon

15.00   Discussion session(s)
                      Working group on solar refrigeration
                      Contributed papers

Thursday, 19 September

9.15 - 10.15   Integrated rural energy systems and agricultural
                      applications of renewable energies II
                      Th. LAWAND (Brace Institute, Canada)

10.15 - 10.45   COFFEE BREAK

10.45 - 11.45   Integrated Energy planning and its data needs
                      A. KHAN (P.A.E.C. Pakistan)

11.45 - 12.45   Qualification and durability tests for photo-
                      voltaic modules
                      G. RIESCH (J.R.C. Ispra, Italy)

Afternoon

14.30   Physics and development programme

15.30   Discussion session(s)
                      Working group on solar refrigeration
                      Contributed papers

Friday, 20 September

9.30 - 12.00   Closing lecture and discussion session

Afternoon

15.00   Working group on solar refrigeration

NB Afternoon sessions will take place at the ADRIATICO

Conference and Workshop on the Physics of Non-Conventional Energy Sources and Material Science for Energy I.C.T.P. 1985 - Trieste.

**WORKSHOP ON THE PHYSICS OF NON CONVENTIONAL ENERGY SOURCES
AND MATERIAL SCIENCE FOR ENERGY**

2 - 20 September 1985

===

F I N A L   L I S T   O F   P A R T I C I P A N T S

===

Name and Institute	Member State

V I S I T O R S

1.   FLECHON Dr. Jean   France
     Universitè de Nancy
     Nancy
     France

2.   GODMEL Dr. Gervais   France
     Universitè de Nancy
     Nancy
     France

Name and Institute	Member State

D I R E C T O R S

1. FURLAN Prof. Giuseppe   ICTP
    ICTP and University of Trieste

2. MANCINI Prof. N.A.   Italy
    Istituto di Fisica
    Università di Catania
    Corso Italia 57
    Catania

3. SAYIGH Prof. Ali A.M.   Kuwait/Iraq
    Energy Resources Department
    OAPEC
    P.O. Box 20501
    Safat
    Kuwait

4. SERAPHIN Prof. Bernard   U.S.A.
    The University of Arizona
    Optical Sciences Center
    Tucson, Arizona 85721
    U.S.A.

Name and Institute	Member State

RESIDENT   SCIENTISTS

1. DATTA Dr. R.L.   India
    Vandana Housing Society
    Flat No. 9, Building A-1
    Thana (Maharashtra)
    India

2. GARG Dr. H.P.   (Associate)   India
    IIT, Delhi
    Hauz Khas, New Delhi 110016
    India

3. KHAN Dr. Arshad   Pakistan
    Pakistan Atomic Energy Commission
    P.O. Box 1114
    Islamabad
    Pakistan

Name and Institute	Member State

L E C T U R E R S

1.   BHATHIA Prof. R.   India
   Institute of Economic Growth
   University of Delhi
   Delhi 110007
   India

2.   CARDINALI Dr. Giancarlo   Italy
   Istituto LAMEL-CNR
   Via Castagnoli 1
   Bologna

3.   DEMICHELIS Prof. F.   Italy
   Dipartimento di Fisica
   Politecnico di Torino
   Corso Duca degli Abruzzi
   Torino

4.   EVANGELISTI Professor F.   Italy
   Dipartimento di Fisica
   Università di Roma, La Sapienza
   Piazziale Aldo Moro 2
   00185 Roma

5.   GRANQVIST Dr. C.   Sweden
   Chalmers University of Technology
   Department of Physics
   Göteborg
   Sweden

5.   GRASSI Dr. G.   Belgium
   Comunità Europea
   Brussels
   Belgium

6.   GRATZEL Prof. M.   Switzerland
   Institut de Chimie Physique
   Ecole Polytechnique Fídírale
   CH-1015 Lausanne
   Switzerland

Name and Institute	Member State

7.   HAMAKAWA Prof. Yoshihiro   Japan
      Osaka University
      Department of Electrical Engineering
      Toyonaka, Osaka
      Japan

8.   KIMURA Dr. Ken-ichi   Japan
      School of Science & Engineering
      Waseda University
      Okubo 3
      Tokyo 160
      Japan

9.   LAMPERT Dr. Carl   U.S.A.
      Lawrence Berkeley Laboratory
      University of California
      MMRD Building 62
      Berkeley CA 94720
      U.S.A.

10.   LAWAND Dr. T.A.   Canada
      Brace Research Institute
      MacDonald College of McGill University
      1 Stewart Park
      Ste Anne de Bellevue
      Quebec
      Canada

11.   LUQUE Prof. Antonio   Spain
      Universidad Politecnica de Madrid
      Instituto de Energia Solar
      ETSI Telecomicacion UPM
      Ciudad Universitaria
      28040 Madrid
      Spain

12.   MEAKIN Prof. John   U.S.A.
      Director
      Institute of Energy Conversion
      University of Delaware
      Newark, Delaware 19716
      U.S.A.

Name and Institute	Member State

13. MESSANA Dr. Calogero   Italy
    ENEA
    Dipartimento di Fonti Alternative
       Rinnovabili e Risparmio Energetico
    S.P. Anguillarese Km 1
    Roma

14. NOBILI Prof. Dario   Italy
    CNR-LAMEL
    Via Castagnoli 1
    40126 Bologna

15. OVSHINSKY Prof. S.   U.S.A.
    President
    Energy Conversion Devices Inc.
    Troy, Michigan 48084
    U.S.A.

16. PALETTA Ing. F.   Italy
    CISE
    Via Reggio Emilia 39
    20090 Segrate (Milano)

17. PEISER Ing. Peter   Italy
    E.N.E.L.
    Piazza G.B. Martini 3
    00198 Roma

18. RIESCH Dr. Gerhard   Italy
    Joint Research Centre
    Commission of the European Communities
    21020 Ispra (VA)

Name and Institute	Member State

19. ROSS Prof. Marc
     The University of Michigan
     The Harrison M. Randall Laboratory
        of Physics
     Ann Arbor, Michigan 48109
     U.S.A.

    U.S.A.

20. SCHNEIDER Prof. Michel
     Directeur du Laboratoire CNRS
     Centre de Recherches
     Batiments Solaires CSTB-CNRS
     B.P. 21
     06562 Valbonne Cedex
     France

    France

21. SCROSATI Dr. B.
     Dipartimento di Chimica
     Università di Roma
     Piazziale Aldo Moro 5
     00100 Roma

    Italy

22. SILVESTRINI Prof. Vittorio
     Istituto di Fisica
     Facultad de Ingenieria
     Università di Napoli
     Piazziale Tecchio
     Naples

    Italy

23. VIVANTI Ms. Giovanna
     LAMEL-CNR
     Via Castagnoli 1
     Bologna

    Italy

Name and Institute	Member State

PARTICIPANTS

1. ABO-NAMOUS Dr. Salem   Kuwait
    Materials Applied Department
    Kuwait Institute for Scientific Research
    P.O. Box 24885
    Safat
    Kuwait

2. ADANU Dr. Komla   Ghana
    Department of Physics
    University of Ghana
    Legon, Accra
    Ghana

3. ADEGBOYEGA Dr. G.   (Associate)   Nigeria
    Department of Electronic
        & Electrical Engineering
    University of Ife
    Ile-Ife
    Nigeria

4. ADJEPONG Dr. Samuel K.   Nigeria/Ghana
    Physics Department
    University of Port Harcourt
    Port Harcourt
    Nigeria

5. AIMIUWU Dr. Victor   Nigeria
    Department of Physics
    University of Benin
    Benin City
    Nigeria

6. AJLOUNI Dr. Ali   Jordan
    Royal Scientific Society
    P.O. Box 6945
    Amman
    Jordan

Name and Institute	Member State

7.   AKORLI Dr. Felix
University of Science & Technology
Kumasi
Ghana   Ghana

8.   ALADLI Dr. Feroda
Ankara Nuclear Research and
    Training Centre
Besevler, Ankara
Turkey   Turkey

9.   AMELIN Dr. Charles
California State University
Pomona
California
U.S.A.   U.S.A.

10.   AMON Dr. Slavko
Fakulterta za Elektrotehniko
Trzaska 25
62000 Ljubljana
Yugoslavia   Yugoslavia

11.   AMORNKITBAMRUNG Dr.   (Affiliate)
Chulalongkorn University
Physics Department
Bangkok 10500
Thailand   Thailand

12.   AMUZU Dr. Jozef Kaku
Department of Physics
University of Ghana
Legon
Ghana   Ghana

13.   APPIAH Dr. Michael K.
Ghana Atomic Energy Commission
P.O. Box 80
Accra, Legon
Ghana   Ghana

Name and Institute	Member State

14. ARCIPIANI Mr. Biagio
     Via A. Manzoni 214/0
     Parco Flory
     80123 Napoli
                                                    Italy

15. ARGIRIOU Dr. Anthanassios
     Universite de Provence
     Centre de St. Jerome
     Dept. d'Heliophysique
     F-13397 Marseille Cedex 13
     France
                                                    France/Greece

16. ARIAS Dr. Manuel   (Associate)
     Faculty of Engineering
     Universidad Central del Ecuador
     Casilla 3972
     Quito
     Ecuador
                                                    Ecuador

17. ATTIA Dr. W.A.   (Affiliate)
     Suez Canal University
     Department of Physics
     Ismailia
     Egypt
                                                    Egypt

18. AWANOU Dr. Cossi Norbert
     Laboratoire de Physique
        du Rayonnement
     F.A.S.T., UNB
     B.P. 526
     Cotonou
     Benin
                                                    Benin

19. BAIMBA Dr. Andrew
     Department of Physics
     Njala University College
     Private Mail Bag
     Freetown
     Sierra Leone
                                                    Sierra Leone

Name and Institute	Member State

20. BALLA Dr. M.Y.   (Affiliate)   Nigeria
     University of Maiduguri
     P.M.B. 1069
     Maiduguri
     Nigeria

21. BAMIRO Professor Olufemi   Nigeria
     Faculty of Technology
     University of Ibadan
     Ibadan
     Nigeria

22. BANTIKASSEGN Mr. W.   Ethiopia
     Physics Department
     Addis Ababa University
     P.O. Box 1176
     Addis Ababa
     Ethiopia

23. BAYOU Dr. Tesfaye   Ethiopia
     Faculty of Technology
     Electrical Engineering Department
     P.O. Box 385
     Addis Ababa
     Ethiopia

24. BELBACHIR Dr. Mohammed   Algeria
     Departement de Chimie
     Institut des Sciences Exactes
     Universite d'Oran
     Es-Senia
     Algeria

25. BHARGAVA Dr. Ashok   India
     Physics Department
     Ramjas College
     Delhi University
     Delhi 110007
     India

Name and Institute	Member State

26. BIRGUL Ms. Gulsen   Turkey
    Ankara Nuclear Research & Training Centre
    Besevler, Ankara
    Turkey

27. BORGES Dr. Jose Carlos   Brazil
    Rua Pacheco Leao 320, Ap. 203
    Rio de Janeiro
    Brazil

28. BOSANAC Dr. Miroslav   Yugoslavia
    Faculty of Electrical Engineering
    R. Boskovica bb
    58000 Split
    Yugoslavia

29. BRACAMONTE Dr. Edwin   Guatemala
    Centro de Investigaçiones de Ingenieria
    Ciudad Universitaria
    Zona 12
    Guatemala

30. BREW-HAMMOND Dr. J.P.   Ghana
    Department of Mechanical Engineering
    U.S.T.
    Kumasi
    Ghana

31. BUCHER Dr. Klaus Günter   Germany F.R.
    Institut für Werkstoffe
        d. Elektrotechnik
    Sekr. 10
    Jebenstr.1
    D-1000 Berlin 12
    Federal Republic of Germany

Name and Institute	Member State

32. BURDULEA Dr. Constantin   Romania
    Inst. Centrale pour Machines-Outils
    Blvd. T
    Vladiumikescu 45-47
    76135 Bucharest
    Romania

33. CAMACHO Ms. Esther   Bolivia
    Instituto de Hidraulica e Hidrologia
    P.O. Box 699
    La Paz
    Bolivia

34. CAPITANI Dr. Jorge   Argentina
    San Nicolas 1526
    1407 Buenos Aires
    Argentina

35. CHAALAN Dr. Ahmed   Lebanon
    Arab Physical Society
    P.O. Box 11-7142
    Beirut
    Lebanon

36. CRNJAK-OREL Ms. Zorica   Yugoslavia
    Boris Kidric Institute
       of Chemistry
    61000 Ljubljana
    Hajdrihova 19
    Yugoslavia

37. DAVID Ms. Adelia   Philippines
    Eulogio Amang Rodriguez Institute
       of Science & Technology
    Nagtahan, Sampaloc
    Manila 2806
    Philippines

Name and Institute	Member State

38. DELGADO Dr. Mario   Portugal
     Rua D. Fuas Roupinho 24-1 Dto
     1900 Lisbon
     Portugal

39. DIAZ Dr. Pedro   Cuba
     Facultad de Fisica
     Universidad de la Habana
     Colina Universitario
     Cuba

40. DUOMARCO Dr. Jose Luis   Uruguay
     Canelones 1055 Ap. 601
     Montevideo
     Uruguay

41. ECEVIT Dr. Ahmet   Turkey
     Department of Physics
     Middle East Technical University
     Ankara
     Turkey

42. EGRICAN Dr. A.N.   (Associate)   Turkey
     ITU Makina Fakultesi
     Termodinamik Anabilimdali
     Taksim
     Istanbul
     Turkey

43. EISSA Dr. N.A.   (Associate)   Egypt
     No.1 Dr. N.A. Hamid Street
     Flat 2 branching from Dr. Azmy St.
     Heliopolis West
     Cairo
     Egypt

44. EKPO ENO Dr. E.   (Affiliate)   Nigeria
     University of Maiduguri
     P.M.B. 1069
     Maiduguri
     Nigeria

Name and Institute	Member State

45. EL-DEGHAIDY Dr. F. (Affiliate)
Department of Physics
Suez Canal University
Ismailia
Egypt

Egypt

46. EL-DESSOUKI Dr. M.S.I.
Physics Department
Faculty of Science
Cairo University
Cairo
Egypt

Egypt

47. EL-MEKKI Dr. Osman
School of Mathematical Sciences
University of Khartoum
P.O. Box 321
Khartoum
Sudan

Sudan

48. ERCELEBI Ms. Cigdem
Department of Physics
Middle East Technical University
Ankara
Turkey

Turkey

49. ESSABOURI Dr. Ahmed
Lab. Spec.
Bat. Physique, 2eme Etage
U.S.T.L.
Place E. Bataillon
34060 Montpellier Cedex
France

France/Morocco

50. FADLY Dr. M.A. (Affiliate)
Department of Physics
Helwan University
Cairo
Egypt

Egypt

Name and Institute	Member State

51.   FAGHIH-HABIBI Dr. M.   Iran
      Tehran Polytechnic
      Hafez Avenue
      Tehran
      Iran

52.   FARIBORZI Dr. Naser   Iran
      Shaid Bahonar University
      Kerman
      Iran

53.   FASULO Dr. Amilcar Jesus   Argentina
      Universidad Nacional de San Luis
      Dep.to de Fisica
      Chacabuco y Perenera
      5700 San Luis
      Argentina

54.   FEGHAHATI Dr. Emad-eddin   Iran
      Tehran University
      Physics Department
      P.O. Box 11365-7693
      Tehran
      Iran

55.   FERNANDEZ Dr. Walter   (Associate)   Costa Rica
      School of Physics
      University of Costa Rica
      San Jose
      Costa Rica

56.   FERREIRA DA SILVA J.   (Affiliate)   Portugal
      Laboratorio de Fisica
      Facultad de Ciencias
      Universidad do Porto
      4000 Porto
      Portugal

Name and Institute	Member State

57. FOFANA Dr. B.   Ivory Coast
    Institut de Recherche sur les
       Energies Nouvelles
    B.P. 34
    Abidjan
    Ivory Coast

58. FUMAGALLI Ms. Simonetta   Italy
    ENEA-FARE
    Ed. 26
    c/o EURATOM
    20120 Ispra (VA)

59. FURLAN Dr. Joseph   Yugoslavia
    University of Ljubljana
    Trzaska 25
    Ljubljana
    Yugoslavia

60. GALAN Dr. O.V.   (Affiliate)   Cuba
    Facultad de Fisica
    Universidad de la Habana
    Colina Universitario
    Cuba

61. GONZALEZ Dr. Fabio   Colombia
    Universidad Nacional de Colombia
    Dep.to de Fisica
    Solar Energy Group
    A.A. 100102
    Bogota
    Colombia

62. GORDILLO Dr. Gerardo   Colombia
    Departamento de Fisica
    Universidad Nacional
    Bogota
    Colombia

Name and Institute	Member State

63. HADJI SAGHATI Dr. A.
Iran University of Science & Technology
Narmak 16
Tehran
Iran

Iran

64. HAN Ms. Daxing
Institute of Physics
Chinese Academy of Sciences
P.O. Box 603
Beijing
People's Republic of China

China

65. HARIDASAN Dr. T.M.   (Associate)
School of Energy Environment
    & Natural Resources
Madurai Kamaraj Univ.
Madurai 625 021
India

India

66. HART Dr. David
Environmental Sciences
University of East Anglia
Norwich NR4 7TJ
U.K.

U.K.

67. HASSAN Dr. A.F.   (Affiliate)
Helwan University
Department of Physics
Cairo
Egypt

Egypt

68. HORN Dr. Manfred   (Associate)
Universidad Nacional de Ingenieria
Facultad de Ciencias
Lima
Peru

Peru/F.R.G.

Name and Institute	Member State

69. HUSAIN Dr. Md. S.
     Physics Department
     Dhaka University
     Dhaka-2
     Bangladesh
     — Bangladesh

70. IBRAHIM Dr. Muhammad   (Associate)
     Physics Department
     Dhaka University
     Dhaka 2
     Bangladesh
     — Bangladesh

71. IBRAHIM Dr. Said
     Department of Mechanical Engineering
     El-Azhar University
     Nasr City
     Cairo
     Egypt
     — Egypt

72. ISLAM Dr. Mohammed Nurul
     Department of Physics
     University of Chittagong
     Chittagong
     Bangladesh
     — Bangladesh

73. ITTYACHEN Dr. M.A.
     TC/12/687
     Barton Hill, Kunnukuzhy
     Trivandrum
     Kerala
     India
     — India

74. JAIN Dr. Prem   (Associate)
     Department of Physics
     University of Lusaka
     Lusaka
     Zambia
     — Zambia/India

Name and Institute	Member State

75. JAMIL Dr. M.
Mühendislik Fakültesi
Erciyes Universitesi
Kayseri
Turkey — Turkey

76. JANUSZEWSKI Dr. Jozef
ul. Potiebni 30/2
51-677 Wroclaw
Poland — Poland

77. JIANG Mr. Shiji
Department of Physics
Chalmers University of Technology
S-412 96 Göteborg
Sweden — Sweden/China

78. JILALI Dr. A.   (Affiliate)
Ecole Normale Superieure de Takaddoum
B.P. 5118
Rabat
Morocco — Morocco

79. JOHRI Dr. G.K.
D.A.B. College
Kanpur
India — India

80. KEITA Dr. Lassana
Ecole Normale Superieure
B.P. 241
Bamako
Mali — Mali

Name and Institute	Member State

81.   KEITA Mr. Mamby   Guinea
      Department of Physics
      Faculty of Sciences
      University of Conakry
      P.O. Box 1147
      Conkary
      Republic of Guinea

82.   KETABI Dr. Gholam H.   Iran
      Iran University of Science & Technology
      Narmak 16
      Tehran
      Iran

83.   KHALEK Dr. F.A.R.   (Affiliate)   Egypt
      Physics Department
      Faculty of Science
      Zagazig University
      Zagazig
      Egypt

84.   KHAN Dr. Mohammad S.R.   Bangladesh
      Department of Applied Physics
      University of Rajshahi
      Rajshahi
      Bangladesh

85.   KHANDANI Dr. S.   Iran
      Mechanical Engineering Department
      Sharif University of Technology
      Tehran
      Iran

86.   KIVAISI Dr. Rogath   Tanzania
      Physics Department
      P.O. Box 35063
      Dar es Salaam
      Tanzania

Name and Institute	Member State

87. KUKU Dr. Titilayo A.
    Department of Electronic and
        Electrical Engineering
    University of Ife
    Ile-Ife
    Nigeria

    Nigeria

88. KULISIC Dr. Petar
    Elektrotehnicki fakultet
    Unska 3
    41000 Zagreb
    Yugoslavia

    Yugoslavia

89. KUNC Dr. Seref
    Faculty of Arts and Science
    Cukurova University
    PK 171
    Balcali, Adana
    Turkey

    Turkey

90. LABOR MONTE Dr. I.W.   (Affiliate)
    University of Sierra Leone
    PMB, Freetown
    Sierra Leone

    Sierra Leone

91. LOPEZ PINEDA Dr. Cesar
    CRE-CASACCIA-CNR
    S.P. 103
    Via Anguillarese 301
    00100 Roma

    Italy/Honduras

92. LUSHIKU Dr. Elias
    Physics Department
    P.O. Box 35063
    Dar es Salaam
    Tanzania

    Tanzania

Name and Institute	Member State

93. MAAFI Mr. Abdelbaki
     Département d'Electronique
     Ecole Nationale Polytechnique
     10 Avenue Pasteur
     El-Harrach-Alger
     Algerie   Algeria

94. MANSOURI Dr. Seyed
     Kerman University
     Engineering Faculty
     Kerman
     Iran   Iran

95. MASSAQUOI Dr. J.G.   (Associate)
     Department of Mechanical Engineering
     Fourah Bay College
     Freetown
     Sierra Leone   Sierra Leone

96. MAYO Mr. Eduardo A.
     Jr. A. Soriano 1059
     Huaraz
     Peru   Peru

97. MBOW Dr. C.M.   (Associate)
     Dakar University
     Physics Department
     Dakar
     Senegal   Senegal

98. MELINTE Ms. Sofia
     Institutul Politechnic Iasi
     Catedra de Fisica
     Iasi
     Romania   Romania

Name and Institute	Member State

99. METAFERIA Dr. Hailu
     Physics Department
     Addis Ababa University
     P.O. Box 1176
     Addis Ababa
     Ethiopia

Ethiopia

100. MHANGO Dr. George
      University of Malawi
      Private Bag 303
      Blantyre 3
      Malawi

Malawi

101. MISKOLCZI Dr. Ferenc M.
      Department of Physics
      University of Calabar
      P.M.B. 1115
      Calabar
      Nigeria

Nigeria/Hungary

102. MOSTAVAN Dr. Aman
      Heliotechnology Laboratory
      Institute of Technology of Bandung
      Bandung
      Indonesia

Indonesia

103. MOUNIR Dr. S.M.   (Affiliate)
      Department of Physics
      Cairo University
      Cairo
      Egypt

Egypt

104. MPAWENAYO Dr. Prosper
      Universite Nationale du Rwanda
      Centre d'Etudes et d'Applications
          de l'Energie
      B.P. 117
      Butare
      Rwanda

Rwanda

Name and Institute	Member State

105. MU Dr. Zaiqin   (Affiliate)
     Institute of Physics
     Chinese Academy of Sciences
     P.O. Box 603
     Beijing
     People's Republic of China

China

106. MUFTI Dr. Atique
     National Institute of Silicon Technology
     15-B Street 44
     Tslamabad
     Pakistan

Pakistan

107. MUHAMAD Dr. M.. Rasat
     Physics Department
     University of Malaya
     Kuala Lumpur 22 11
     Malaysia

Malaysia

108. NABELEK Dr. Bohumil
     Institute of Physics
     Czechoslovak Academy of Sciences
     Na Slovance 2
     Praha 8, Liben
     Czechoslovakia

Czechoslovakia

109. NAIR Dr. Padmanabhan
     Department of Physics
     University of Jos
     P.M.B. 2084
     Jos
     Nigeria

Nigeria/India

110. NAJAFABADI Dr. Bijan   (Affiliate)
     Sharif University of Technology
     P.O. Box 11365-8639
     Tehran
     Iran

Iran

Name and Institute	Member State

111. NAQVI Ms. Farzana   Pakistan
      Applied Systems Analysis Group
      Pakistan Atomic Energy Commission
      P.O. Box 1114
      Islamabad
      Pakistan

112. NEGRETE Dr. P.   (Associate)   Venezuela
      Departamento de Fisica
      Facultad de Ciencias
      Universidad Central
      Caracas
      Venezuela

113. NSABIMANA Dr. Marcel   Rwanda
      C.E.A.E.R.
      Universite National du Rwanda
      B.P. 117
      Butare
      Rwanda

114. ODUKWE Professor Okay A.   Nigeria
      Faculty of Engineering
      University of Nigeria
      Nsukka
      Nigeria

115. OGANA Dr. Wandera   Kenya
      Department of Mathematics
      University of Nairobi
      P.O. Box 30197
      Nairobi
      Kenya

116. OKTIK Dr. Sever   Turkey
      University of Selcuk
      Konya
      Turkey

Name and Institute	Member State

117. OMAR Dr. Tarek   (Affiliate)   Egypt
      Department of Physics
      Faculty of Science
      Cairo University
      Cairo
      Egypt

118. ONYANGO-OTIENO Dr. V.   (Associate)   Kenya
      Department of Mathematics
      University of Nairobi
      P.O. Box 30197
      Nairobi
      Kenya

119. OSMAN Dr. M.A.   (Affiliate)   Egypt
      Assiut University
      Department of Physics
      Assiut
      Egypt

120. OTITI Dr. Tom   Uganda
      National Research Council
      Ministry of Planning and
          Economic Development
      P.O. Box 6884
      Kampala
      Uganda

121. OUAIDA Dr. Mohamed Bassam   Lebanon
      National Council for Scientific Research
      Solar Energy Group
      P.O. Box 11-8281
      Beirut
      Lebanon

122. OUMBA Dr. Marie-Therese   Congo
      Universite Marien Ngouabi
      INSSED
      B.P. 237
      Brazzaville
      Congo

Name and Institute	Member State

123. PANDE Dr. Dinkar
Department of Physics
Nagpur University
Nagpur
India

India

124. PEER Dr. Mushtaq
P.G. Department of Physics
University of Kashmir
Srinagar 190006
India

India

125. PIETRUSZKO Dr. Stanislaw
Institute of Electron Technology
Faculty of Electronics
University of Warsaw
Koszykowa 75
Warsaw
Poland

Poland

126. POVOLO Dr. Francisco
Comision Nacional de Energia Atomica
Depto de Materiales
Av. del Libertador 8250
1429 Buenos Aires
Argentina

Argentina

127. PRAKASH Dr. J.
Dipartimento di Energetica
Universita degli Studi di Firenze
Via di S. Marta 3
50139 Firenze

Italy/India

128. PRESUTTO Ms. Milena
ENEA-FARE
Ed. 26
c/o EURATOM
20120 Ispra (VA)

Italy

Name and Institute	Member State

129. PURICA Dr. Ionut.   Romania
     IMGB-FECNE
     Sos. Berceni Nr. 104, Sec. 4
     78004 Bucharest
     Romania

130. PYARE LAL Dr.   Zambia/India
     Department of Physics
     University of Zambia
     P.O. Box 32379
     Lusaka
     Zambia

131. QUINTINO Dr. I.   (Affiliate)   Cuba
     Facultad de Fisica
     Universidad de la Habana
     Colina Universitario
     Cuba

132. RAI Dr. B.   Nigeria
     Department of Physics
     Bayero University
     P.M.B. 3011
     Kano
     Nigeria

133. RAKSHANI Dr. Ali   Kuwait
     Kuwait University
     P.O. Box 5659
     Safat
     Kuwait

134. RAMADAN Dr. A.   (Affiliate)   Egypt
     Menia University
     Department of Physics
     Egypt

Name and Institute	Member State

135. RAMADAN Dr. Muhammad   Egypt
      Physics Department
      Faculty of Science
      Tanta University
      Tanta
      Egypt

136. RATHANANTHAMMPAN Dr.   (Affiliate)   Thailand
      Chulalongkorn University
      Physics Department
      Bangkok 10500
      Thailand

137. ROMAN Dr. Roberto Latorre   Chile
      Depto. de Ingenieria Mecanica
      Universidad de Chile
      Casilla 2777
      Santiago ·
      Chile

138. RU GUANG Dr. Cheng   China
      Shanghai Institute of Ceramics
      Academia Sinica
      Shanghai
      People's Republic of China

139. RUSU Dr. Gheorghe   Romania
      Scientific Research & Technical
          Engineering Inst.
      ICPE
      Str. Parcului No. 7
      Bucharest
      Romania

140. SADEGHIPOUR Dr. Mohammad   Iran
      School of Mechanical Engineering
      Sharif University of Technology
      P.O. Box 8639
      Tehran 11365
      Iran

Name and Institute	Member State

141. SADER Dr. E.M.   (Affiliate)   West Bank
     Birzeit University
     Physics Department
     P.O. Box 14
     Birzeit, West Bank, via Israel

142. SALEHI Dr. Mahmoud   (Affiliate)   Iran
     Sharif University of Technology
     P.O. Box 11365
     Tehran
     Iran

143. SALEHPOOR Dr. Behrooz   Iran
     Chancellor's Office
     University of Tabriz
     Tabriz
     Iran

144. SALMON Ms. Elaine   Jamaica
     Irvine Hall University of
         the West Indies
     Mona, Kingston
     Jamaica

145. SAMBO Dr. A.S.   (Affiliate)   Nigeria
     Department of Physics
     Bayero University
     Nigeria

146. SAMUEL Dr. T.   (Associate)   Sri Lanka
     Faculty of Engineering
     University of Peradeniya
     Peradeniya
     Sri Lanka

147. SEGHAYER Dr. Khalifa   Italy/Libya
     ENEA/FARE
     S.P. Anguillarese 301
     00100 Roma

Name and Institute	Member State

148. SHARMA Dr. Ashok K.
S.T.R.L.
Centre of Energy Studies
I.I.T., Delhi
New Delhi 110016
India
   India

149. SHARMA Dr. Kamalesh Kumar
Department of Physics
R.K. College
Shamli 247776
Muzaffarnagar U.P.
India
   India

150. SHOUSHA Dr. Abdel Halim
Electronics Department
Faculty of Engineering
Cairo University
Cairo
Egypt
   Egypt

151. SHRESHTA Dr. V.M.
Tribhuvan University
Research Centre for Applied
    Science and Technology
P.O. Box 1036
Kathmandu
Nepal
   Nepal

152. SILAWATSHANANAI Dr. C.
Physics Department
Faculty of Science
P.O. Box 3
Kohong, Hatyai
Thailand
   Thailand

Name and Institute	Member State

153. SINGH Dr. K.M.
Department of Physics
Nagpur University
University Campus
Nagpur 440 010
India

India

154. SINGH K.   (Affiliate)
University of Science & Technology
Kumasi
Ghana

Ghana/India

155. STAPINSKI Dr. Tomasz
Academy of Mining & Metallurgy
al. Mickiewicza 30
30-059 Krakow
Poland

Poland

156. SUPARNO Dr. Satira
Physics Department
ITB
Jalan Ganesha 10
Bandung
Indonesia

Malaysia

157. TAIBI Dr. M.   (Affiliate)
Ecole Normale Superieure de Takaddoum
B.P. 5118
Rabat
Morocco

Morocco

158. TANER Dr. Kemal
Arifiye Mah
Bakim Sokak Karabacokoglu
Apto. no. 1
Eskisehir
Turkey

Turkey

Name and Institute	Member State

159. TASEVSKI Dr. Milan
    Institute za Elektronike
      in Vakuumsko Tehniko
    P.O.B. 1
    Ljubljana
    Yugoslavia   Yugoslavia

160. TOFIGHI Dr. Aliassghar
    2 rue des Abeilles
    31000 Toulouse
    France   France/Iran

161. TORRES Dr. Juan
    Universidad Nacional San Antonio
      Abad del Cusco
    Dep.to de Fisica
    Av. de la Cultura s/n
    Cusco
    Peru   Peru

162. TRESSO Ms. Elena
    Dipartimento di Fisica
    Politecnico di Torino
    C.so Duca degli Abruzzi
    Torino   Italy

163. TUNSIRI Dr. P.
    Physics Department
    Faculty of Science
    Chulalongkorn University
    Bangkok 10500
    Thailand   Thailand

164. TURKOVIC Ms. Aleksandra
    Institute Rudjer Boskovic
    Bijenicka 54
    41001 Zagreb
    Yugoslavia   Yugoslavia

Name and Institute	Member State

165. VERCIN Dr. Abdullah   (Affiliate)   Turkey
      Ankara University
      Faculty of Science
      Ankara
      Turkey

166. VIRK Dr. H.S.   India
      Department of Physics
      Guru Nanak Dev University
      Amritsar 143005
      India

167. VISITSERNGRAKUL Dr. Supan   Thailand
      School of Energy & Materials
      King Mongkut's Institute of Technology
      Thonburi,
      Bangkok 10140
      Thailand

168. XI Dr. Yiming   Italy/China
      Dipartimento di Fisica
      Via G, Campi 213/A
      41100 Modena

169. YAGHOUBI Dr. Mahmood A.   Iran
      Mechanical Engineering Department
      Engineering School
      Shiraz University
      Shiraz
      Iran

170. YIANOULIS Dr. Panayiotis   Greece
      Physics Laboratory II
      University of Patras
      Patras
      Greece

Name and Institute	Member State

171. ZAOUK Dr. Aref
   National Council for Scientific Research
   Solar Energy Group
   P.O. Box 11-8281
   Beirut
   Lebanon

   Lebanon

172. ZHU Mr. Huanlinag
   Guangzhou Institute of Energy Conversion
   Chinese Academy of Sciences
   81 Central Martyrs' Road
   Guanzhou
   People's Republic of China

   China

ENERGY MANAGEMENT

ROLES FOR THE PHYSICIST OR ENGINEER

IN MEASUREMENT, ANALYSIS AND MANAGEMENT OF

INDUSTRIAL ENERGY USE

Marc Ross
Physics Department
University of Michigan
Ann Arbor, Michigan 48109, USA

OUTLINE

I. ENERGY MEASUREMENT & ANALYSIS TO SUPPORT ENERGY CONSERVATION IN INDUSTRY

   A. Introduction

   B. Availability (Available Work) Analysis

   C. Energy Measurement & Analysis

      1) Aggregate Energy Intensity
      2) Part-Load Analysis, Energy-Use by Stage of Production
      3) Parametric Studies of Process Units
      4) Economic Analysis of Capital Projects

II. ENERGY CONSERVATION IN MANUFACTURING

   A. Types of Conservation Categorized by Level of Investment

   B. Major Change in Manufacturing Processes

      1) The Movement of Some Energy-Intensive Manufacturing to
         Developing Countries
      2) Radically Improving Energy-Intensive Manufacturing Processes
      3) Example: Potential New Processes for the Steel Industry
      4) Decision Making About Manufacturing Processes in Developing
         Countries

4

I. ENERGY MANAGEMENT & ANALYSIS TO SUPPORT ENERGY CONSERVATION IN INDUSTRY

A. Introduction

The technical effort needed to support energy conservation, or the achievement of increased energy efficiency by cost-effective means, involves three activities: systematic analysis of energy-using systems, observations (where needed), and technical support for implementation of agreed actions. The appropriate action is typically either changes in operations or investment in new equipment.

An energy service is a desideratum measured in terms of the service: for example pumped water in terms of liters/min., light on a task, or a manufactured material suitable for a specific purpose. Energy service is not measured in kwh electricity, liters of fuel oil, etc.

Almost all energy services can, with technical effort and capital investment, be provided with much less energy, with reduced side effects and typically at lower cost. Typically the cost of saved energy:

$$\$/(Gcal)_{saved} \ll \$/(Gcal)_{added supply}$$

That is, the investment to save a unit of energy is typically much less than an investment in new fuel wells, solar or other facilities to produce the energy. The supply curve for a fully developed program of conservation projects will typically look like Fig. 1. Specific examples of such supply curves will be given below.

B. Availability (Available Work) Analysis:

Consider the capacity of a system S, in contact with the environment whose temperature and pressure are T_O and P_O, to do work on another system S'. The availability of S is the maximum work.

FIGURE 1

SUPPLY CURVE FOR SAVED ENERGY

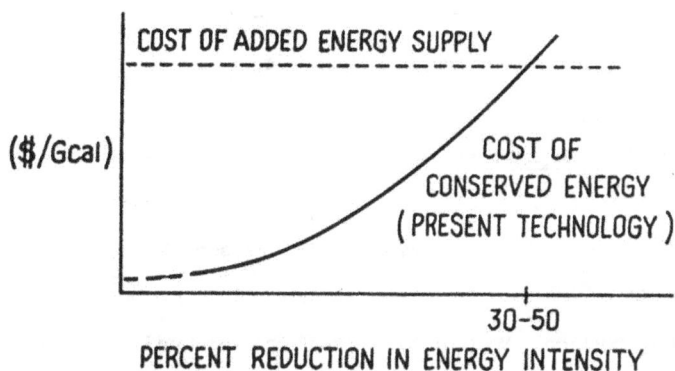

PERCENT REDUCTION IN ENERGY INTENSITY

FIGURE 2

AVAILABILITY OF HEAT FROM A RESEVOIR

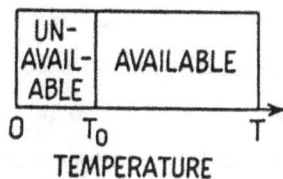

A very simple example is the availability of a quantity of heat ΔQ taken from a large reservoir at temperature T. According to Carnot, the availability is $\Delta B = \Delta Q \; (1-(T_0 /T))$. See Fig. 2. Somewhat more generally:[2]

$$\Delta B = \Delta H^\circ - T_0 \Delta S = (U - U_0) + P_0(V - V_0) - T_0(S - S_0)$$

where U_0, V_0, S_0 are properties of S when in equilibrium with the environment.

Application to crude oil separation. In Fig. 3 energy use in a crude distillation unit at a petroleum refinery is shown from first-law (enthalpy) and second-law (availability) perspectives. The second-law analysis shows that about 30% of the availability of a fuel is lost in the (irreversible) process of combustion. Most of the rest of the availability is lost in the thermal degradation of heat in distillation. That is the essence of the crude distillation process: the material is raised to a high temperature by direct heating and then various components decline in temperature as they rise through the tower. A moderate amount of availability (8%) is lost with the hot combustion gases up the heater stack. Most of the enthalpy (66%) but relatively little availability (9%) is lost in cooling the product streams. This discrepancy results from the relation of the availability, B (the work available in principle from the heat Q), and Q. In the case at issue where the reservoir is finite (assuming constant heat capacity)

$$B = C\!\int_{T_0}^{T} dT' \; (T' - T_0)/T' = Q \; (1 - \frac{T_0}{T - T_0} \ln(T/T_0))$$

If for example the "dead state" is at 38°C then B/Q = .145, corresponding to the result shown for heat rejected by air and water coolers in Fig. 3.

The implication of this is that heat recovery from stack gas tends to be economically justified because the temperature is high. Heat recovered at high temperature can, for example, be transformed into steam and used

FIGURE 3
ENTHALPY BALANCE

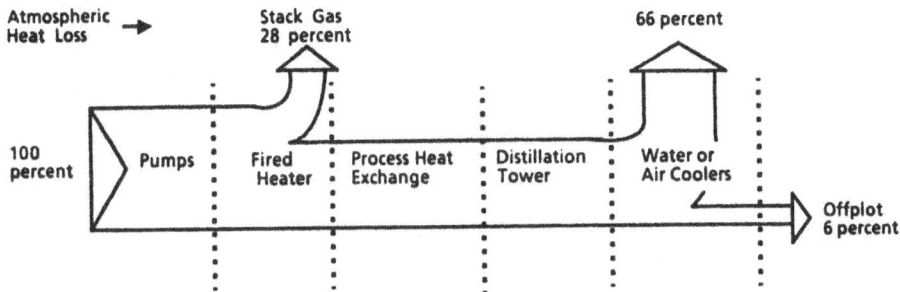

AVAILABLE WORK BALANCE

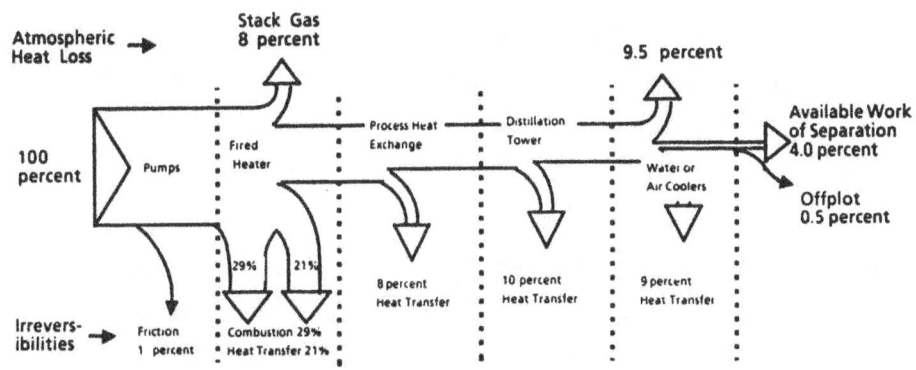

Enthalpy and Available Work Balance for a Crude Separation Unit

elsewhere. On the other hand, if there is no use for low-temperature heat very nearby, it probably doesn't pay to recover it.

This example suggests that there are two approaches to conservation, both of which are needed: (1) Given the process, reduce the lost work in waste streams. The large amount of energy rejected at low temperature by coolers is probably the least attractive target. (2) Change the process-specific availability requirements. This may consist of fine tuning or through radical process change.

Minimum availability requirements across manufacturing. The laws of physics show that change other than endothermic chemical reactions typically requires very little available work in a thermodynamically ideal process. Thus macroscopic physical change and physical separations such as drying have very small minimum thermodynamic requirements. Of common industrial transformations only the reduction of metal ores and some other endothermic chemical reactions have high availability requirements in principle. (See Fig. 4.) One other type of manufacturing has a high fuel requirement in principle, the manufacture of organic materials where fuel is used as the feedstock.

The use of energy in practice is thus many orders of magnitude more than the minimum availability requirement in shaping, forming, joining and assembling processes. For some of the energy-intensive industries with endothermic chemical reactions the situation is completely different. Actual use is as little as 3 to 4 times the minimum in the primary metals industries, and closer for a specific process like smelting of aluminum. (See Fig. 5.)

Among physicists engaged in energy-efficiency improvement there is a belief (which I share) that technological improvement in the future will

FIGURE 4

MINIMUM AVAILABILITY REQUIREMENTS ACROSS MANUFACTURING

	High Specific Requirement	Modest Requirement	Negligable or Negative Requirement

---

Chemical Change	reduction of ores		exothermic reactions
	other endothermic reactions		
Micro-physical Change		molecular separation drying internal structure of solids	
Macro-physical Change			shaping forming joining assembling

FIGURE 5

ENERGY INTENSITIES FOR SELECTED BASIC-MATERIALS*

	Energy Intensity 1980(U.S.) (Gcal/tonne)		Thermodynamic Minimum[c] (Gcal/tonne)
	primary energy[a]	carrier energy[b]	
Paper	7.3[d]	5.6[d]	—[e]
Steel	9.4	7.9	1.7
Aluminum	46[f]	21[f]	7.0[h]
Petroleum Refinging	1.17	1.08	0.1
Cement	1.7	1.4	0.2

[a]with purchased electricity evaluated at about 2.9 Mcal/kwh (11,500 Btu/kwh)
[b]with electricity evaluated at 0.86 Mcal/kwh (3413 btu/kwh).
[c]Gyftopoulos et al., Ref. 8.
[d]wood derived fuels not included.
[e]For paper the absolute value of the minimum is small and its sign depends on
accounting conventions and product.
[f]The energy intensities are per tonne of shipped product. If the base is
taken to be tonnes of primary plus secondary metal the energy intensities are
17% higher and if the base is tonnes of primary metal it is 37% higher.
[h]per tonne of primary metal.

*1Gcal/tonne = 4.18 GJ/tonne
           = 1.16 Mwh/tonne
           = 3.60 Million Btu/short ton

enable efficiency improvement to be made at very little, if any, additional cost. The structure of this cost-per-unit-of-energy-service vs energy efficiency relationship will depend however on whether one is approaching, in practice, the theoretical availability limit. (See Fig. 6.) The figure suggests that, for energy services with low or negative minumum-availability-requirement (Fig. 6a), simple cost considerations will not really determine an optimal energy efficiency. Other criteria, such as goals set by the society, are thus essential.

C. Energy, Measurement & Analysis[3]

1) Aggregate Energy Intensity

The aggregate energy intensity of an entire production process or plant is :

$$e = \frac{\text{total energy use}}{\text{production}} = \frac{E}{P}$$

The purpose of determining such a quantity is to facilitate communication about energy use, the setting of goals, and temporal monitoring of energy use. A number of issues must be settled in defining it. For the numerator, one must decide whether to combine different forms of energy used by adding, e.g., the Joules of the energy carriers. I prefer to add primary energies, i.e. to multiply electrical energy by a factor, before adding, so that it approximately represents fuel use at the generating plant. For the numerator one must also decide whether to include biomass fuels (e.g. in the paper and food processing industries). In addition, should one adjust for changes in time of input materials, output mix, pollution control measures, and new production facilities?

FIGURE 6

COST OF SERVICE VS ENERGY INTENSITY
FOR SMALL AND FOR LARGE
MINIMUM AVAILABILITY REQUIREMENT (SCHEMATIC)

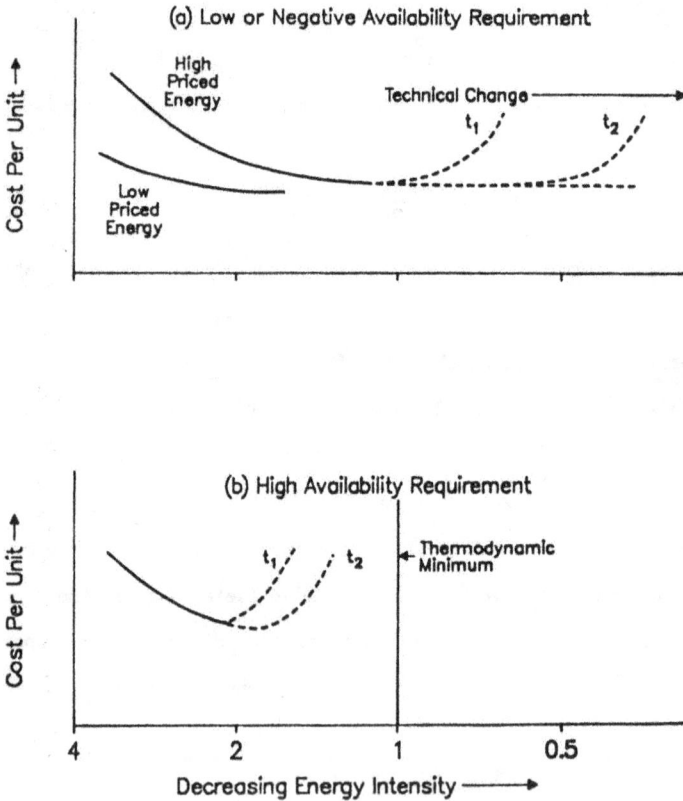

For the denominator there is little problem at a factory with a single product like cement; quote production in tons. With multiple products there are two options: a) Add the measures, e.g. tons, of different products even where some are intermediates. b) Construct an index

$$I(t) = \sum_i W_i P_i(t)$$

where each W_i is a fixed weight and is the estimated energy used in the production process to make a ton of product i in a base year.

2) Part-Load Analysis, Energy Use by Stage of Production

Energy use tends to have a large production-independent part because motors are left on, high temperature furnaces are left on, etc., even when production rates are low or when there is no production. Thus, approximately,

$$E = A + BP$$

where A and B are constants. The energy intensity is:

$$e = B + \frac{A}{P}$$

Month-to-month data for E and P scatter, as might be expected, but roughly describe this curve. Regression analysis of data for whole industries shows the A term to be comparable to the B term suggesting a major target for energy conservation in industries which do not operate 24 hours a day or which operate well below production capacity.

If fuel use at each furnace and electricity use in each section of a factory is metered, energy use can be allocated by stage of production. The factory must be systematically divided into subsystems and mass/product flows measured.[4] Then for each stage of production, $E_i = A_i + B_i P_i$. (Such an equation should also include on the right-hand side other parameters known to be important to energy use in the particular process, such as degree days.)

The purpose of carrying out this more detailed analysis is to enable a) cleaner interpretation of energy-intensity time series, b) sufficiently definitive information for cost accounting and incentive programs, c) specific identification of opportunities for conservation, e.g. through scheduling. "Knowledge is power".

3) Parametric Study of Process Units

Quantitative relationships are determined between energy (and other costs) and operational parameters of a process, like temperatures, pressures, feed rate and compositions. The purposes are: development of targets for energy performance, optimization of day-to-day operations, creation of a basis for possible on-line automatic control, and creation of a basis for analysis of possible capital projects. The method is to develop a theoretical model on the basis of the process design and installed equipment, to conduct systematic experimental tests, and then to adjust the theory and check the experiments to bring them into conformity. This effort typically requires installing submetering.

Kenney[5] details an example of such a parametric study, a study of a distillation unit for separation of C_4 and C_5 compounds at an ethylene plant, in the context of an overall energy management program. I recommend Kenney's book enthusiastically.

4) Economic Analysis of Capital Projects

The central problem for this analysis is to put the initial expense and the annual cash flows on the same basis. See Fig. 7. Here the bottom row, the discounted cash flow (DCF) defines a present value for a cash flow in the future. For a cash flow B, n years in the future,

$$present\ value = B/(1 + r)^n$$

FIGURE 7

DISCOUNTED CASH FLOWS ASSOCIATED WITH A CAPITAL PROJECT[a]

Year	0	1	2	...	T[b]
Capital-related flows[c]	$-K$	K_1	K_2	...	K_T
Net operating benefits[d]	0	B_1	B_2	...	B_T
Cash flow	$-K$	$K_1 + B_1$	$K_2 + B_2$	...	$K_T + B_T$
DCF	$-K$	$\dfrac{K_1 + B_1}{1 + r}$	$\dfrac{K_2 + B_2}{(1 + r)^2}$	...	$\dfrac{K_T + B_T}{(1 + r)^T}$

[a]In the example shown the entire construction phase lasted one year or less.
[b]The economic life of the project is T years. No associated cash flows occur after that time.
[c]Initial capital cost is K. The K_i are net tax benefits associated with the capital.
[d]The difference between annual operating benefits, e.g. from forgone energy use, and any operating and maintenace cost increases.

where   r is a "discount rate".   In many situations the numerical value of the discount rate may not be well defined, but it can be a useful concept.   One popular method to develop an economic ranking for possible investments is to

FIGURE 9

SAMPLE REFINERY ENERGY CONSERVATION PLAN:

CAPITAL COST, ENERGY SAVINGS & ECONOMIC PERFORMANCE

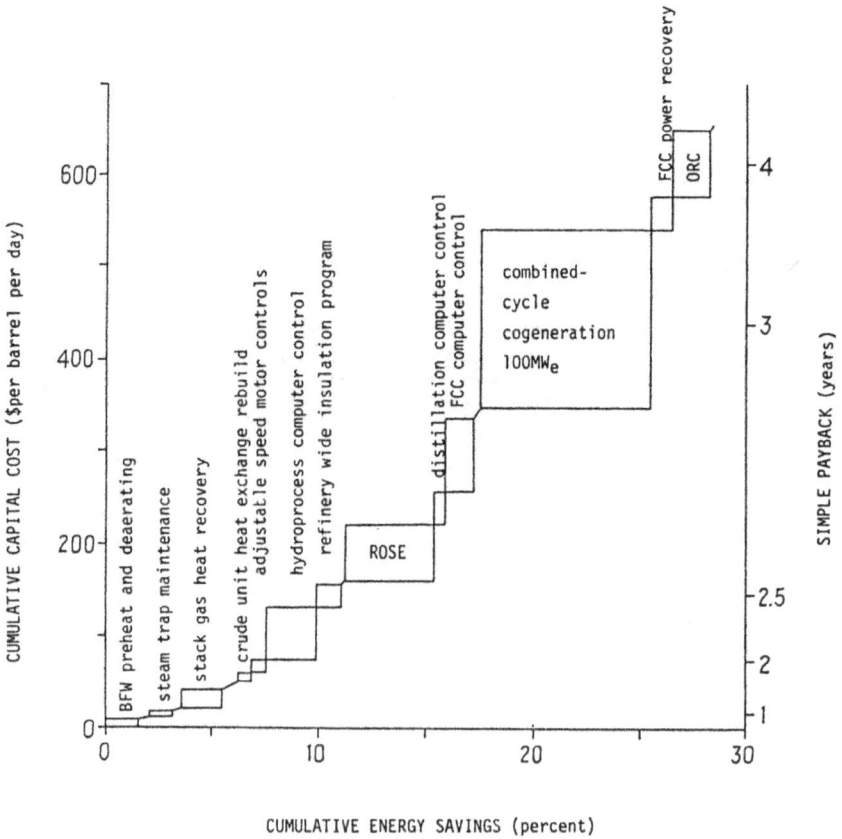

CUMULATIVE ENERGY SAVINGS (percent)

FIGURE 8

SAMPLE STEEL MILL ENERGY CONSERVATION PLAN

CAPITAL COST, ENERGY SAVINGS & ECONOMIC PERFORMANCE

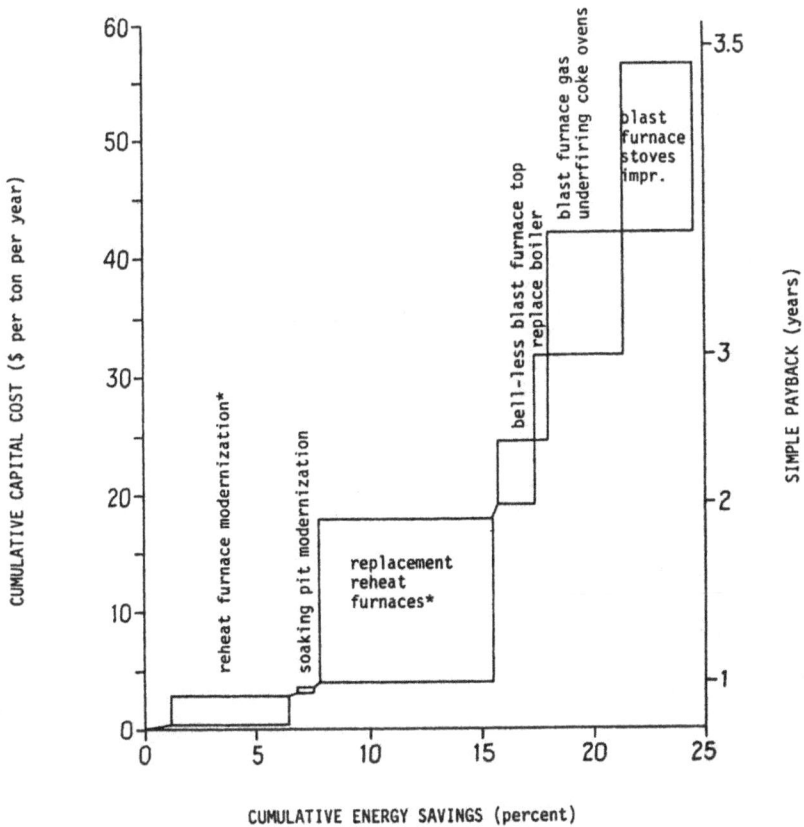

*These projects substantially overlap each other. Net total
energy savings for the plan are 20%.

FIGURE 9

SAMPLE REFINERY ENERGY CONSERVATION PLAN:

CAPITAL COST, ENERGY SAVINGS & ECONOMIC PERFORMANCE

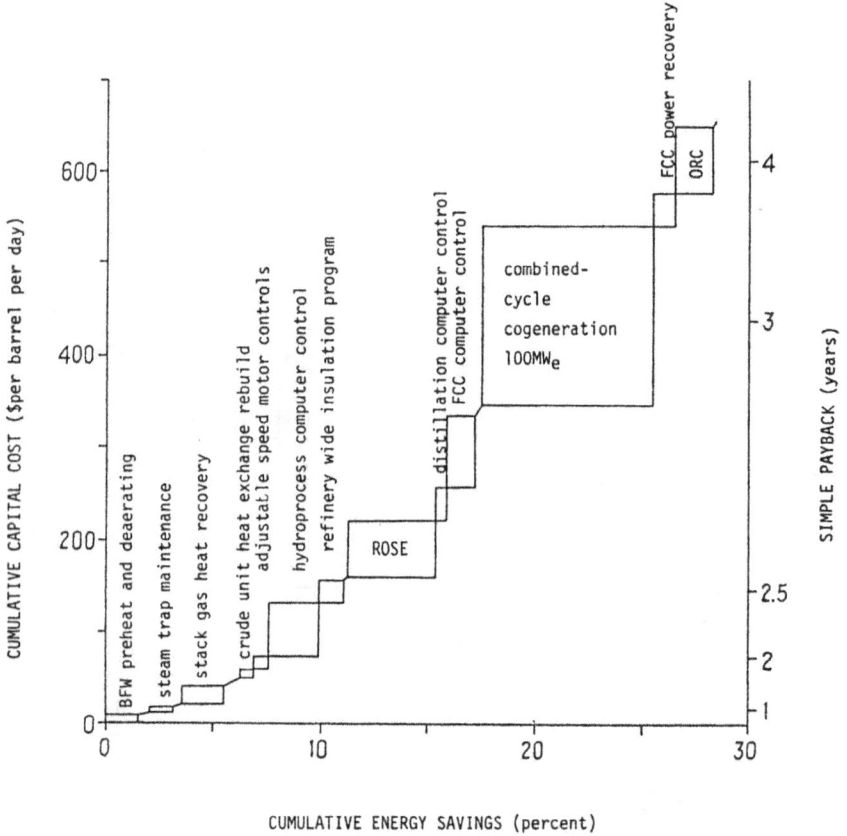

CUMULATIVE ENERGY SAVINGS (percent)

II.   ENERGY CONSERVATION IN MANUFACTURING

A.   Types of Conservation Actions, Categorized by Level of Investment

There are three major categories of technological actions which are taken to improve energy efficiancy:

1) Improve operations and maintenance practices with little or no investment (< $50,000 per project).

2) Replace energy-intensive equipment or install add-on conservation equipment (< $10 million per project).

3) Build a new facility with components optimized to realistic energy costs; or adopt a major process change. The latter may or may not require a new manufacturing facility.

Examples of operations/maintenance improvements are: inspections, training, scheduling, systematic maintenance and accounting procedures. Of course, improved maintenance will lead to added expenditures at first. In addition, improved maintenance may call for modern inspection equipment such as portable meters, an infrared scanner, etc. Accounting procedures refers, for example, to charging the separate production groups for their use of energy. This would require sub metering.

Categories of add-on conservation equipment are: heat and power recovery, utility system improvements & cogeneration, combustion controls, motor controls, motor and motor-drive improvements, and advanced automatic process controls. These constitute a very large subject which I will not address here. Instead I will go on to emphasize process improvement through research and development.

B. Major Change in Manufacturing Process

1) The Movement of Some Energy-Intensive Manufacturing to Developing Countries

As background for the rest of my discussion I will present a brief explanation for the decline of certain energy-intensive industries in highly industrialized nations and for the growth of those sectors in developing nations.[6]

One factor is the saturation of demand for bulk manufactured materials in developed countries even as economic growth continues, while demand for materials is growing in developing countries. Figure 10 shows two measures of steel consumption during the past century in the United States. The steel/GNP ratio peaked over 60 years ago and per capita steel consumption leveled off some 2 or 3 decades ago. The picture for European countries is similar, while consumption is, of course, low but growing in developing countries (Fig. 11).

A second factor is that there are major cost advantages at certain sites or regions for hard-to-transport raw materials like natural gas, ethane and hydropower. Some examples are shown in Fig. 12. For one of the raw materials, wood, the United States has an advantage. For organic chemicals and aluminum, countries with underdeveloped gas or hydropower resources have strong advantages.

A third factor favoring some developing countries is that there can be technological advantages in creation of a growth industry from a relatively small base. In addition to lower wages, national governments may take a special interest, and, most important, decision makers are not wedded to old facilities and old ways of doing things.

FIGURE 10

A CENTURY OF STEEL CONSUMPTION IN THE U.S.

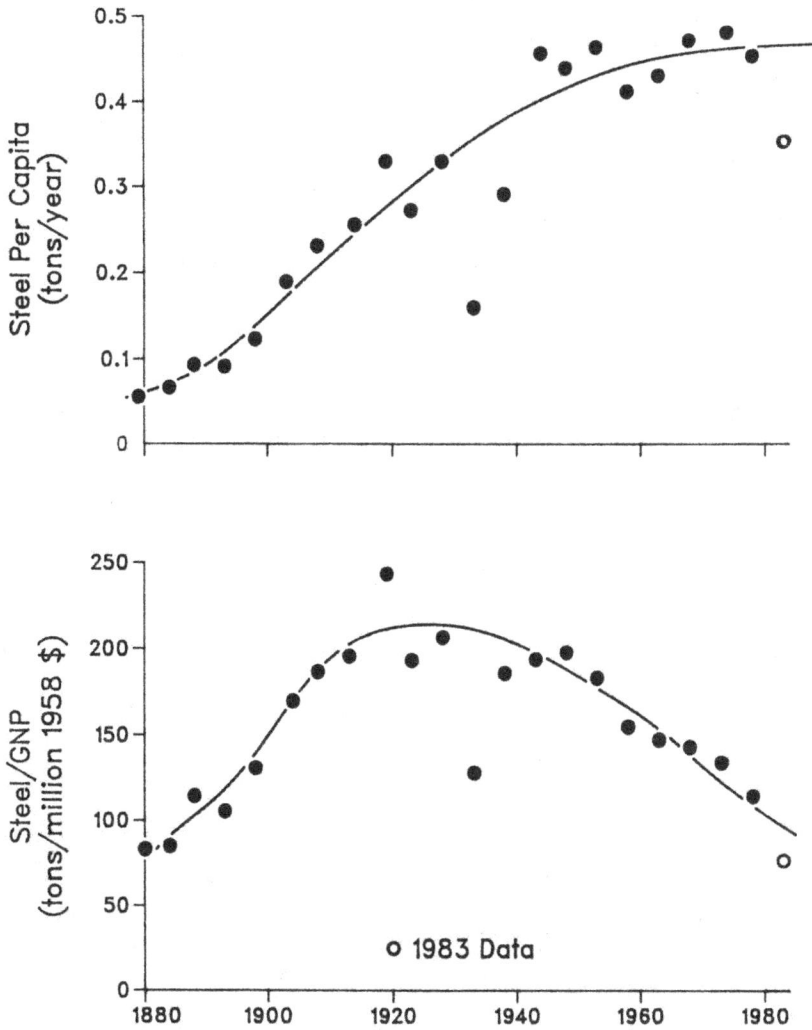

o 1983 Data

FIGURE 11

APPARENT STEEL CONSUMPTION (kg per capita)

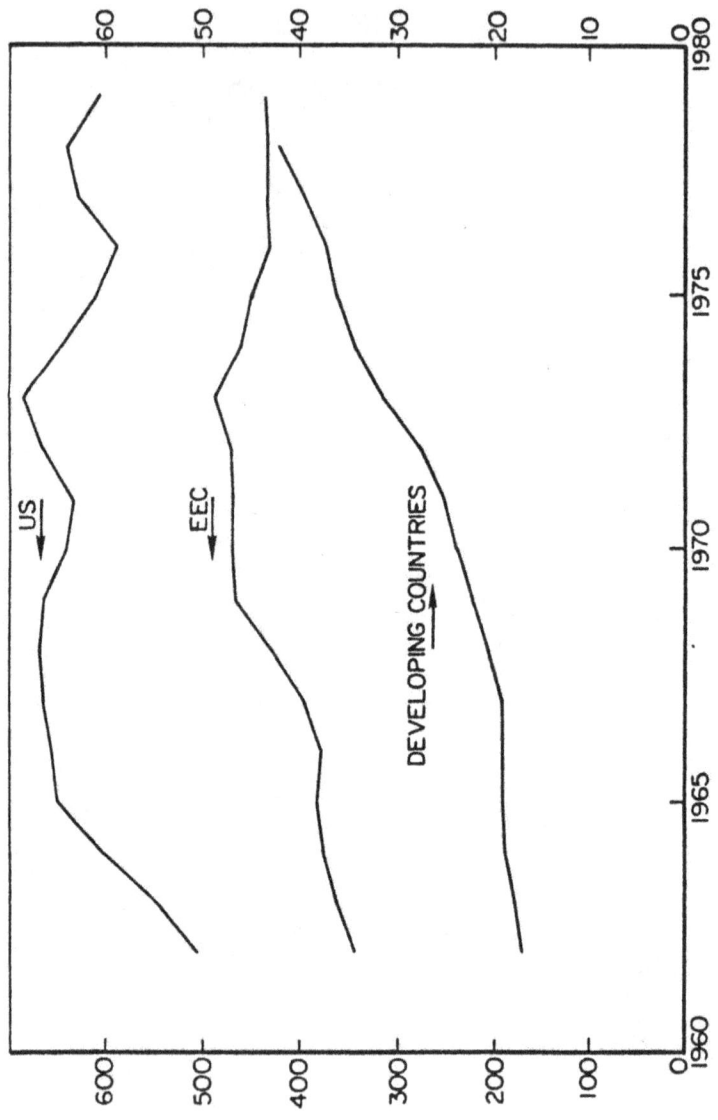

FIGURE 12

COST POSITION OF EXPORTERS AND POTENTIAL EXPORTERS TO THE UNITED STATES
(PARTIAL LIST)

	Cost Advantage[*]	Rationale
STEEL[a]		
Japan	significant	high technology
East-Asia Rim	major	low-cost labor[b]
Canada, South Africa	marginal	
Brazil	significant	low-cost labor
Europe	uncertain	subsidies
Mexico, Venezuela	uncertain	low-cost labor, gas
PAPER		
Canada	similar costs[c]	
Scandinavia	disadvantage	limited resources
Brazil	significant[d]	natural resources
USSR	uncertain	natural resources
ETHYLENE		
Canada	significant	natural resources
Mid-East	significant	natural resources
Mexico, Venezuela	significant	natural resources
East-Asia Rim	uncertain	
Indonesia, Australia	uncertain	natural resources
ALUMINUM		
Canada	major	hydropower
Brazil	major	hydropower
Australia	significant	coal,[e] gas
USSR	uncertain	hydropower
Other	uncertain	hydropower, flared gas

[*]Relative to the U.S. The present high value of the dollar also provides an advantage.
[a]All these countries have some or many modern, more productive, mills than in the U.S., although Europe also has many antiquated mills.
[b]Relatively skilled labor and good infrastructure provide a major advantage.
[c]The potential for expansion is more limited than in the U.S.
[d]Expansion will be slow and only hardwood pulp is involved.
[e]Low-quality coal unsuited for export. Some analysts see significant cost advantages in Australia. Others see the opportunity dependent on help, e.g. capital subsidies.

2) Radically Improving Energy-Intensive Manufacturing Processes

Technical advances of recent decades have created the opportunity to radically improve old manufacturing processes and products. Energy requirements can be substantially reduced; material requirements, side effects, unit capital costs and, often, the minimum scale of production can also be reduced. Through research and development one seeks processes, in materials manufacturing (which dominates industrial energy use), which are

　　-much nearer the thermodynamic ideal,

　　-offer easy and complete control, and

　　-are rapid and continuous.

3) Example: Potential New Processes for the Steel Industry

Iron and Steel mill products are made in two broad stages: a) iron and steel making in which the ore is reduced and the material purified and alloyed to produce the desired liquid steel, and b) shaping and treating to obtain the desired shape and internal structure.

Iron and Steel making. The main opportunity for improvement over the present process is not reduction of energy intensity, but reduced unit capital cost, reduced scale and elimination of coke making with its problems. Elements of the present process are sketched in Fig. 13a. These are all bacth processes.

One proposal for a new process seems particularly promising to me.[7] The reduction of the ore and removal of most impurities would be accomplished in one vessel, continuosly and rapidly. The concept is to carry out the process in a liquid iron bath as shown in Fig. 13b. The vessel would be something like the basic oxygen converter, with powdered ore, coal and flux and oxygen

FIGURE 13a

IRON- & STEEL-MAKING TODAY

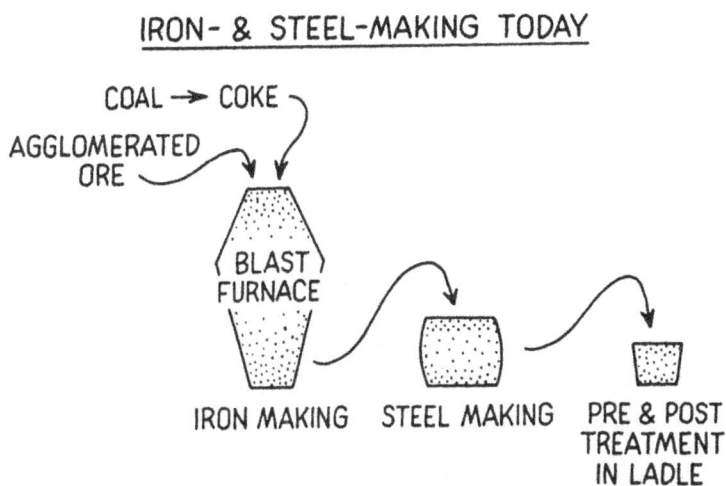

FIGURE 13b

PROPOSED PROCESS FOR REDUCING IRON ORE

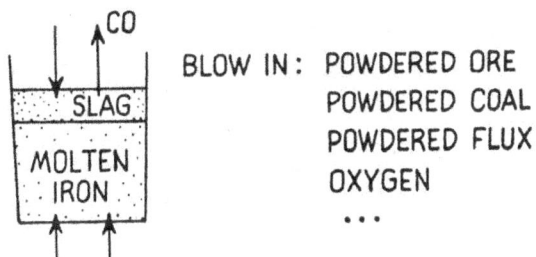

BLOW IN: POWDERED ORE
         POWDERED COAL
         POWDERED FLUX
         OXYGEN
         ...

being blown in, and CO and, perhaps, H_2, produced, as well as steel and slag. With continuous operation, a trickle of only 50 kg/s would provide for a moderatly large 1 million t/year production. The main problems to be solved are: a) The reduction reaction is endotermic, but the carbon in the coal tends to be oxidized to CO rather than CO_2 so that relatively little heat is supplied. In this situation great quantities of coal are used and CO produced. b) The refractories need to be protected.

An approach which might be developed through appropriate research would be to accurately model the flows in the bath and attempt to develop a design in which the flows are controlled so that carbon would be oxidized to CO_2 in one zone (creating adequate heating), and reduction in another zone, while appropriate conditions are maintained at walls to protect refractories.

Shaping. The casting of thick shapes (e.g. ten or more centimeters) as at present does not permit control of solidification. The desired internal structure, now obtained by extensive rolling and heat treating, might be obtained, in large part, by fully controlled solidification. This might be possible if continuous thin casting at the millimeter scale could be developed. If so, the energy savings, through avoidance of reheating and rolling would be enormous.

A possibility which I find especially interesting with controlled solidification is that it might enable the development of high-quality steels even where they contain impurities like copper which, using present shaping and treating processes, render steel unsatisfactory for many purposes. Such a development would enable much more effective and economic use of scrap steel from consumer products than is now customary. For example, cars might be recycled into cars.

4) Decision Making About Manufactoring Processes in Developing
   Countries.

The decision to adopt a particular process technology for
energy-intensive manufacturing has enormous ramifications for a developing
country. The requirements for energy and raw materials, the potential for
pollution and the potential for local manufacturing of economically-beneficial
products downstream in the production process, and other factors affecting the
control of employment and markets, are all intimately dependent on the
technology selected. Even if it might be thought a priori that the prospects
for a significant R & D program relating to a particular manufacturing sector
are not favorable in country X, if there are serious prospects for expanded
production in that sector, individuals and organizations should interest
themselves in research on process technologies for that sector. The potential
for poor decisions on process technology by governments is great. Technical
people are needed who can bring their knowledge to bear, within or outside of
official channels, on these important decisions.

REFERENCES

1. See Efficient Use of Energy, Part I, Chap. 2, American Institute of
    Physics Conference Series vol. 25 (1985); also thermodynamics texts for
    engineers which discuss availability, available work, lost work,
    second-law efficiency, etc.

2. For greater generality, see AIP, reference 1.

3. See, for example, articles by Aird (p. 223) and Strecker (p. 261) in
    Energy Economics and Management, A. Reis, et al., eds. Pergamon Press
    (1985).

4. I. Bousted and G. F. Hancock, Handbook of Industrial Energy Analysis,
    Halstead Press, John Wiley (1979).

5. W. F. Kenney, Energy Conservation in the Process Industries, Academic
    Press (1984).

6. M. Ross, E. Larson and R. Williams, "Energy Demand and Materials Flows
    in the Economy", Center for Energy and Environmental Analysis, Princeton
    University report No. 193 (1985), Princeton, NJ 08544.

7. S. Eketorp, "Energy Considerations of Classical and New Iron and
    Steelmaking Technology", to be published in Energy the International
    Journal, 1986 or 1987.

8. Elias P. Gyftopoulos, Lazaros J. Lazarides, and Thomas F. Widmer,
    Potential Fuel Effectiveness in Industry, a report to the the Energy
    Policy Project of the Ford Foundation, Ballinger (1974).

PROSPECTS OF ENERGY DEMAND AND
SUPPLY IN DEVELOPING COUNTRIES

Arshad Muhammad Khan
Pakistan Atomic Energy Commission
Islamabad - Pakistan

1.   INTRODUCTION

The intimate relationship between the level of energy
consumption in a country and its stage of socio-economic
development is now a well established phenomenon. It is on
this account that the developing countries* with three quarters
of the world population at present account for a mere 1/5th of
the global consumption of commercial energy. While the increasing scarcity of global energy supplies together with sharp rise in
energy prices during the 1970s has adversely affected the
economies of almost all countries, the plight of the resource-
poor developing countries is particularly appalling. These
countries, being at the lower rungs of the economic development
ladder, have correspondingly low levels of energy consumption
but would need large inputs of energy in the coming years
in order to support their much needed industrial and other
socio-economic development activities and for satisfying the
growing aspirations of their rapidly increasing population.

---

* The developed countries in the context of this paper are
the market economy industrialised countries and the nonmarket
economy countries of Eastern Europe. The developing countries
comprise Low-income countries, Middle-income countires and High-
income oil exporters; Low-income countries are those with GNP/Cap
in 1981 less than U.S. $ 410, and High-income oil exporters
consist of Bahrain, Brunei, Kuwait, Libya, Oman, Qatar, Saudi
Arabia and United Arab Emirates. Taiwan and South Africa are
not covered in any group.

Most of them are acutely short of indigenous energy resources, lack investment capital as well as technological know-how for expanding domestic energy resources base and have become heavily dependent on imported energy. Due to a ten fold increase in the price of imported oil over the last one decade, the cost of energy imports is now putting a crippling burden on their economies; the energy import bill in several of these countries now exceeds 50%, and in some cases even 100%, of their total earnings from merchandise exports. As a result, a number of developing countries have, in order to finance their developmental activities, been forced to borrow quite heavily from international sources, generally at high interest rates. With their 1982 debt standing at 550 billion dollars (equivalent to 23% of their annual GNP), the low- and middle-income developing countries are spending about 5% of their combined GNP or 21% of their export earnings on debt servicing alone. These financial difficulties have seriously affected their economic growth, which now barely exceeds their annual population growth rate.

There are glaring socio-economic disparities between the developed and the developing countries which, if left unattended, may have serious implications for future global security and world peace. In order to narrow the gap between the "haves" and the "have-nots" it will be necessary to increase significantly the pace of industrial and economic development in the developing countries, which in turn will

call for much expanded reliable supplies of commercial energy
to these countries at reasonable costs. Difficult though it
may appear, it is not an impossible task. The world has
sufficiently large stock of conventional and unconventional
energy resources which, if used judiciously in combination
with new energy technologies, may adequately meet the future
energy requirements of both the developed and developing
countries in a sustainable way. What is realy needed is an
environment of goodwill and co-operation between the developed
and developing countries leading to a mutual sharing of
both resources and technologies in an equitable manner with
the welfare of mankind as a whole being the foremost objective.

This paper reviews the current energy situation in the
developing countries with respect to both the level of
consumption and resource availability, and compares it with
that in the developed countries. It then looks into the cost
of energy imports in the resource-deficient developing
countries. After giving a brief resume on the expected growth
of energy demand in the developing countries, the paper
analyses the prospects of future energy supply in these
countries based on different energy options. It presents an
overview of the financial difficulties of the low- and middle-
income developing countries and recommends possible measures
of enhancing international co-operation in energy investments
so as to provide enhnaced energy security for both the
developed and developing countries.

2. ENERGY SITUATION OF DEVELOPING COUNTRIES

2.1. Consumption

The level of commercial energy consumption in the
developing countries is very small compared to that in the
developed countries. A large fraction (75%) of their population
still lives in rural areas and relies heavily on use of
noncommercial fuels (firewood, bagasse, dung etc.) for household
needs and draft power for farming and irrigation activities.
Inspite of massive effort on rural electrification and provision
of electricity to urban poor less than 20% of the population
actually has electricity connections in their households. In
most of the developing countries, industrial activity is
extremely rudimentary and public transport system abominably
weak while private cars are still a luxury afforded by less than
1% of the population. Table 1 compares the levels of per capita
consumption of total commercial energy and electricity in
different groups of countries. Also listed in the Table are the
corresponding figures for population and per capita GNP. The
large imbalance in the global energy consumption pattern is
clearly visible. The average annual per capita consumption of
commercial energy and electricity in the developing countries
is only about one quarter of the world average, one-tenth of the
average for the developed countries and one-twentieth of that of
North America. The average figures for low-income developing
countries, which constitute 2/3rd of the developing country
population, are only 1/3rd of those for the developing countries
as a whole. Thus, as seen in Table 2, the developing countries
constituting about three quarters of the world population are

responsible for only 20 percent of world commercial energy
consumption, roughly in proportion to their share in gross
world product. Judging from the past experience of the developed
countries as well as that of the developing countries themselves
in the recent past, the Third World countries will have to
substantially increase their level of energy consumption if
they are to achieve a reasonable economic growth in step with
their urgent development needs. Still, as indicated by various
international studies [1-3],the large disparity in per capita
consumption of energy in the developed versus the developing
countries is likely to continue well into the next century.

2.2.   Resources

There is a widespread misconception that the developing
countries are richly endowed with conventional energy resources.
The fact is that the energy resource base in most of the developing countries is extremely limited. This becomes quite evident
from Table 3, which summarises the present status of proven
fossil fuel and uranium reserves and hydro potential in various
groups of countries. The developing countries, which represent
about three-fourths of the world population, account for only
36% of the proven global reserves of fossil fuels, 25% of those
of uranium and 65% of the technically exploitable hydro potential.
If such known occurrences of coal, heavy oil and oil shales, that
at some time in the future may acquire an economic value, were
also to be taken into account, the known fossil fuel resource
potential in the developed countries would amount to 10,000
billion tons of coal equivalent (TCE) as against 2500 billion
TCE in the developing countries [4] . The energy resource
situation of both the oil importing and the low- and middle-income

oil exporting developing countries is particularly precarious.
These two groups of countries with 41% and 34% share in global
population respectively hold only 6% and 22% of the proven world
reserves of fossil fuels. The per capita availability of fossil
fuel reserves in the same two groups of countries is only 1/16th
and 1/4th of that in the developed countries. Contrary to the
general belief, even the fossil fuel reserves of high-income oil
exporting countries are also limited amounting to only 8% of the
world total reserves. The situation of the developing countries
with respect to per capita availability of uranium reserves and
hydro potential is likewise much worse than that of the developed
countries.

It may be pointed out that a number of oil exporting
developing countries are now depleting their oil reserves at a
very high rate. If such high rates of production are continued
some of them (e.g. Bahrain, Ecuador, Egypt, Gabon, Indonesia,
Peru, Qatar) will nearly exhaust their known oil reserves within
a matter of next 20 years, with some of them even becoming net
oil importers.

2.3.   Imports
It is worth noting that the developing countries as a
whole are net exporters of energy, with their annual production
of 3600 million TCE in 1980 (oil : 2560 million TCE, natural
gas : 210 million TCE, coal : 650 million TCE, primary electricity : 180 million TCE) being almost twice as large as their
own domestic consumption for the same year (oil : 900 million
TCE, natural gas : 175 million TCE, coal : 665 million TCE,
primary electricity : 180 million TCE). About half of total

fossil fuels, in particular two-third of oil, produced in these
countries during 1980 was exported to developed countries
meeting some 25% of their total fossil fuel requirments and
75% of those of liquid fuel in the market economy developed
countries. However, only 28 out of the 130 odd developing
countries are net exporters of energy (26 being net exporters of
oil), while the rest meet their requirements to a varying extent
through import of fossil fuels, mainly in the form of crude oil
and refined oil products. In fact imported oil is responsible
for meeting over three-fourth of commercial energy demand in
two-third (as many as 68) of the Oil Importing Developing
Countries (OIDCs). In 1980 the net oil imports of OIDCs were
360 million TCE equivalent to about 80% of their oil demand or
45% of their commercial energy consumption. Imported coal
accounted for an additional 2% of the 1980 commercial energy
consumption in OIDCs.

3. PROSPECTS FOR ENERGY DEMAND AND SUPPLY

3.1. Growth of Demand

The developing countries, in general, are now passing
through an energy intensive phase of socio-economic development,
a phase in which they have to mechanise agriculture, establish
industrial infrastructure, augment transportation and other
service facilities, and provide electricity and other environmentally cogent fuels at adequate levels to their rapidly growing
and increasingly urbanising population. The present day developed
countries passed through this phase at a time when energy was
cheap and abundantly available with prices of fossil fuels
falling in real terms. But now that the developing countries are

striving to climb on the development ladder, energy is becoming
a scarce and increasingly expensive commodity. Yet energy being
an essential input for practically all socio-economic development
activities, its demand in the developing countries will continue
to grow quite significantly in the foreseeable future, inspite
of price-induced conservation measures, and restructuring of the
economy and lifestyles, if a reasonable pace of economic development is to be maintained for countries of the Third World.

Until early 1970 the demand of commercial energy in the
developing countries was growing at about 7% per annum as
compared to 5% in the developed countries. Following the oil
price hikes of 1973/74 the growth rates were slowed down in
all but a few oil-rich countries due to the ensuing economic
recession and conservation measures adopted by various countries.
The average annual growth rates of commercial energy consumption
during 1970-80 were: Developed countries: 2.4%; Developing
countries: 6.1%; World: 3.0%. The overall income elasticity
of energy demand during this period was 0.73 for the developed
countries and 1.13 for the developing countries.

The future requirements of energy in both the developed
and developing countries would depend upon their economic
growth as well as the effectiveness of conservation measures
that are adopted in the wake of rising costs of energy.
Estimates made by international agencies [1-3] indicate that
the average income elasticity of commercial energy demand in
the next two decades or so will be in the range 0.5-0.7 for
the developed countries and 0.9-1.1 for the developing countries,
with prices of oil increasing in real terms at 0-1.5% per annum.
Taking median values of these elasticity coefficients and

assuming, in general agreement with the above studies, average
annual GNP growth rates of 3.5% and 5% per annum, respectively,
for developed and developing countries over the period 1980-
2000, the demand of commercial energy will increase by the year
2000 to:

Developed Countries	:	11.0 billion TCE
Developing Countries	:	5.1 billion TCE
World		16.1 billion TCE

The share of developing countries in global demand will thus
increase from 21% in 1980 to 32% in the year 2000. About 45%
of the global increase in commercial energy consumption by the
year 2000 will thus take place in the developing countries.
The developing country demand in 1980 was distributed among
different groups as:

Low-income oil importers :	10%
Low-income oil exporter (China, only) :	33%
Middle-income oil importers :	33%
Middle-income oil exporters :	20%
High-income oil exporters :	4%

Only minor changes in this distribution are to be expected
over the next 20 year period.

Two other aspects deserve a mention here. First, in the
light of the past experience all over the world, the demand
of electricity as end-use energy is expected to grow in both
the developed and developing countries, much faster than
that of fossil fuels used as such. Assuming then, in line with
recent IIASA and World Bank studies [1, 3], the income elasticity

of electricity demand to be about one and a half times as high
as the elasticity for overall commercial energy demand, the
share of commercial energy demand originating from the requirements of electricity generation will increase respectively from
26% and 39% in the developed and developing countries in 1980
to roughly 40% and 48% by the year 2000. Thus electricity
generation, already a major consumer of commercial energy, will
become increasingly important in the coming years. Second,
whereas the demand of fossil fuels used in thermal electricity
generation and in supplying process heat to households, industry
etc., is intersubstitutable among oil, gas and coal, a certain
fraction of the fossil fuels demand also corresponds to specific
fuels e.g. liquid fuel for transportation, agricultural tractors,
and feedstocks used in petrochemical industries; coal as coke
for steel industry. It is estimated [1] that such specific uses
accounted for only half of the global oil consumption in 1975,
with the share of oil in global commercial energy consumption
in that year being about 47%. Thus, in view of the growing
scarcity of oil relative to that of other fossil fuels, it
should be possible for both the developed and developing
countries to reduce the share of oil in their commercial energy
consumption by substituting alternate fossil fuels for oil
wherever practically feasible.

The total amount of electricity generated in these
countries in 1980 was about 1200 billion KWH supplied by
306,000 MW installed power generation capacity comprising 32%
based on oil, 8% on gas, 21% on coal, 38% on hydro, 0.6% on
nuclear and 0.2% on geothermal energy. The distribution of
installed power generation capacity in terms of energy sources

among different groups of developing countries is shown in
Table 4. Also shown in the Table is the fraction of total
technically exploitable hydroelectric potential in each group
that had been developed by 1980.

As discussed earlier, the income elasticity of
electricity demand in the developing countries over the next
two decades or so is expected to be around 1.5. Then, for a
medium pace of 5 percent per annum average overall economic
growth in these countries, their requirements of electricity
generation capacity will increase by factors of 2.1 and 4.3
during the 1980s and 90s respectively, necessitating total
installed power generation capacity of about 650×10^3 MW in
1990 and 1300×10^3 MW in the year 2000.

3.2 Supply Options

The main commercial energy supply options for the developing countries during the next two decades will continue to be
fossil fuels for transportation and process heat, and these
fuels together with hydro and nuclear power for electricity
generation. Organised supply of renewables other than centralised
hydropower generation (e.g. biogas, charcoal from plantations
and harvesting of natural forests, alcohol production from
biomass, solar energy, geo-thermal energy etc.) will be able
to play only a minor role in the foreseeable future. The
prospects of energy supply in the developing countries from
these various options are briefly discussed below.

3.2.1. Renewables

The only significant applications, so far, of renewable sources of energy in the developing countries have been centralized hydropower generation and use of noncommercial fuels derived from disorganized cutting of forests and from agricultural and animal wastes. Recently some countries (most notably China and India) have started promoting the use of biogas plants in rural areas, while Brazil has embarked on a program of production of alcohol from sugarcane for use as fuel. Other applications such as those of windmills, small hydropower units, soft solar devices, and plantation schemes, are lagging further behing and are still in the exploratory stages.

In this section we identify the soft/decentralized (S/D) technologies based on renewables, which appear promising for wide scale application in developing countries, and show what role may be realistically expected of them in meeting the future energy requirements of developing countries in case an all out effort is made to support their use over the next several decades. In the context of this discussion the term "soft" refers to simple technologies such as harvesting of wood from forests and plantations and use of small scale hydropower, whereas the term"decentralized" implies localised systems e.g. windmills, small hydropower units, biogas plants etc. Large windmills are not considered as soft technologies but are included as decentralized technologies, whereas wood harvesting may involve a very large organised effort but is still taken here as a soft technology. The technologies considered here are not necessarily "soft" and "decentralized" at the same time but they belong to at least one of the categories.

Since the various soft/decentralized technologies have not been commercialized it is not possible to make firm estimates of their investment requirements or fuel production costs. In addition both the investment requirements per unit of installed capacity and the total cost per unit of energy produced will vary consider- ably for each technology, depending on the geography, environmental considerations and indigenous industrial capability. Nevertheless some rough estimates are necessary in order to identify the technologies that hold promise of large-scale utilization in areas where resource conditions are favourable.

Table 5 presents some estimates (in 1983 US dollars) of the capital costs (per unit capacity) and the average energy production costs of S/D renewables for electricity generation. The cost of electricity production has been calculated by assuming a 10% fixed annual capital charge, and neglecting operating and maintenance costs and by using the appropriate duty cycle. For comparison the specific capital costs and electricity production costs of coal-and oil-fied plants are also listed. The cost of generated electricity in these cases includes the fuel cost.

As centralized systems of electricity supply also entail large investments in transmission networks and have associated maintenance and distribution expenses, the actual energy supply costs from centralized systems would be some 50-100% higher than those in Table 5. Thus under favourable conditions the supply from individual windmills and small hydropower units may be more economical. However, windmills and small hydropower units are irregular souces of energy. This shortcoming makes them unsuitable for supplying regulated power to large cities, major industries,

and electrified transportation systems. Nevertheless, this unsteady
nature would not pose much of a problem in meeting irrigation water-
pumping requirements. Similarly, villages and small towns might
tolerate to a considerable extent an irregular electricity supply
and could meet part of their requirements from diesel-operated
systems or central grids. However, use of these S/D renewables
would call for considerable investments, which may be difficult for
individulas or small groups to afford without government finances.

Table 6 lists the capital costs (per unit thermal capacity)
and the average energy production costs of some solar devices,
biogas plants and alcohol production plants. The energy costs in
column 4 (expressed per unit of useful energy for solar devices
and of energy content of fuel for other plants) are, again, based
simply on a 10% fixed annual capital charge and the assumed duty
cycle in column 3. For comparison the energy costs of coal and of
oil are also shown. The actual price paid by the user for
transportable fuel (fuelwood, charcoal, alcohol, coal and oil)
will, in general, be much higher when taxes, profits, and
transportation are taken into account.

To the extent that one can rely on the estimates in
Table 6, renewable energy supply schemes based on biogas plants,
energy plantations, and harvesting of natural forests appear
attractive when their energy production costs are compared with
the prices of coal and oil. Alcohol production may become more
economical in coming years as coal and oil prices rise. The
energy costs for solar devices are relatively high but may fall
as a result of the current R&D effort and the possible introduction
of mass production.

Biogas generation offers a very efficient and convenient form of fuel but its application is limited to rural areas. The supply of wood/charcoal from natural forests and energy plantations and production of alcohol from crops, in different developing countries will depend on the relative availability of forests, marginal land and surplus arable land respectively. Although wood can be used directly as fuel it is not as convenient to handle and transport as charcoal and burns with a lower efficiency (for supplying useful energy). It is therefore expected that most of the available wood from forests and plantations will be converted to charcoal for use in industry and urban households.

In view of the above, only the following four soft/decentralized technologies based on renewables are considered as promising for wide application to developing world regions over the next 50 years.

(i)   Windmills, small hydropower units: for irrigation water pumping and supplying electricity to villages and small towns.

(ii) Charcoal: for industry, households, and the service sector, mainly as a source of thermal energy.

(iii) Biogas: for rural areas of Africa and Asia, where the handling of animal wastes is traditionally and culturally acceptable.

(iv) Solar heat: mainly for supplying hot water/steam to industries and hot water to households and services; of limited use for space heating (in rich households and the service sector where the availability of capital is not a problem).

Many factors will determine the extent to which soft/decentralized technologies based on renewable energy sources may be invoked to meet the future energy demand. The most important are listed below:

the cost economics of S/D renewables as compared with those of conventional forms of energy;

- the magnitude of the domestic resources of conventional fuels;
- the production potential of renewables close to demand centres;
- the convenience of use and social preferences;
- the access of different sections of the population to central power grids;
- the investment potential of individuals and small groups for financing independent installations;
- the government loan and investment policies for funding decentralized supply sources in preference to centralized facilities; and
- the problems of institutional changes and management.

All these factors will vary from region to region and much more so from country to country. A detailed analysis that acknowledges so many factors with their inherent uncertainties may prove to be a formidable task. Therefore, some simplifying assumptions need to be made in order to estimate the possible share of S/D renewables in meeting the future energy demands of the developing countries. Khan [11] has made one such estimate by considering the demand for various sectoral activities separately for "cities" (large urban agglomerations each with 100,000 or more inhabitants), towns (urban agglomerations having upto 100,000 inhabitants) and "villages" (covering all rural agglomerations) and assuming the likely extent of S/D renewables penetration, under favourable conditions, in the supply of electric and nonelectric commercial energy for broad sectoral activities at City, Town and Village levels. His major assumptions about the penetration of S/D renewables in electricity supply and in supply of nonelectric energy by the year 2000 and 2030 are listed in Tables 7 and 8. The analysis shows that by the year 2030 these renewables might be able to meet about 7% of the electricity demand and 17% of

non-electric commercial energy demand in the developing
countries as a whole.   The corresponding figures for
the year 2000 are only 1.5% and 11% respectively.

3.2.2   Fossil Fuels

The energy resources of fossil fuels are very unevenly
distributed among the developing countries, as may be seen
from Table 3. The high-income oil exporters (8 countries) are
richly endowed with petroleum resources and are not likely to
have any problems in meeting their domestic requirements of
energy in the coming decades. In fact most of these countries
will in all probability, continue to export for the next
several decades their surplus production of oil, and possibly
natural gas (through pipelines and as LNG) to meet the energy
déficits in other developing countries and the industrialized
world. It is, however, in their interest to increase the share
of natural gas in their domestic energy consumption, particularly by utilising gas which is still being flared due to
lack of collection and distribution facilties. The use of gas
in place of oil, wherever possible, will release some additional
supplies of oil which are likely to fetch higher revenues
through exports than the export of equivalent quantities of
natural gas (whether as gas or liquid).

The middle-income oil exporting developing countries
(17 in all) were able to export more than three quarters of
their oil production in 1980 and, as a group, are likely to

continue their oil exporter status over the next two or three
decades. However, 11 of these countries have at present oil
reserves-to-production ratio [3] of less than 30:1, with 4 of
them having this ratio even less than 15:1. It is mainly
because of a few countries (e.g. Iran, Iraq, Mexico) with large
reserves and relatively low depletion rates that the overall
oil reserves-to-production ratio for these countries was 43:1
in 1980. Unless sufficient effort is made on exploration of
new oil in these countries, their oil export potential may
become stagnant or even decline in the coming decades.
Countries such as Egypt, Gabon, Indonesia, Peru and Trinidad
and Tobago are particularly vulnerable to becoming oil importers
for want of adequate effort on exploration and development of
new oil fields and secondary oil recovery in the near future.
The natural gas potential in the middle-income oil exporters
is still very much underutilised. Their present reserves-to-
production ratio for gas is close to 300:1 while the share of
natural gas in their commercial energy consumption in 1980 was
only 22%. These countries need to invest more in the development
of natural gas both for domestic consumption and export purposes.
None of the middle-income oil exporting countries has significant reserves of coal; their total coal reserves are less
than 4 billion TCE with about one third of these in Indonesia.
Indonesia, which is particularly experiencing rapid depletion
of its oil reserves and does not have large gas reserves
(total gas reserves: 800 million TCE), may find it useful to
develop its coal fields for domestic market.

China is the only oil exporting country among the low-
income countries. Its present oil production exceeds the

country's domestic consumption by only 20%. With present oil
reserves-to-production ratio of about 25:1 it can hardly be
considered as a potential exporter of oil in the coming decades.
Its natural gas reserves are inadequate to allow significant
increase in the share of gas in domestic energy consumption,
which in 1980 stood at about 3%. The bulk of commercial energy
consumption (about 72% in 1980) in China is based on coal. The
country is one of the three giant holders of global coal
reserves. Its proven coal reserves are about 100 billion TCE
and the present production (all meant for domestic consumption)
stands at a level of 450 million TCE. China would need to
further develop its coal resources both for domestic consumption and export.

Among the oil importing developing countries, (more than
100 in number) several are believed to have reasonable prospects for economic petroleum reserves. However, the exploration
activity in these countries has been extremely low due to
shortage of capital and necessary technical know-how as well as
lack of interest by International Oil Companies (IOCs). The
total number of exploratory wells drilled per year in these
countries has stayed at a level of 3-4% of the world total [3]
throughout 1970s. Whatever little activity there exists is
also concentrated in a few oil producing countries, most notably
Argentina, Brazil and India. With their present oil reserves-
to-production ratio being close to 15:1, production can not be
increased significantly in the OIDCs unless sizable new
reserves are discovered. Their natural gas reserves are,
however, still underutilised due to lack of both capital and
interest by IOCs. The share of natural gas in overall commercial energy consumption of these countries is still less than

4% while the gas reserves-to-production ratio is close to 80:1.

Large coal reserves (about 34 billion TCE) exist in India
with sizable deposits (3-8 billion TCE) also being in Brazil,
Colombia and Yugoslavia. Coal had a significant share in commercial energy consumption in 1980 in India (55%), Republic of
Korea (31%), Yugosalvia (38%) and Turkey (21%), with its share
in the overall commercial energy consumption of the rest of
OIDCs being only 14%. The OIDCs with large coal reserves are,
in general, finding it difficult to enhance production due to
capital shortage, inadequacy of transportation facilities
and lack of interest by users due to inconvenience of its use
and insufficinet economic margin over alternative fuels.

There is a bleak future for those OIDCs with little or
uneconomic domestic resources of coal and natural gas. They
will not only have to continue importing oil for transportation
and similar other specific uses but will also have to import
both oil and coal for meeting the bulk of their requirements of
process heat. At present their import of coal is only 1/20th
that of oil and more than 80% of it is due to just three
countries: Brazil, Republic of Korea and Yugosalvia. Substituting
imported coal for imported oil will not be an easy transition
for OIDCs; it will call for heavy investments in establishing/
extending the necessary infrastructure (port handling, storage,
internal transportation etc). These countries therefore need,
more than any one else, to develop non-fossil energy supply
alternatives in parallel with intensification of their effort
on fossil fuel exploration and development.

3.2.3. Hydro

Hydro is a particularly attractive source for power
generation since its fuel cost is zero and operation and

maintenance cost is minimal.   However, this advantage is
mitigated in terms of overall cost of electricity generation
if the capital cost gets too high. At present international
prices of coal and oil, hydropower's economic limit of capital
cost is roughly $ 2000 to $ 3000 per KW of installed capacity
[3] . As shown in Table 4, so far only about 8% of the technically exploitable hydropower potential in the developing
countries has been put to use. There are considerable difficulties
both in further expansion of hydropower and in continuing
heavy reliance on this source. First, the most attractive
sites have already been developed and the cost of construction
of new dams is increasing with increasing complexity of dams at
less favourable sites. Second, in most cases new sites are far
away from demand centers, thereby necessitating huge additional
investment in transmission lines and, still having their
generated electricity subject to substantial transmission losses.
Third, the water levels in storage dams undergo large seasonal
variation which markedly reduces the generation capability in
low reservoir level months. And fourth, large dams in the
developing countries are often multipurpose (power generation,
supply of irrigation water, flood control) -- that is how the
heavy investment gets economically justified. In general, the
water release pattern from such dams in low reservoir level
periods is dictated by the irrigation requirements rather than
by the needs of power generation. These last two factors, for
example, in the case of Pakistan are responsible for reducing
the generation capability of the hydro system in April to less
than half of that in August.

     In view of the above factors, rapid development of the
technically exploitable hydropower potential in the developing

countries can not be expected in the coming decades. The
World Bank has recently estimated [3] that by 1995 the installed
hydropower capacity in these countries will be less than 15%
of the technical potential. Thus hydro may be expected at best
to contribute about 300×10^3 MW generation capacity by the year
2000 to the electric supply system of the developing countries,
thereby meeting some 23% of their envisaged requirements.

3.2.4. Nuclear

Nuclear power is now a technologically well established
and economically proven option which holds considerable promise
for providing electricity in an economic way, independent of
fossil fuel prices, to both developed and developing countries.
Already by the end of 1982 there were 297 nuclear power reactors
with a total capacity of 173,000 MW, operating in 25 countries
and meeting about 11% of the global requirements of electricity
[6]. The share of nuclear in electricity generation in seven
developed countries was more than 25% (e.g. Finland: 42.4%,
France 38.7%, Sweden: 38.6%) and in some it is expected to
exceed 50% by 1990 (e.g. France: 70%, Belgium: more than 50%).
The nuclear power now has more than 3000 reactor-years of
operating experience with a safety record better than that of
any other large industry. Inspite of some set-backs in a few
large countries, due mainly to economic and political reasons,
nuclear power is expected to continue growing and would be able
to supply [7] about 18% of global electricity in 1990 (22-23%
in the developed countries) and 21-26% in the year 2000 (28-35%
in the developed countries). In addition to the 25 countries with
operating power reactors in 1982 there were 12 others with
power reactors under construction or in advanced planning stages [8].

Despite the increasing role of nuclear power world-wide, and particularly in energy resource-poor developed countries, the developing countries have been rather slow in making use of this technology. By the end of 1982 only 6 developing countries had operating power reactors with total capacity of about 3700 MW and contributing 1.6% to electricity generation of developing countries. On the basis of reactors under construction or in advanced planning stages it is expected that 3 additional developing countries will have operating power reactors by 1990 and perhaps 7 more in the years shortly thereafter. Table 9 lists the developing countries with operating, under construction and planned power reactors (as of December, 1982), the number of reactors and their capacity. Based on the recent IAEA projections it is to be expected that the nuclear generation capacity in the developing countries will increase by the year 2000 to 30,000 - 75,000 MW and contribute 5-8% to their total electricity generation in that year [7].

Various international studies have shown that nuclear power has economic advantage over power generation based on coal and oil fired plants except perhaps in countries having large reserves of good quality coal but with low production and transportation costs. Table 10 based on a recent OECD study [2] compares the overall cost economics of power generation based on nuclear and coal/oil fired plants in the U.S., Western Europe and Japan. It is seen that, inspite of the per KW capital cost of nuclear plant being 1.5 and 2 times as high as those of coal and oil fired plants, respectively, nuclear generated electricity is considerably cheaper in all cases except the U.S., where coal-fired power generation has marginal economic advantage

over nuclear. A World Bank study [4] also concluded that both
large and medium sized nuclear plants would have significantly
lower electricity generation costs than oil fired plants of
comparable size in the OIDCs, but the margin would be relatively
small compared to coal fired plants. A similar conclusion has
been reached in a recent analysis by IAEA for plants of 600 MW or more
[6] . As seen in Table 10, the fuel cost of nuclear plant is a
small component (25%) of the generation cost in the case of
nuclear plant but accounts for 40-80% of generation cost in the
case of fossil fuel fired plants. As a result the higher capital
investment in the nuclear plant will be offset by its lower
fuelling cost in the first few years of plant operation and its
cummulative life-time discounted cost will turn out to be much
lower than that of an equivalent fossil fuel fired plant.

At present indigenous nuclear power plant construction
capability exists only in one developing country: India, which
is building plants of about 220 MW capacity, but plans to build
larger plants of 500 MW capacity in the near future. Other
developing countries have been importing nuclear plants from
the developed countries which now are being supplied in unit
capacities of 440 MW (USSR design) and 600-1400 MW (designed by
USSR, Western Eruope and North American countries). Small units
in the range of 100-200 MW (e.g. KANUPP in Pakistan and TARAPUR
in India) were supplied by the developed countries in 1960s and
early 70s but those designs are not commercially available any
more. In fact the current nuclear power industry in the developed
countries is geared to construction of only large sized plants
which fit well into their own comparatively much larger grids.

Although many developing countries are interested in
making use of nuclear power to diversify their power generating
system and reduce fossil fuel consumption, they can not plan
for it simply because the cost economics of small sized nuclear
power plants, now being offered by some manufacturers, is extremely unfavourable and the larger more economic plants would not
fit into their grids in the next decade or so. If only smaller
nuclear plants, 100-300 MW, may be made economically competitive
and offered to developing countries, a very large number of them
could benefit from this relatively abundant source of energy.
On the basis of the criterion that the capacity of an individual
unit should not be more than 10% of the total installed capacity
in the national grid, by the year 2000 as many as 71 developing
countries would be able to accomodate in their national grids
(assuming that the grid is integrated in each country) nuclear
units of 100 MW or more, 46 countries: 300 MW or more, 33
countries: 600 MW or more and 26 countries: 900 MW or more.
Work is reported to be in progress in several developed countries
(U.K., France, Japan, F.R.G., Sweden, Canada, U.S.S.R. and the
U.S.) on the development of small and medium power reactors
(SMPRs) intended for electricity and/or process heat generation
[9]. The IAEA has also been taking some interest -- although it
does not seem to be adequate -- in promoting development and
information exchange on small plants [6]. There is a need that
concerted effort be made to open up the vistas of nuclear power
to developing countries with relatively small grids by making
small and medium power reactors economically competitive.

From the preceding discussion it follows that, as
seen by energy analysts to-day, hydro and nuclear power are

expected to contribute less than 25% and 10%, respectively, to
the installed electricity generation capacity in the developing
countries by the year 2000. The brunt of the load will, therefore, have to be borne by fossil fuels until after the turn of
the century. In view of the growing scarcity of oil its share
in thermal power generation is expected to decrease in almost
all countries except perhaps those with capacity requirements
of less than a few hundred MW. The likely sources of substitution are: domestic coal in countries with significant coal
reserves, natural gas in countries where gas reserves are still
very much underutilised or gas is being flared for want of
gathering/distribution facilities, and imported coal, mainly
from developed countries, in those developing countries which
are short of all fossil fuels resources. Again, this renewed
dependency on import of yet another fossil fuel will hardly help
solve the financial problems of the resource-deficient OIDCs or
provide them with any degree of energy security. The real
solution for them lies in rapid expansion of their domestic
production capacity of fossil fuels through intensive exploration and development effort, together with use of nuclear power,
hydro and other renewable sources of energy to the maximum
feasible extent. All this calls for large capital investments,
full mobilisation of domestic resources and last, but not
the least, a high degree of international co-operation

4.   ENERGY INVESTMENTS: NEED FOR INTERNATIONAL CO-OPERATION

The development activities in the energy sector are
becoming increasingly more expensive. In the low- and middle-

income developing countries the share of energy investments in GDP has already increased from a level of 1-2% in the early 1970s to 2-3% now, and is likely to increase further to 4-6% over the next two decades [1,3]. According to the World Bank estimates [3] these countries will need to spend on energy investments over the next 10 years an average of $ 130 billion a year (in 1982 prices) of which about half ($ 64 billion a year) will have to be in foreign exchange. Still the energy import dependence of OIDCs would not be eliminated -- with the aforementioned investments, if they were to take place, it is projected to decrease modestly from about 43% in 1980 to 32% in 1995, while the energy export capability of the low- and middle-income oil exporters is projected to decrease from about 48% to 44% of their production over the same period. The particularly difficult situation will be that of OIDCs, which will not only have to find large investments for energy sector development but will also have to **Pay heavily for their energy imports**.

The sharp increase in oil prices during the last one decade has given a serious blow to the economies of OIDCs by causing several fold increase in their energy import bills, leading to large current account deficits (see Figure 1), heavy borrowing from international sources -- frequently at high interest rates -- high debt service ratios and acute shortage of investment capital for financing the much needed socio-economic and domestic energy resource development programmes. A number of these countries, which were spending 5-10% of their export earnings on import of oil during the 1960s and early 70s are now forced to divert more than 50% of such earnings to pay for their oil imports. For some countries e.g. Panama, Turkey, Yemen Arab Republic, the cost of oil imports even exceeds their earnings from merchandise exports. As estimated by the World Bank [5], the net fuel import bill of OIDCs in 1980 was $ 74 billion equivalent to 5.3% of their combined gross domestic product (GDP), compared to 2.8% of GDP in 1978 and less than 1% of GDP in early 1970s. The same study projected that the bill will rise further to 6.2% of GDP by 1990. The net fuel imports of OIDCs in 1980 corresponded to 27% of their total merchandise imports and 36% of their total merchandise exports in that year.

The irony is that most of the OIDCs are exporters of primary commodities and the price index of non-oil commodities has since 1973 been decreasing not only with respect to index of petroleum prices but also relative to that of prices of manufactures imported by the developing countries. It decreased (as estimated by the World Bank [5] for 33 non-petroleum commodities) by 85% during 1972-82 relative to petroleum prices (see also Table 11) and by 43% between 1974 and 1982 relative to prices of manufactures imported by developing countries. The oil importing developing countries are under a double squeeze : they have to pay higher not only for oil imports but also for the manufactured goods imported mainly from the developed countries, while the worth of their own exports is decreasing. Thus during 1970-80 inspite of more than 80% increase in the volume of non-fuel primary exports by the groups of low-income and middle-income OIDCs, each, the net purchasing power of such exports increased merely by about 15% in the case of the former group and hardly 30% in the case of the latter [5] . Due to worldwide slowing down of economic growth since 1980 and consequent reduction in

international trade, the exports of non-fuel primary products
and of manufactures from developing countries have grown (in
volume) hardly at annual rates of 0.6% and 4% respectively
during 1980-82. This has further cut down the purchasing power
of merchandise exports by OIDCs which, more than half, consist
of non-fuel primary commodities in the case of low-income OIDCs
and more than 35% in the case of middle-income oil importers.

The decrease in purchasing power of exports by OIDCs
coupled with their increasing requirements of imports of both
fuel and manufactured goods, necessary for supporting the socio-
economic development activities, have resulted in increasing
their current account deficits in alarming proportions. The
current account deficit of these countries increased from 9
billion US dollars in 1970 to $ 72 billion in 1980 and $ 80
billion in 1981 but is estimated [5] to decrease to $ 69 billion
in 1982, due partly to recent reduction in oil prices and partly
to impact of conservation measures following the second oil
price hike during 1979-80. Since the Official Development
Assistance (ODA) available to OIDCs can hardly cope with a
quarter to one-third of their current account deficit, the
OIDCs have by 1982 accumulated about $ 400 billion of official
and private debts and are paying 23% of their export earnings
towards debt service alone.

The middle-income oil exporters are also faced with
financial difficulties. Their gross domestic product has grown
during the last two years at about 1.7% per annum -- only
slightly better than the 1.1% growth rate for middle-income oil
importers for the corresponding period. In order to finance
their industrial investment programmes leading to higher economic

growth and gradual independence from heavy reliance on income
from their diminishing oil reserves, these countries have also
been borrowing heavily from international sources on the strength
of their present petroleum reserves. Their current account
deficit, outstanding debt and debt service ratio in 1982 were
respectively about $ 50 billion, $ 150 billion and 19% of
export earnings.

Table 12 summarises the overall current account deficit,
outstanding debt, debt servicing, interest payment and ODA
position of the developing countries (excluding the eight
high-income oil exporters) in 1982 and compares it with that
in 1970. It is seen that there is a deterioration all round
except in the case of ODA which has also remained stagnant at
a level of 1% of the GNP of the recipients.

The financial problems of the developing countries
(excluding high-income oil exporters) are due essentially to the
following reasons:

    i)   Decreasing prices of non-petroleum primary commodities
        relative to prices of oil and those of manufactures
        imported by developing countries. This trend has
        continued inspite of efforts by UNCTAD, GATT and the
        Group of 77 to rationalise the prices of non-petroleum
        primary commodities.

   ii)   Undue barriers such as import restrictions, quota
        system, heavy import duties etc., imposed by the
        developed countries on import of manufactured and semi-
        processed goods from the developing countries. Again,
        efforts made by relevant U.N. bodies and developing
        countries themselves to seek removal/relaxation of
        these barriers have not so far found much success.

  iii)   Small size of ODA that is loans and grants made at

concessional financial terms by the Development
Assistance Committee (DAC) of OECD, members of OPEC,
and nonmarket developed countries. Although we are now
approaching the midway mark of the 3rd UN Development
Decade, the ODA contribution of a number of DCA
countries is well below the bench mark (0.7% of GNP
of the donor country) set by the UN for the 2nd UN
Development Decade. In fact the overall average of ODA
from the DAC countries has since 1970 remained close
to half of the benchmark level. The OPEC countries,
as a whole, have been relatively more generous; their
contribution as a fraction of their combined GNP,
although reducing since 1979, has still remained at
or above 1.5% of GNP throughout since 1975. The
contribution of nonmarket developed countries, on the
other hand, has been even smaller than that of DAC
countries.

iv) The very limited availability of loans from official
sources necessitating increased borrowing from private
sources at relatively much higher interest rates. The
share of private loans in outstanding debt of developing countries increased from one-half in 1970 to two-
third in 1980 and consequently the average effective
interest rate on total developing country debt increased from 6.3 percent in 1970 to 8.9 percent in 1980.
The London Inter Bank Offer Rate (LIBOR) for six-month $
deposits, which determines interest payments on the
bulk of private bank loans, averaged 16.6 percent in
1981 and 13.5 percent in 1982 [5].

v) Difficulty in obtaining long-term loans (official as
well as private) necessitating increased short-term
borrowing or running down of reserves. Increasing
dependence on short-term loans makes developing
countries subject to rising interest rates and
vulnerable to sudden withdrawl of support by commercial
banks.

vi) Increasing difficulty of oil importing developing
countries, particularly low-income countries, in

obtaining private loans due to their conceived
doubtful creditworthiness. The low-income countries
have had to rely on ODA for more than 75% of their
external capital throughout 1970s.

All these financial difficulties of the developing
countries have been affecting their energy sector development
programmes in the past and are likely to continue doing so,
perhaps much more severely, in the coming decades. The developing countries, both oil importing and middle-income oil
exporting, have in recent years faced considerable difficulties in financing their energy investments which acquired particular importance in the wake of oil price increases of 1973/74.
The total amount of publicly guaranteed external borrowing (both
official and private) for energy investments in developing
countries in 1982 was, in terms of 1982 US dollars, $ 10 billion
in 1975, $ 16 billion in 1980 and is estimated to be less than
$ 25 billion in 1982 [3] . The share of private loans was about
75% during the period 1975-80. Even in the case of borrowing for
the energy sector, the low-income countries have to be essentially content with multilateral and concessional bilateral loans,
which have have contributed about 80% to their external energy
borrowing between 1975 and 1980 as against 22% and 17% for
middle-income oil importers and middle-income oil exporters,
respectively, and 25% for developing countries as a whole.

The external loans in 1982 (estimated at $ 25 billion),
meant specifically for energy sector development in the develop-

ing countries correspond to about 30% of their total borrowing and 1.0% of their GNP for the year. The external energy sector loans to OIDCs during the last three years were hardly sufficient even to offset 15-20% of their oil import bill. Thus the external flows have so far been extremely inadequate to help the developing countries overcome their investment difficulties in the energy sector.

In view of the fact that future development prospects of OIDCs are heavily dependent on their ability to rapidly increase domestic energy production and cut down import of energy and those of the low- and middle-income energy exporting countries on expanding their energy export revenues while maintaining a high reserves-to-production ratio and a stable price of oil at a level conducive to world economy, it is necessary that:

i) The level of official loans (both on concessional and nonconcessional terms) be increased, in general, and that for the energy sector, in particular. The donor countries now contributing below the 0.7% of GNP level recommended by the UN, should reach this target without delay, while those with higher contributions should maintain their higher level. The IDA commitments to low- and lower middle-income countries have been decreasing, even in terms of current dollars, for the last three years; they decreased from $ 3.8 billion in 1980 to $ 3.5 billion in 1981 and $ 2.7 billion in 1982 [10]. This trend should not only be reversed but also the IDA commitments increased substantially in real terms.

ii) Loans from private banks and financial institutions for energy sector development, should be increased and terms made easier for the developing countries. International financial agencies should encourage such lending to low-income or less creditworthy countries by providing the necessary guarantees. The present practice of co-financing by the World Bank and similar other international organisations should be considerably strengthened, particularly for the energy projects, in order to attract much more external capital.

iii) International Finance Corporation, an affiliate of the World Bank, should substantialy increase its equity participation programme in energy sector investments in the developing countries to attract much more direct private investment by international energy industry, particularly international oil companies. Other international organisations should also likewise launch equity investment programmes in the developing countries.

Unless these steps are taken by the international agencies and the developed/high-income countries, there is little hope that the low- and middle-income developing countries will be able to invest in the development of the energy sector to the required level. Consequently their problems will get much worse and prospects of their economic growth considerably reduced. The

slowed down pace of economic development and increasing
scarcity of energy supply in the developing countries will
not only be harmful to them but will also adversely affect
the developed countries through loss of market for their
manufactures and reduced availability of energy to cover
their own deficits. It is, therefore, in the interest of the
developed countries to help boost economic development and
energy supply in the developing countries through transfer
of advanced technology and technological skills and by
providing necessary funds for critical investments.

REFERENCES

1. International Institute for Applied Systems Analysis,
   "Energy in a Finite World", Report by Energy Systems
   Program Group, W. Häfele, ed. (Cambridge, Mass.:
   Ballinger, 1981).

2. International Energy Agency "World Energy Outlook",
   (Paris: OECD/IEA, 1982).

3. "The Energy Transition in Developing Countries",
   Washington, D.C., World Bank, 1983.

4. "Energy in Developing Countries", Washington, D. C.,
   World Bank, 1980.

5. "World Development Reports 1981, 1982, 1983",
   Washington, D.C., World Bank, 1981,1982,1983.

6. "Status and Trends of Nuclear Power Worldwide",
   Report by the Division of Nuclear Power (Vienna:
   International Atomic Energy Agency, September 1983).

7. "Energy, Electricity and Nuclear Power Estimates for
   the period up to 2000", IAEA Reference Data Series
   No. 1 (Vienna: International Atomic Energy Agency,
   September 1983).

8. "Nuclear Power Reactors in the World", IAEA Reference
   Data Series No. 2 (Vienna: International Atomic Energy
   Agency, September 1983).

9. J. R. Egan, Small Reactors: Facts or Fancy? Conference
   on Nuclear-Electric Power in the Asia-Pacific Region,
   East-West Resource Systems Institute, Honolulu,
   Hawaii, January 1983.

10. The World Bank, "IDA in Retrospect", (Oxford:
    Oxford University Press, 1982).

11. Khan, A. M. (1981) The Possible Share of Soft/Decentralized
    Renewables in Meeting the Future Energy Demands of Developing
    Regions. RR-81-18, Laxenburg, Austria: International Institute
    for Applied Systems Analysis.

12. United Nations (1981) Preparatory Committee for the UN
    Conference on New and Renewable sources of Energy.
    A/CONF.100/PC/41 and 42, New York NY: UN

13. "Yearbook of World Energy Statistics, 1980",
    (New York: United Nations, 1981).

14. "Uranium Resources, Production and Demand", A joint
    Report by OECD Nuclear Energy Agency and the
    International Atomic Energy Agency (Paris: OECD, 1982).

15. South (London), p. 79 (March, 1984).

Table 1

Population, per capita GNP, and per capita consumption
of commercial energy and electricity in 1980.

| | Population (billion) | GNP/Capita ($-1980) | Per Capita Consumption | |
			Commer-cial Energy (TCE)	Elect-ricity (KWH)
Developing Countries	3.27	680	0.58	370
Low-income				
Oil Importers	1.20	237	0.17	130
Oil Exporters	0.98	290	0.64	308
Middle-income				
Oil Importers	0.58	1,544	1.11	908
Oil Exporters	0.50	1,164	0.80	395
High-income Oil Exporters	0.01	12,630	3.45	2,068
Developed Countries	1.09	8,304	6.62	6,268
Market Economy	0.71	10,330	7.08	7,180
(North America)	(0.25)	(11,243)	(11.33)	(10,822)
Nonmarket Economy	0.38	4,478	5.75	4,543
World	4.36	2,594	2.07	1,850

Source: Based on Ref.[3,5,13].

Table 2

Shares of developing and developed countries
in global population, GNP, and commercial
energy consumption in 1980 (in percent)

	Population	Gross World Product	Commercial Energy Consumption
Developing Countries	74.9	19.6	20.9
Low-income			
Oil Importers	27.5	2.5	2.2
Oil Exporters	22.4	2.5	6.8
Middle-income			
Oil Importers	13.2	7.9	6.9
Oil Exporters	11.5	5.1	4.2
High-income Oil Exporters	0.3	1.6	0.8
Developed Countries	25.1	80.4	79.1
Market Economy	16.4	65.4	55.4
Nonmarket Economy	8.7	15.0	23.7
World	100	100	100

Sources: See Table 1.

Table 3

Proven reserves of fossil fuels and uranium,
and technically exploitable hydropower potential.

| | Fossil Fuels | | | Hydro |
	Petroleum $(10^9$ TCE)	Total $(10^9$ TCE)	Uranium $(10^3$ Tons)	Potential $(10^3$ MW)
Developing Countries	148	297	448	1,498
Low-income				
Oil Importers	2	26	218	291
Oil Exporters	5	104	N.A	379
Middle-income				
Oil Importers	3	25	179	590
Oil Exporters	69	73	51	237
High-income Oil Exporters	69	69	-	1
Developed Countries	77	521	1,355	778
Market Economy	28	326	1,355	533
Economy	49	196	N.A	245
World	255	818	1,803	2,276

Sources: Based on Ref. $\left[3,4,13,14\right]$.

Table 4

Installed power generation capacity and fraction of hydro
potential exploited till 1980 in developing countries.

| Country Group | Installed Power Capacity | | | | | Fraction of Hydro Potential Exploited (%) |
	Total 10^3 MW	Thermal (%)	Hydro (%)	Nuclear (%)	G. Thermal (%)	
Low-income						
Oil Importers	46	54	44	2.2	-	7
Oil Exporters	66	69	31	-	-	5
Middle-income						
Oil Importers	130	51	48	0.7	0.4	3
Oil Exporters	54	69	31	-	0.3	26
High-income Oil Exporters	9	100	-	-	-	-
All Developing Countries	305	60	39	0.6	0.2	8

Source: Based on Ref. [3,4,13].

Table 5

Electricity supply from S/D renewables: estimates of capital costs and electricity production costs, and comparison with centralized systems. (1983 US $)

Technology	Capacity	Capital Cost ($/kW(e))	Assumed duty cycle (hr/yr)	Electricity cost ($/kWh)
Windmills	< 1kW(e)	5000-10000	2500	0.20 - 0.40
	5-15kW(e)	1500- 3500	2500	0.06 - 0.14
	3 MW(e)	800	2500	0.03
Mini-Hydro	0.5-10kW(e)	2000-12000	4000	0.05 - 0.30
Photovoltaic devices	< 1kW(e)	20000-30000	2000	1.0 - 1.5
For Comparison				
Large Scale Hydro	250 MW(e)	1500- 3000	4000	0.04 - 0.08
Coal-Fired (coal @ $50/Ton)	300 MW(e)	900	4000	0.04
Oil-Fired (oil @ $30/bbl)	300 MW(e)	700	4000	0.07

Sources: Based on references [4,11,12]

Table 6

Thermal energy supply from S/D renewables: estimates of capital costs and energy production costs (1983 US$)

Technology	Capital cost ($/kW)	Assumed duty cycle (hr/yr)	Energy cost (10^{-3}/kWh)
Solar water heating	500-1000	2000	25 - 50
Solar space heating	700-1400	500	140 - 280
Solar cooking	300- 600	500	60 - 120
Biogas:			
one-family units	1000	Continuous	12
community plants	500	Continuous	6
Alcohol from sugarcane	1600	4000	40
Fluewood production:			
natural forests	-	-	1
energy plantations	-	-	1 - 2
Charcoal production:			
natural forests	-	-	3
energy plantations	-	-	4 - 6
For Comparison			
Coal @ $50/Ton	-	-	5
Oil @ $30/bbl	-	-	17

Sources: Based on references [4,11,12]

Table 7

Potential markets and projected shares
of S/D renewables in electricity supply

Demand Sector	Potential Market	Projected Share (%) in Potential Market	
		2000	2030
Households			
Villages	Half	10	60
Towns	Half	5	30
Cities	-	-	-
Service Sector			
Towns	Half	5	30
Cities	-	-	-
Manufacturing			
Towns	Half	5	30
Cities	-	-	-
Agriculture	Half	15	80
Other Sectors (Transport, Mining, Construction)	-	-	-

Table 8

Projected shares of renewables in
nonelectric energy supply

Sector	Type of Demand	Projected Share (%)	
		2000	2030
Households	cooking, space and water heating		
Villages		75	90
Towns		60	80
Cities		40	60
Service Sector	-do-		
Towns		30	60
Cities		20	40
Manufacturing			
	L.T. steam/hot water	40	80
	H.T. steam	30	60
	furnace	6	12
	coke	30	60
	feedstocks (oil replacement)	-	-
Other Sectors (Transport, Agriculture, Construction, Mining)	mainly liquid fuel demand	-	-

Table 9

Status of nuclear power reactors in
developing countries, as of December 1982.

	Operating Reactors		Reactors under Construction		Planned Reactors	
	Number	Capacity (MW)	Number	Capacity (MW)	Number	Capacity (MW)
1. Argentina	1	335	2	1,292	3	1,800
2. Brazil	1	626	2	2,490	6	7,470
3. China	-	-	-	-	?	?
4. Cuba	-	-	1	408	1	408
5. Egypt	-	-	-	-	8	8,400
6. India	4	809	4	880	6	1,320
7. Iraq	-	-	-	-	1	600
8. Israel	-	-	-	-	1	900
9. Korea, Rep.	2	1,193	7	6,227	4	3,456
10. Libya	-	-	-	-	2	816
11. Mexico	-	-	2	1,308	-	-
12. Pakistan	1	125	-	-	1	937
13. Philippines	-	-	1	620	-	-
14. Thailand	-	-	-	-	1	900
15. Turkey	-	-	-	-	1	1,000
16. Yugoslavia	1	632	-	-	1	1,000
Total	10	3,720	19	13,225	36	29,007

Source: Based on Ref. [6,8].

Table 10

Indicative cost estimates for electricity generation
by fuel (in 1981 US $, capacity factor : 65%).

| | Oil 2x600 MW | | Nuclear | Coal with FGD* 2x600 MW | | |
	Low sulphur	High sulphur with FGD*	PWR 2 x 1100 MW	US	Europe	Japan
Capital Investment ($/KW)	577	692	1331	920	920	956
Fuel Cost ($/TCE)	$ 33/bbl (157)	$ 27/bbl (129)	(27)	$ 40/t (40)	$ 65/t (67)	$ 65/t (67)
Generation Cost (Mills/KWH)	67.9	64.7	39.0	38.2	48.2	48.9
Capital Cost	15.9%	19.9%	63.8%	44.7%	35.5%	36.4%
Operating Cost	3.7%	6.5%	10.7%	13.4%	10.6%	12.4%
Fuel Cost	80.4%	73.6%	25.5%	41.9%	53.9%	53.2%

* Flue Gas Desulphurization.

Source: Based on Ref. [2].

Table 11

Erosion of purchasing power of primary commodities for oil
(1975 - 1984)

Commodity	Tons of Commodity Needed for Purchase of 1000 Barrels of Oil	
	1975	Jan. 1984
Copper	8.7	21.1
Lead	26.0	73.2
Tin	1.6	2.4
Coffee	3.5	9.5
Cotton	8.4	17.4
Suger	23.9	189.4
Maize	94.3	246.3

Source: Based on Ref. [15].

Table 12

Deterioration in financial status of low- and middle-income developing countries between 1970 and 1982.

	1970	1982
Current Account Deficit (U.S. $)	12.0	118.2
O.D.A.* (U.S. $)	4.7	23.9
Debt Outstanding (U.S. $)	69.4	548.0
Interest Payments on Loans (U.S. $)	2.7	49.5
Current Account Deficit as % of GNP	2.3	5.0
Current Account Deficit as % of Exports	17.3	22.0
O.D.A.* as % of GNP	0.9	1.0
Debt Outstanding as % of GNP	13.3	23.2
Debt Service as % of GNP	1.8	4.7
Debt Service as % of Exports	13.5	20.7
Interest Payments as % of GNP	0.5	2.1

* Net Official Development Assistance defined as net
  disbursements of concessional official loans plus
  net official transfers.

Source: Based on Ref [5].

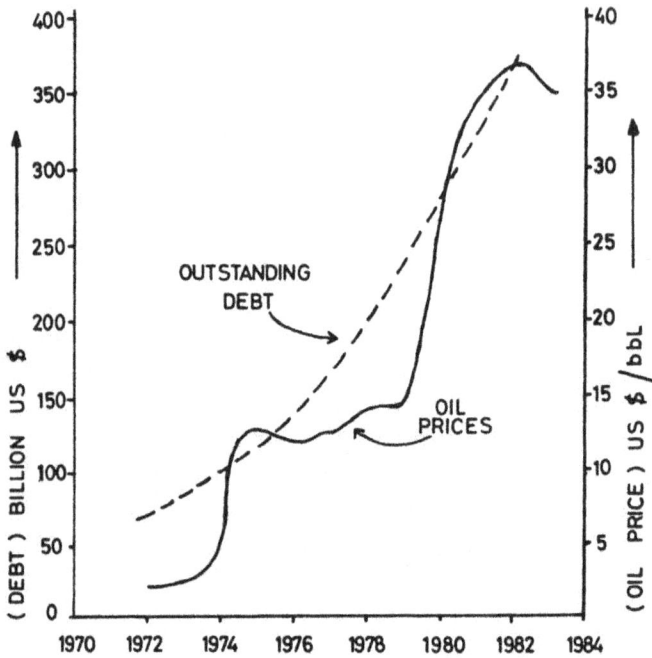

Figure 1.   Variation of (i) outstanding debt of oil
             importing developing countries (OIDCs) and
             (ii) oil prices, between 1972 and 1983

SOLAR MATERIAL AND PHOTOVOLTAIC TECHNOLOGY

RECENT PROGRESS OF AMORPHOUS SILICON CELL TECHNOLOGY

Yoshihiro Hamakawa
Faculty of Engineering Science, Osaka University
Toyonaka, Osaka 560, Japan

INTRODUCTION

Since the first announcement of solar cells using amorphous silicon (a-Si) by Carlson and Wronski[1] in 1976, remarkable progress has been seen in both the device physics and technological development of this new electronic material.[2] Its properties, such as excellent photo-conductivity and high optical-absorption coefficient for visible light, are well matched with the current need for development of a low-cost solar photovoltaic system. The advantages of a-Si alloys as solar-cell materials are:

a) high absorption coefficient and high photoconductivity,

b) low cost,

c) large-area and non-epitaxial growth on any substrate material, and

d) large-scale production and mass reproducibility.

With the aid of national promotion and support for photovoltaic projects due to their potential for renewable energy conversion, great progress has occurred in fields from film-growth technology to cell-fabrication processes for hetero-junction- and multi-junction-structure solar cells which utilize new materials such as a-SiC:H and micro-crystalline Si.[3] As a result, conversion efficiencies of greater than 12% have been obtained for small-area laboratory-phase cells with a-Si/

poly c-Si stacked junction structures.[4] The size of the market for and industrial use of a-Si solar cells are steadily growing.

Significant evidence of this progress is to be found in the fact that there has been a reduction in solar-cell prices of more than one order of magnitude during the last ten years. In 1974, module costs were more than US$80/$W_p$. These have now come down to $5-8/$W_p$.[5] The main reasons for this are the large-scale production of solar cells and a 100-fold increase in annual production in recent years. In the near future, with the advent of technological innovations using polycrystalline, ribbon crystal, amorphous silicon and other thin-film solar cells, examples of which are shown in Figure 1, the cost of solar-cell modules can be expected to drop by a factor of ten. As seen in Figure 2,[6] it is forecasted that, through these technological innovations, the electricity cost from photovoltaics might be competitive with that of present utility-generated electricity by 1990 and that the photovoltaics industry might be as developed as the electronics industry by then.[7]

(a) ACRO-Solar 0.1-m² (1-ft²) glass module.

(b) Kyocera ceramic-substrate module.

(c) ECD ultra-light a-Si solar cell compared with a stainless substrate 0.1-m² (1-ft²) module

(d) Kanaka flexible solar cell.

Fig. 1. a-Si solar-cell modules used in power-system applications.

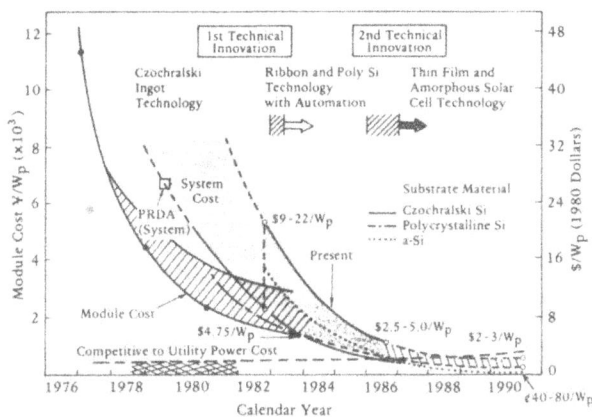

Fig. 2. Changes in cost of innovative solar-cell modules and photovoltaic systems with good potential.

FILM-QUALITY IMPROVEMENTS AND HIGH DEPOSITION RATES

Major efforts to understand the plasma-deposition mechanism and improve film quality have been made using several different approaches such as the cross-field method,[8] application of magnetic fields,[9] and plasma-emission spectroscopy.[10] Although a massive amount of experimental data exists,[11] the plasma-decomposition mechanism is still not completely understood. Recently, systematic investigations of the relationship between solar-cell performance and plasma-decomposition configurations[12] and also plasma frequency[13] were carried out. From an analysis of these investigations, three main processes, that is, plasma decomposition, transport of radicals and compiling decomposed species into an a-Si network to control the quality of the deposited film, were identified.[14]

Subsequently, several new types of plasma-deposition furnace were designed, for example, proximity dc plasma furnaces,[15] multi-chamber horizontal glow furnaces[16] and consecutive separated reaction chambers.[17] Figures 3(a) and (b) show schematic illustrations of two typical mass-production processes for a-Si solar-cell modules.[17-21]

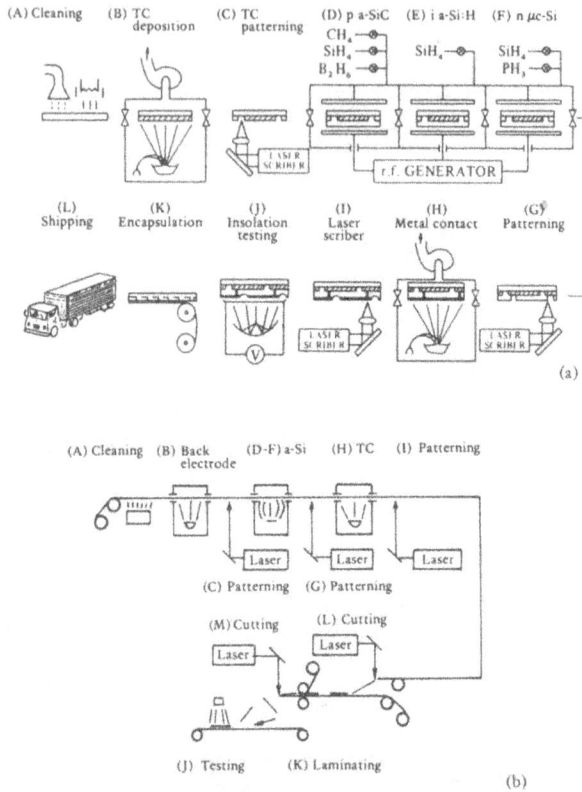

(A) Cleaning (B) TC deposition (C) TC patterning (D) p a-SiC (E) i a-Si:H (F) n μc-Si

(L) Shipping (K) Encapsulation (J) Insolation testing (I) Laser scriber (H) Metal contact (G) Patterning

(a)

(A) Cleaning (B) Back electrode (D-F) a-Si (H) TC (I) Patterning

(C) Patterning (G) Patterning

(M) Cutting (L) Cutting

(J) Testing (K) Laminating

(b)

Fig. 3. Two typical mass-production processes for a-Si solar cells: (a) a budge-type production sequence, and (b) a roll-to-roll mass-production sequence.

The quite low deposition rate $(1-5 \text{ Å/s})$ in plasma chemical-vapour deposition (CVD) is one of the limiting factors of the mass-producibility of a-Si solar cells. In attempts to circumvent this, a wide variety of experimental trials using the following have been made in recent years, including the following:

a) reactive sputtering and post hydrogenation,[22]

b) sputter-assisted plasma CVD,[23]

c) CVD of higher silanes by Chronar, USA,[24]

d) Plasma CVD of Si_2H_6 by the Electrotechnical Laboratory (ETL), Japan,[25-27]

e) photo-CVD of Si_2H_6,[28]

f) plasma-confinement CVD,[29] and

g) electron cyclotron resonance CVD

Recently, the Hitachi-Mitsui Toatsu group reported about 30-Å/s growth rates using a 100% Si_2H_6 source gas. A 7.1% efficiency solar cell was fabricated using this system. A mercury-sensitized photo-CVD reaction chamber was used by Konagai *et al.*[28] to produce a 9.64% efficiency a-SiC/a-Si hetero-junction solar cell. Recently, Hamazaki *et al.*[29] reported deposition rates of greater than 40-Å/s by the plasma-confinement CVD process.

MICRO-FABRICATION TECHNOLOGIES

The laser-scribing and dry-etching processes have become the most important processes for the fabrication of not only a-Si solar cells but also a-Si thin-film transistors and imaging devices. Figure 4 shows a magnified view of the electrode-coupling portion of an integrated a-Si solar cell. Transparent conductor (TC) cutting, a-Si deposition, and the cell isolation and interconnection processes are made in three scribing steps, indicated as steps C, G and I in Figure 3(a), while the back electrode is formed by the evaporation of aluminium, indicated as

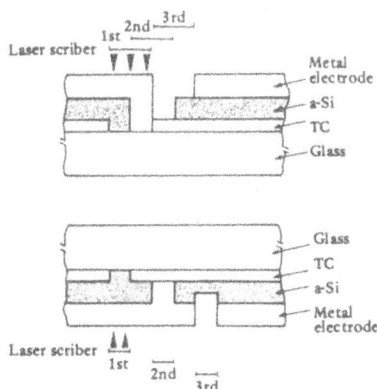

Fig. 4. Cross-sections of the laser-scribing sequence for the integrated a-Si solar cell. Transparent conductor isolation, cell isolation and side-contact interconnection are made in three laser-scribing steps.

step H. The YAG laser is now widely employed for a-Si micro-fabrication
due to the absorption provided by the light of wavelength 0.53 μm, the
second harmonic of the 1.06-μm wavelength light, that it produces.
Figure 5 shows a schematic representation of a YAG laser scriber. a-Si
solar-cell roofing tiles have been fabricated by Sanyo Electric Co.,
Japan, by a combination of laser scribing and ordinal mask processing.

Dry etching as an alternative to conventional wet-chemical methods
has become an essential technique for large-scale integrated-circuit
fabrication. This is because the dry process provides both high-fidelity
pattern transfer from the resist mask to the underlying substrate layer
and better control of the etching process. The use of dry etching is
also increasing in amorphous-semiconductor technology, particularly in
lithographic applications, better known as inorganic resist technology.

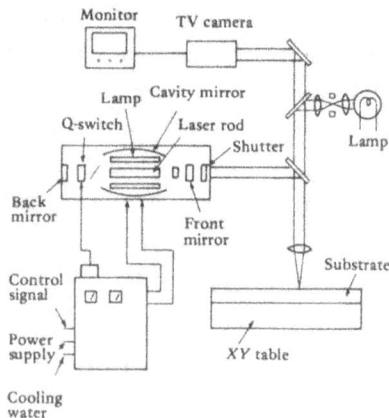

Fig. 5. A schematic diagram of the laser-scribing system for
integrated a-Si solar cells.

For the past several years, one of the major aims of resist
research has been to facilitate dry development of the resist. Dry
development was first achieved in inorganic resist systems, using high
etching selectively between Ag-photodoped and undoped chalcogenide films

in CF_4^{30-32} or $SF_6^{32,33}$ gas plasma. Since previous experiments were performed predominantly under isotropic etching conditions using a barrel-type plasma reactor, undercutting was more or less unavoidable. This was because the Ag-photodoped layer was relatively thin, typically several hundred angstroms in the Ag/SeGe system, and the underlying undoped layer was etched isotropically. Recently, laser-scribed dry etching and photo-scribed CVD of a-Si have been tried in other areas. These techniques may well be developed further in the next few years.

HETERO-JUNCTION SOLAR CELLS WITH NEW a-Si ALLOYS

There has been noticeable progress in recent research and development efforts with the introduction of amorphous mixed alloys such as the new materials, a-SiC, a-SiGe and a-SiSn. All these materials permit good valency control by the hydrogenation passivation of free bonds and the doping of substitution impurities using a suitable gas mixture technique. Since 1978, studies on valency-electron control of amorphous mixed alloys have been made at Osaka University, Japan. In applying the results, an a-SiC/a-Si hetero-junction solar cell having an efficiency greater than 8% was developed in 1980.[34] The $J_{SC} - V_{OC}$ (short-circuit current density vs. open-circuit voltage) curve for this cell is shown in Figure 6(a). Comparing this with the J-V characteristics of an ordinary a-Si solar cell shows that not only J_{SC} but also V_{OC} are clearly improved by utilizing p-type a-SiC:H as the window junction material. Recent developments in material-synthesis technology, in which the carbon fraction in a-Si$_{1-x}$C$_x$[35] is controlled, and in optimum design theory[36] based upon the concept of the drift-type photovoltaic process[37] have led to rapid improvements in the efficiency of this type of hetero-junction solar cell. In 1982, Catalano *et al.*[38] of the RCA group achieved an efficiency of 10.1% with an a-SiC/a-Si hetero-junction solar cell having an active area of 1.01 cm^2. The highest efficiencies to date for single p-i-n junction a-SiC/a-Si hetero-junction solar cells are 11.5% by Sanyo[39] and 11.1% by Fuji.[40] Figure 6(b) shows the area dependence of efficiency for hetero-junction cells.

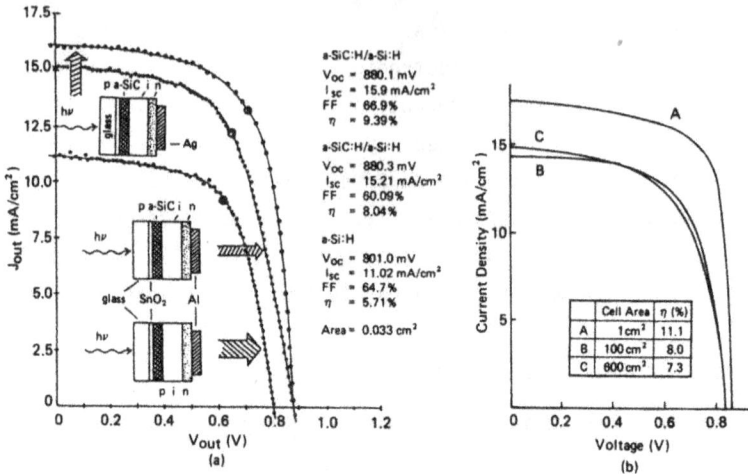

Fig. 6.   Efficiency increases using the p-type a-SiC window.
(a) Clear evidence of the wide-gap window effect
    demonstrated in 1980 (ref. 34).
(b) In 1984, the efficiency was improved to greater
    than 11% (ref. 40).

Figure 7 indicates the progress made in cell efficiency for
various types of a-Si solar cells since 1976.   As can be seen from this
figure, a step-like increase in cell efficiency occurred about 1981.
The increase in efficiency of a-Si solar cells prior to 1981 corresponds
mainly to improvements in film quality and routine cell-fabrication
processes.   In 1981, the development of hetero-junction solar cells with
a-SiC:H [35] and a-SiGe:H [41] led to a significant jump.   By examining
the wide-gap window effect, the built-in potential and minority-carrier
mirror effect of a-SiC/a-Si hetero-junctions were intensively studied
by Tawada et al.,[35] Nonomura et al.[42] and Okamoto et al.[43] Due to the
previously mentioned advances since 1981 in material-synthesis technology[44] and optimum-design theory,[36] the efficiency of this type of
hetero-junction solar-cell has continued to improve.   Recently, efficiencies of greater than 10% were obtained by RCA,[38] with Sanyo,[39]
Fuji,[45] Komatsu,[46] TKD-SEL,[47] ARCO[48] and ECD[49] also achieving high
efficiencies for this type of a-SiC/a-Si hetero-junction solar cell.

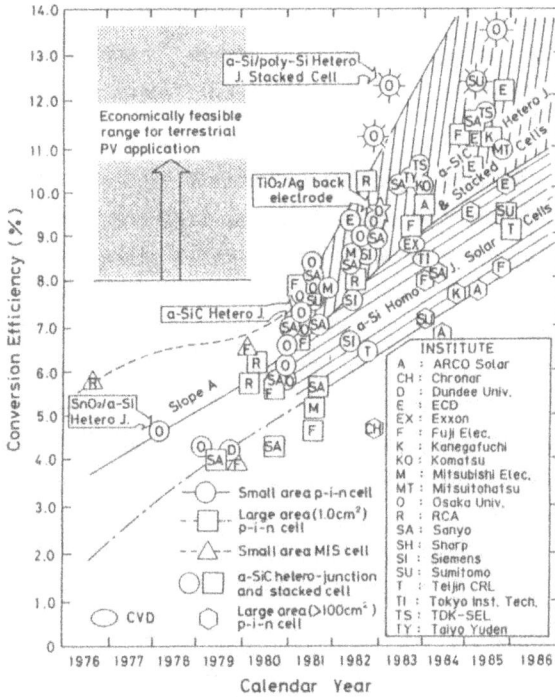

Fig. 7. Recent progress in cell efficiencies of a-Si
solar cells.

STACKED AND TRIPLE SOLAR-CELL STRUCTURES

One important area in which there remains scope for further
improvement of a-Si solar cells is the collection efficiency of low-
energy photons just above the band edge of a-Si. This is because the
penetration depth of, for example, 1.8-eV photons is of the order of
5 μm while the thickness of a-Si solar cells is only 0.6 μm. The
concept of efficient collection of long-wavelength photons by a highly
reflective random surface was first proposed by Böer et al. in 1981,[50]
and its theoretical basis was established by the Exxon group.[51] This
concept has been extended to include the more efficient utilization of
optical and carrier confinement in multi-layered hetero-junctions.[52]
Recently, Fujimoto et al.[53] developed a practical process for the
fabrication of cells of ITO/n microcrystalline Si/i-p a-Si/TiO_2/Ag-

plated semi-textured stainless steel and having an efficiency of 9.17%. Quite recently, the Taiyo-Yuden and ETL groups[54] reported an efficiency of 10.26% for cells which use the optical-confinement effect of milky transparent glass.

Another way to collect longer-wavelength photons is by absorption using stacked junctions of lower-energy-gap semiconductors. The concept of efficiency improvement using hetero-junction stacked cells is shown in Figure 8. Because the energy gap in a-Si of 1.7-1.8 eV is higher than that of crystalline solar-cell semiconductors, for example, 0.66 eV for Ge, 1.1 eV for Si and 1.43 eV for GaAs, the carrier-collection efficiency for solar radiation in a-Si solar cells is considerably lower than that of crystalline-based solar cells. However, the fabrication of large-area polycrystalline thin films using techniques such as CVD, MOCVD, molecular-beam epitaxy, sputtering and ion plating is already established for a wide variety of classic semiconductors. By combining these well-developed techniques with low-temperature a-Si solar-cell deposition technology, it is possible to make more efficient low-cost a-Si-based solar cells. Quite recently, a new type of stacked a-Si solar cell deposited on p-type polycrystalline silicon[55] was developed, for which an efficiency of greater than 12.5% was obtained. The structure of this cell and its J-V characteristics are shown in Figure 9. Although material-selection and economic feasibility studies are now in progress, an efficiency of greater than 15% could be obtained with stacked a-Si solar cells on only 1 μm of polycrystalline GaAs or a Ge thin film deposited on a stainless-steel substrate. In the case of polycrystalline silicon films, a thickness of about 20-30 μm is required for conventional CVD or photo-CVD from SiH_4 or $SiHCl_3$.[56] Some of the more remarkable achievements in this field are summarized in Table I.

RECENT APPLICATIONS OF SOLAR CELLS

A wide variety of solar-cell applications have been developed in recent years. In particular, consumer electronics applications are expanding very rapidly. For example, about 5 million a-Si-operated pocket calculators were fabricated each month in 1984 in Japan alone.

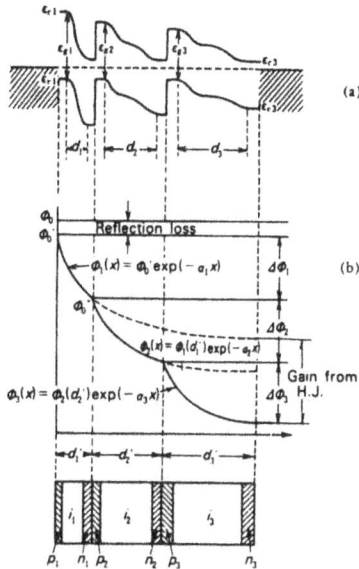

Fig. 8. A band-diagram representation of (a) the multi-band-gap stacked solar cell and (b) its photon-energy collection.

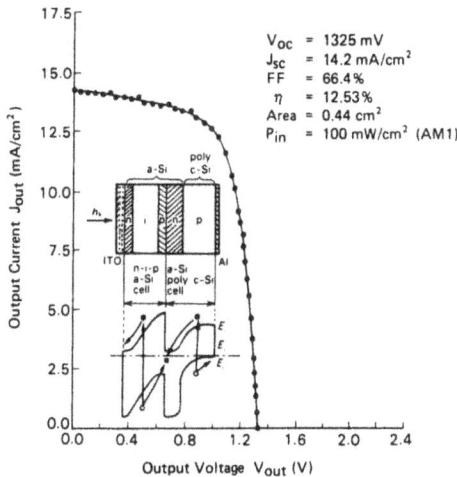

Fig. 9. The J-V characteristics and junction structure of the a-Si/polycrystalline Si stacked solar cell (ref. 4).

Table I. A summary of notable a-Si solar cells developed recently.

Type	Configuration	Area (cm^2)	Eff. (%)	V_{OC} (V)	J_{SC} (mA/cm^2)	FF	Year	Institute
SJ	ITO/nipn a-Si/p poly Si/Al	0.082	12.7	1.380	14.18	0.65	1984	Sumitomo
SJ	ITO/nipn a-Si/p poly Si/Al	0.44	12.5	1.325	14.2	0.66	1982	Osaka Univ.
HJ	Ag/ni a-Si/p a-SiC/text. TCO/glass	1.0	11.5	0.869	18.9	0.70	1984	Sanyo
HJ	Me/ni a-Si/p a-SiC/text. SnO$_2$/ITO/glass	1.05	11.63	0.850	18.7	0.732	1985	TDK-SEL
TJ	ITO/nipnipn a-Si/i a-SiGe/ p a-Si/ stainless steel	1.0	11.2	.	.	.	1985	ECD
HJ	Me/ni a-Si/p a-SiC/TCO/glass	1.0	11.1	0.864	17.6	0.73	1984	Fuji
HJ	Ag/ni a-Si/p a-SiC/text. SnO$_2$/glass	1.0	11.0	0.860	21.5	0.6	1985	Kanegafuchi
HJ	Me/ni a-Si/p a-SiC/text. TCO/glass	0.32	10.7	0.840	18.8	0.68	1984	Komatsu
HJ	Ag/ITO/ni a-Si/p a-SiC/MTG	0.045	10.26	0.802	22.32	0.57	1984	ETL & Taiyo
HJ	Al/ni a-Si/p a-SiC/text. SnO$_2$/glass	4.15	10.2	0.865	16.1	0.73	1984	ARCO
HJ	Ag/ni a-Si/p a-SiC/SnO$_2$/glass	1.09	10.1	0.840	17.8	0.676	1982	RCA
HJ	Ag/p a-SiC/in a-Si/SnO$_2$/glass	0.084	9.64	0.848	17.1	0.664	1984	TIT
SJ	ITO/nipn a-Si/i a-SiGe/ p a-Si/ stainless steel	929	9.42	.	.	.	1985	ECD
TJ	ITO/nipnipn a-Si/i a-SiGe/ p a-Si/ stainless steel	0.09	8.5	2.200	6.74	0.57	1982	Mitsubishi
Mod	Me/ni a-Si/p a-SiC/TCO/glass	100	8.1	11.96	15.6	0.61	1984	Sanyo
HJ	Me/ni a-Si/p a-SiC/TCO/glass	100	8.0	0.850	14.4	0.654	1984	Fuji
SJ	Al/nipni a-Si/p a-SiC/ SnO$_2$/glass	4.15	7.7	1.710	6.23	0.71	1984	ARCO
HJ	Al/ni a-Si/p a-SiC/SnO$_2$/glass	1.0	7.7	0.880	14.1	0.62	1982	Osaka Univ.
Mod	Al/ni a-Si/p a-SiC/SnO$_2$/glass	400	7.5	47.8	114 mA	0.55	1985	Kanegafuchi
Mod	Me/ni a-Si/p a-SiC/SnO$_2$/glass	600	7.0	16.0	12.5	0.63	1984	Sumitomo
Mod	Me/ni a-Si/p a-SiC/SnO$_2$/glass	3200	6.7	48.7	890 mA	0.50	1985	Kanegafuchi

SJ: Double-stacked junction.　　HJ: Hetero-junction.　　TJ: Triple-stacked junction.　　Mod: Module.

Power applications, however, are still in the experimental phase. Some a-Si power modules developed recently are shown in Figure 1. a-Si solar-cell experimental plants installed in Japan include power and distillation applications.

REFERENCES

(1)   D.E. Carlson and C.R. Wronski. Applied Physics Letters, vol. 28, p.671 (1976).

(2)   Reviewed in Japan Annual Reviews in Electronics, Computers and Telecommunications (JARECT), vol. 2, Amorphous Semiconductor Technologies and Devices, p.1-7. Y. Hamakawa (ed.), Ohmsha and North-Holland (1982).

(3)   Y. Hamakawa, H. Okamoto and Y. Nitta. Applied Physics Letters, vol. 35, p.187 (1979).

(4)   K. Okuda, H. Okamoto and Y. Hamakawa. Japanese Journal of Applied Physics, vol. 22, p.L605 (1983).

(5)   Reviewed in Japan Annual Reviews in Electronics, Computers and Telecommunications (JARECT), vol. 16, Amorphous Semiconductor Technologies and Devices, p.1-9. Y. Hamakawa (ed.), Ohmsha and North-Holland (1984).

(6)   P. Maycock. P.81 of Proceedings of the Fourth New Energy Industrial Symposium (held Tokyo, Japan, 1984).

(7)   Y. Hamakawa. Business Japan, vol. 29, p.36 (1984).

(8)   H. Okamoto, T. Yamaguchi and Y. Hamakawa. Journal of Non-Crystalline Solids, vol. 35-36, p.313 (1980).

(9)   M. Taniguchi, M. Hirose, T. Hamasaki and Y. Osaka. Applied Physics Letters, vol. 37, p.787 (1980).

(10) A. Matsuda, K. Nakagawa, K. Tanaka, M. Matsumura, S. Yamasaki, H. Ohkushi and S. Iijima. Journal of Non-Crystalline Solids, vol. 35-36, p.183 (1980).

(11) See Part II, preparation session of Proceedings of the Ninth International Conference on Amorphous and Liquid Semiconductors (held Grenoble, France, 1981).

(12) S. Hotta, Y. Tawada, H. Okamoto and Y. Hamakawa. Journal de Physique, vol. 42, sup. 10, C-4, p.631 (1981).

(13) R.R. Gay, D.L. Morel, D.P. Tanner, D. Kanani and H.S. Ullal. P.714 of Proceedings of the Fourth European Community Photovoltaic Solar Energy Conference (held Stresa, Italy, 10-14 May 1982). 1982.

(14) S. Hotta, N. Nishimoto, Y. Tawada, H. Okamoto and Y. Hamakawa. Journal of Applied Physics, vol. 21, p.289 (1982).

(15) D.E. Carlson and C.R. Wronski. Chapter 10 of Amorphous Semicon-
      ductors. M.H. Brodsky (ed.), Springer-Verlag (1979).

(16) Y. Kashima, H. Kida, H. Okamoto and Y. Hamakawa. Journal of Non-
      Crystalline Solids, vol. 59-60, p.755 (1983).

(17) Y. Kuwano, M. Ohnishi and H. Shibuya. Japanese Journal of Applied
      Physics, vol. 20, p.157 (1981).

(18) Y. Hamakawa. Journal of Non-Crystalline Solids, vol. 59-60, p.1265
      (1983).

(19) Y. Kashima, S. Nonomura, H. Kida, H. Okamoto and Y. Hamakawa.
      P.793 of Proceedings of the Fifth European Community Solar Energy
      Conference (held Athens, Greece, 1983). (1983).

(20) M. Ohnishi, H. Nishiwaki, K. Enomoto, Y. Nakashima, S. Tsuda and
      Y. Kuwano. Journal of Non-Crystalline Solids, vol. 59-60, p.1107
      (1983).

(21) H. Okaniwa, K. Nakatani and I. Ohuchi. Proceedings of the Annual
      Amorphous Silicon Contractor Meeting. Sunshine Project, 20 Apr.
      1984.

(22) T.D. Moustakas. P.698 of Proceedings of the Fifth European
      Community Photovoltaic Solar Energy Conference (held Athens,
      Greece, 1983). (1983).

(23) H. Hitotsuyanagi. Proceedings of the 1983 Annual Amorphous Silicon
      Contractor Meeting. Sunshine Project, 20 Apr. 1984.

(24) V.L. Dalal, M. Akhpar and A. Dalahoy. P.1385 of Proceedings of the
      16th Photovoltaics Specialists Conference (held San Diego, CA, USA,
      1982). (1982).

(25) A. Matsuda, T. Kaga, H. Tanaka, L. Malhotra and K. Tanaka.
      Japanese Journal of Applied Physics, vol. 22, p.L115 (1980).

(26) N. Fukuda. Proceedings of the 1983 Annual Amorphous Silicon
      Contractor Meeting. Sunshine Project, 20 Apr. 1984.

(27) J. Shimada. Proceedings of the 1983 Annual Amorphous Silicon
      Contractor Meeting. Sunshine Project, 20 Apr. 1984.

(28) M. Konagai and K. Takahashi. Proceedings of 1985 Solar Energy
      Research Institute Workshop (1985).

(29) T. Hamasaki, M. Ueda, A. Chayahara, M. Hirose and Y. Osaka.
      Applied Physics Letters, vol. 44, p.600 (1984).

(30) M.S. Chang and J.T. Chen. Applied Physics Letters, vol. 33,
      p.892 (1978).

(31) A. Yoshikawa, O. Ochi and Y. Mizushima. Applied Physics Letters,
      vol. 36, p.107 (1980).

(32) S.A. Lis, J.M. Lavine and J.I. Masters. P.275-84 of Proceedings
      of Microcircuit Engineering 82. (1982).

(33) P.G. Huggett, K.F. Rick and H.W. Lehmann. Applied Physics
      Letters, vol. 42, p.592 (1983).

(34) K. Okuda, K. Fujimoto, Y. Kashima, H. Okamoto and Y. Hamakawa. P.189-92 of Extended Abstracts of the 15th Conference on Solid State Devices and Materials (held Tokyo, Japan, 1983). (1983).

(35) Y. Tawada, K. Tsuge, M. Kondo, H. Okamoto and Y. Hamakawa. Journal of Applied Physics, vol. 53, p.5273 (1982) and Y. Hamakawa Late News of Proceedings of the 17th IEEE Photovoltaics Specialists Conference (held Orlando, FL, USA, May 1984). (1984).

(36) H. Okamoto, H. Kida and Y. Hamakawa. Solar Cells, vol. 8, p.97 (1983).

(37) H. Okamoto, T. Yamaguchi and Y. Hamakawa. Journal of Physics Society of Japan, vol. 49, p.1213 (1980).

(38) T. Catalano, A. Friester and B. Fanghman. P.1421 of Proceedings of the 16th IEEE Photovoltaics Specialists Conference (held San Diego, CA, USA, 1982). (1982).

(39) S. Nakano, H. Kawada, T. Matsuoka, S. Kiyama, S. Sakai, K. Murata, H. Shibuya, Y. Kishi, I. Nagaoka and Y. Kuwano. P.583 of Technical Digest of the International Photovoltaic Science and Engineering Conference (held Kobe, Japan, 1984). (1984).

(40) H. Sakai, K. Maruyama, T. Yoshida, Y. Ichikawa, T. Hama, M. Ueno, M. Kamiyama and Y. Uchida. P.591 of Technical Digest of the First International Photovoltaic Science and Engineering Conference (held Kobe, Japan, 1984). (1984).

(41) G. Nakamura, M. Kato, H. Kondo, Y. Yukimoto and K. Shirahata. Journal of Physics, vol. 42, p.C4-483 (1981).

(42) S. Nonomura, H. Okamoto, K. Fukumoto and Y. Hamakawa. Journal of Non-Crystalline Solids, vol. 59-60, p.1099 (1983).

(43) H. Okamoto, H. Kida, K. Fukumoto and Y. Hamakawa. Journal of Non-Crystalline Solids, vol. 59-60, p.1103 (1983).

(44) Y. Tawada, K. Tsuge, K. Nishimura, H. Okamoto and Y. Hamakawa. Japanese Journal of Applied Physics, vol. 21, sup. 21-2, p.291 (1982).

(45) Y. Uchida, H. Sakai, M. Nishiura, M. Miyagi and K. Maruyama. Vol. 1, p.669 of Proceedings of the Eighth International Vacuum Congress (held Cannes, France, 1980). (1981).

(46) G. Kagaya. Photovoltaic Insider News, vol. 3, no. 1, p.4 (1984).

(47) S. Yamazaki, K. Itoh, S. Watanabe et al. Proceedings of the 17th IEEE Photovoltaics Specialists Conference (held Orlando, Florida, USA, May 1984). (1984). P1-1B-1.

(48) K.W. Mitchell. Annual Review of Energy, vol. 10 (1985). Annual Reviews Inc., USA.

(49) S. Ovshinsky, private communication.

(50) W. den Böer and R.M. Van Strijp. P.764 of Proceedings of the Fourth European Community Photovoltaic Solar Energy Conference (held Stresa, Italy, 1982). (1982).

(51) E. Yablonovitchi and G.D. Cody. IEEE Transactions, Electron Devices, vol. ED-29, p.300, (1982).

(52) Y. Hamakawa, Y. Tawada, K. Nishimura, K. Tsuge, M. Kondo, K. Fujimoto, S. Nonomura and H. Okamoto. P.679 of Proceedings of the 16th IEEE Photovoltaics Specialists Conference (held San Diego, CA, USA, 1982). (1982).

(53) K. Fujimoto, H. Kawai, H. Okamoto and Y. Hamakawa. Solar Cells, vol. 11, p.357 (1984).

(54) H. Iida, T. Miyado and Y. Hayashi. Proceedings of the 44th Fall Meeting of Society of Japan Applied Physics (held Sendai, Japan, Sept. 1983). (1983). 25p-L-2 and 3.

(55) Y. Hamakawa, K. Okuda, H. Takakura and H. Okamoto. P.1386 of Proceedings of the 17th IEEE Photovoltaics Specialists Conference (held Orlando, Florida, USA, 1984). (1984).

(56) Y. Hamakawa and H. Okamoto. Japan Annual Reviews in Electronics, Computers and Telecommunications (JARECT), vol. 16, Amorphous Semiconductor Technologies and Devices, p.200. Ohmsha and North-Holland (1984).

The author wishes to express his sincere thanks to his research colleagues throughout the world who kindly supplied the latest information in this field. Due to limited space, not all details could be cited. The author also gratefully acknowledges the advice and assistance of Dr. H. Takakura and Dr. H. Okamoto in the preparation of this review.

REFINED COMPUTER SIMULATION OF AMORPHOUS SOLAR CELLS

F. Demichelis[*], E. Mezzetti, A. Tagliaferro and E. Tresso
Dipartimento di Fisica, Politecnico di Torino, Italy
* Gruppo Nazionale di Struttura della Materia, U.R., 24

Abstract

A model based on the theory of electromagnetic radiation through a multilayer stratified medium for determination and optimization of the fundamental parameters defining the quality of amorphous solar cells is developed.
Optical , electrical and recombination properties are taken into account and it is possible to deal with a cell composed by as many active layers as are to be considered, such as p-i-n, mixture, amorphous/poly c and amorphous μc/poly c.

REFINED COMPUTER SIMULATION OF AMORPHOUS SOLAR CELLS

F. Demichelis[*], E. Mezzetti, A. Tagliaferro and E. Tresso
Dipartimento di Fisica, Politecnico di Torino, Italy
* Gruppo Nazionale di Struttura della Materia del C.N.R., U.R., 24

In the last few years a great effort has been done towards the increase of amorphous solar cells efficiency, not only through the improvement of material quality, but also by means of the optimization of the solar cells layout. Simulation by computer is indeed an important tool to achieve such an optimization, in order to save efforts and time, which are wasted when an empirical optimization is attempted.

To meet these requirements, we developed a model based on the theory of electromagnetic radiation through a solar cell considered as a multilayer stratified medium: the resulting expression for absorptance is used for carrier generation evaluation [1].

By means of this model it is possible to deal with several kinds of cells, such as p-i-n, mixture, amorphous/poly c and amorphous/μc/poly c.

THEORY

BASIC EQUATIONS

The photogenerated carrier continuity equations in one dimensional device, under steady-state conditions, can be written as

$$\begin{cases} D_n \dfrac{d^2 \Delta n}{dx^2} + \mu_n \dfrac{d(\Delta n \cdot E)}{dx} + G_n(x) - R_n(x) = 0 \\[3mm] D_p \dfrac{d^2 \Delta p}{dx^2} - \mu_p \dfrac{d(\Delta p \cdot E)}{dx} + G_p(x) - R_p(x) = 0 \end{cases} \qquad (1)$$

where $D_n(D_p)$, $\mu_n(\mu_p)$, $G_n(G_p)$, $R_n(R_p)$ are respectively diffusion coefficient, mobility, generation rate and recombination rate for electrons (holes); $\Delta n(\Delta p)$ is the photogenerated carrier density of electrons (holes).

Since R(x), in the Shockley-Read model, for amorphous semiconductors, where generation rate is larger then intrinsic carrier density is given by[2]

$$R(x) = \frac{1}{\tau_n(x)/\Delta n(x) + \tau_p(x)/\Delta p(x)} \qquad (2)$$

where $\tau_n(\tau_p)$ is the electron (holes) lifetime, as a consequence $R_n(x) = R_p(x)$ and, moreover, the recombination rate is fixed by the minority carriers, when the two densities are very different.

SOURCE OF CARRIERS

The solar cell system is considered as a multilayer composed by different materials. To evaluate the photon absorption in the doped and undoped layers of the cell we calculated the Poynting vector:

$$\vec{P} = \frac{1}{2} \, Re \, (\vec{E} \wedge \vec{H})$$

The ratio of the power absorbed in a generical j-th layer to the incident power is given by:

$$A_{abs_j} = \frac{|P_{inc}^{(j)}| - |P_{inc}^{(j+1)}|}{P_0^t}$$

The number of photons absorbed in the j-th layer, L_j thick, is

$$N_{abs}(\lambda, L_j) = A_{abs}(\lambda, L_j) \cdot AM1(\lambda) \cdot \lambda/hc$$

for AM1 illumination. The total number of photons absorbed which, in the present theory, is taken as the source of carrier generation G(x), is given by:

$$G(x) = N_{tot} = \int_{\lambda_0}^{\lambda_g} A_{abs_j}(\lambda, L_j) \, AM1(\lambda) \frac{\lambda}{hc} \, d\lambda$$

where λ_0 is the shortest wavelength of incident light and λ_g the band-gap wavelength respectively. In our model the local generation rates for electrons and holes are always taken equals.

DERIVATION OF CURRENTS

a) p-i-n cell

In a p-i-n cell the three layers have to be separately considered, since in doped layers recombination is controlled by minority carriers, while in i layer a single minority-carrier type does not exist:

doped layers

Since the two layers are very close in behaviour we examine in detail only p-one.
Since conditions $\Delta m / \tau_m \ll \Delta p / \tau_p$ and E=0 hold, system (1) reduces to

$$\begin{cases} D_m \dfrac{d^2 \Delta m}{dx^2} + G_m(x) - \dfrac{\Delta m}{\tau_m} = 0 \\[4mm] D_p \dfrac{d^2 \Delta p}{dx^2} + G_p(x) - \dfrac{\Delta m}{\tau_m} = 0 \end{cases} \qquad (3)$$

Setting proper boundary conditions, that is $\Delta m = 0$ for x=0 (fig.1), indicating the sweeping of carriers at the p-i interface towards the i layer, and $D_m \, d(\Delta m)/dx = S_m \Delta m$ for x=x_p, due to surface recombination (S_n is, in fact, the surface recombination velocity), the first of the (3) stands alone and can be solved. The solution is analytical if $G_n(x)$ is an analytical function; assuming a constant $G_0 (G_0 = \frac{1}{|x_p|} \int_{x_p}^{0} G_m(x)dx)$ and bearing in mind that $D_{p,m} = \frac{KT}{q} \mu_{p,n}$ and $L_{m,p} = \sqrt{D_{m,p} \tau_{m,p}}$ we can write

$$\Delta m = A e^{-x/L_m} + B e^{x/L_m} + G_0 \tau_m \quad \text{with} \begin{cases} A = -B - G_0\tau_m \\ B = G_0\tau_m \dfrac{D_m/L_m + G_m (1 - e^{-x_p/L_m})}{G_m (2 e^{x_p/L_m} - 1) - D_m/L_m (e^{2x_p/L_m} - 1)} \end{cases} \quad (4)$$

Similar considerations hold for the hole density in n-layer Δp_n, but, since such a layer is thicker two types of generation rate profiles have been considered, that is constant and linear generations.

The current densities coming from the doped layers are given by (fig.1)

$$\begin{cases} J_{m_p}(0) = q \, D_{m_p} \left. \dfrac{d(\Delta m)}{dx}\right|_0 \\ J_{p_m}(L) = -q \, D_{p_m} \left. \dfrac{d(\Delta p)}{dx}\right|_L \end{cases} \quad (5)$$

i-layer

Here no one of the above semplifications holds, so that the two equations of system (1) are coupled. For this reason the solution is find through an iterative method based either on the Green functions or on the finite differences. Without entering in the details, let us say that

a) the Green function method is physically the most significant, since it tells how many carriers generated in one point will be able to reach any other fixed point in the layer

b) the (a) method is greatly simplified when E in i-layer can be taken as a constant, as it is in short circuit conditions

c) the finite differences method requires a shorter calculation time.

In short circuit conditions the internal electric field is high ($\simeq 10^4$ V/cm) so that the diffusion term in the continuity equation can be neglected. Appropriate boundary conditions are:

$$J_{p_i}(L) = J_{p_m}(L) + J_p^*(L) \qquad \text{n-i interface}$$

$$J_{m_i}(0) = J_{m_p}(0) + J_m^*(0) \qquad \text{p-i interface}$$

$$(6)$$

where J^* are surface recombination currents.

As an example, $J_{ni}(L)$ turns out to be (Green function method, short circuit conditions)

$$J_{n\ell}(L) = qE \frac{\mu_m}{D_m} A_m \int_0^L [G_m(\xi) - R_m(\xi)] d\xi$$

where

$$A_m = - \frac{J_{mp}(o)}{q S_{pi}} \int_0^L G_q(\nu) e^{-\nu/|\alpha_m|} d\nu \qquad (7)$$

and

$$|\alpha_m| = \frac{2kT}{qE} \quad ; \quad G_q(\nu) = \frac{G_p(\nu) - R_p(\nu)}{D_p} e^{-q V(x)/2kT}$$

The pair recombination at the interfaces imposes

$$|J_{pp}(o)| = |J_{pi}(o)| - |J_n^*(o)|$$

and similar relationship for the holes current at p-i interface

doped layers again

Once known the majority photocarriers current entering the layer, the remaining continuity equation is easily solved, leading to the current coming from the cell. Taking into account doped layer-electrode interface we get

$$J_{ph} = |J_{pp}(x_p)| - |J_n^*(x_p)|$$

i.e. the photocurrent produced by the amorphous cell.

b) mixture cells

The only difference in this case from the one discussed above is that i-layer is a mixture of silicon and germanium: a-$Si_x Ge_{1-x}$:H. Other than the layer thickness, also the best x value has to be determined.

c) stacked cells

When two or more cells are stacked, different configurations can stand:
(1) pin amorphous-crystalline
(2) pin amorphous-amorphous/polyc junction
(3) pin amorphous-mixture
A general rule is that the back cell has to show an energy gap lower than the front one, in order to achieve a better use of the solar radiation.
The calculus is done separately for the two cells, with the appropriate generation rates, and the two photocurrents are used in the solution of the equivalent electric circuit.

EFFICICIENCY DETERMINATION

The efficiency is evaluated by solving the equivalent circuit of the given cell. As an example, let us consider the case of a tandem cell (fig.2), the equations read:

$$
\begin{cases}
J = \frac{1}{2}\left\{ f\left(J_{ph_1} + J_{ph_2}\right) - \left[J_{o1}\left(\exp\left(q\frac{V_I}{mkT}\right) - 1\right) + J_{o2}\left(\exp\left(q\frac{V_{II}}{mkT}\right) - 1\right) + \frac{V_I}{R_{sR_1}} + \frac{V_{II}}{R_{sR_2}}\right]\right\} \\
V = J\left(R_c + R_{s1} + R_{s2}\right) + V_I + V_{II} \qquad f = \left(\text{lighted}/\text{total}\right)\text{ cell area}
\end{cases}
\tag{8}
$$

where for the dark current value usual expressions have been used.

Once fixed a desired V value, the procedure to solve (8) needs to be iterative, since V_I and V_{II} (voltage of the two cells) values fix, through the relationship $E = [V_{bi} - V_I]/L$ (V_{bi} is the built in potential), the electric field in the i layer of the pin cell and, in a similar way, the electric field in the junction zone of the poly-c one. Since the photogenerated current on its turn depends on the electric field value, the need of an iterative approach is clear. Once obtained the J-V characteristic, evaluation of typical parameters, such as efficiency , fill factor FF, short circuit current J_{sc} and open circuit voltage V_{oc} are straightforward obtainable, since, being P_{max} the maximum power density given by the cell, P_{inc} the incident solar radiation power density, the following relationships hold:

$$\eta = P_{max}/P_{inc} \qquad\qquad FF = P_{max}/V_{oc} J_{sc}$$

OPTIMIZATION METHOD

The value of any solar cell parameter (V_{oc}, J_{sc}, η,...) is a function of several material characteristics, such as absorption coefficients, carrier diffusion lengths, and of the layers thicknesses. Once the materials composing a solar cell have been chosen, the only left variables are the thicknesses, so that

$$J_{sc} = J_{sc} (L_1, L_2, \dots L_m)$$

We used as a relevant characteristic to be optimized the short circuit current instead of the efficiency since the computing time is sensibly lower, due to the fact that the knowledge of the J-V characteristic is not needed and the differences in the final layout are negligible.

We are dealing with N+1 dimensional space (N independent variables $L_1,...L_n$ and the dependent one J_{sc}), so that the gradient of J_{sc} in a given point $\vec{L_i}$ is given by

$$\overrightarrow{grad}\, J_{sc} (\vec{L_i}) = \begin{vmatrix} \partial J_{sc}/\partial L_1 \\ \partial J_{sc}/\partial L_2 \\ \partial J_{sc}/\partial L_N \end{vmatrix}$$

Starting from the point P_0 we move along the gradient direction, up to a maximum (nul gradient). At each step an increase $\vec{\Delta L}$, parallel to the gradient, is taken:

$$\vec{\Delta L_i} = const \cdot \overrightarrow{grad}\, J_{sc} (\vec{L_i})$$

The new thickness vector turns out to be $\qquad \vec{L_{i+1}} = \vec{L_i} + \vec{\Delta L_i}$

and the maximum is reached, as stated above, when all gradient components become negligible.

NUMERICAL PROCEDURE

The procedure followed to solve the problem, here outlined for a pin-cell , is:

a) choice of the cell structure (pin, tandem, Schottky, stacked...)

b) insertion of the physical parameters of the chosen materials, such as absorption coefficients, refractive indices, diffusion lengths, interface recombination velocities,...

c) choice of a starting thicknesses configuration

d) evaluation of the source term

e) evaluation of electrons density in p layer and holes density in n layer: $J_{pn}(L)$ and $J_{np}(0)$ are evaluated

f) iterative solution of the coupled continuity equations in the i layer: $J_{ni}(L)$ and $J_{pi}(0)$ are found

g) determination of majority carriers density in doped layers, leading to J_{ph} evaluation

h) repetition of (d)-(g) steps until the maximum J_{ph} is found

i) evaluation of J-V characteristic and relevant cell parameters (η, V_{oc}, FF,..)

RESULTS

As a test for our method, because of the relevant importance of the Green function method developed, our choice was to focus on pin structure.

The characteristics of the optimized cells, either with a-Si:H i layer or a-SiGe:H i layer, are reported in fig.3 together with an experimental characteristic[3]. The relevant photovoltaic parameters relative to the above mentioned cells are reported in Table I. As it can be seen, a careful layout of the cell allows, using the same materials, a large efficiency improvement (from 5.7% to 7.6%).

CONCLUSIONS

We developed a detailed model of the solar cells behaviour, taking into account recombination and generation, applicable to several kind of solar cells, and tested it for pin a-Si:H and pin a-SiGe:H structures. The results have shown that a careful layout allows to achieve higher efficiency in shorter time, since the time indeed to deposit a cell is longer by far of the time required for the computer simulation of its behaviour. The optimum use of this model is, then, to give a good starting point to obtain a good efficiency solar cells, avoiding the waste of time connected to the empirical attempt to reach the best results.

ACKNOWLEDGEMENTS

This work was partially supported by the Research Project PF2 sponsored by the Consiglio Nazionale delle Ricerche (C.N.R.)

REFERENCES

(1) F. Demichelis, A. Tagliaferro and E. Tresso - Solar Cells, 11, 4, 375, (1984)

(2) W. Shockley and W.T. Read - Phys. Rev. 87, 835, (1952)

(3) Y. Hamakawa - Proc. 5-th E.C. Conference on Photovoltaic Solar Energy, Athens, 690, (1983).

T A B L E I

Structure of the cell	Thickness of the layer (μm)	J_{sc} (mA cm^{-2})	V_{oc} (V)	FF	η %	References
AR1 (MgF$_2$)	0.20	12.60	0.86	0.70	7.6	Present work
AR2 (ZnS)	0.05					
a-Si:H	0.40					
Ag						
AR1 (MgF$_2$)	0.04	16.10	0.81	0.67	8.7	Present work
AR2 (ZnS)	0.05					
a-Si$_{0.48}$:H/	0.50					
a-Ge$_{0.52}$:H						
AR (SnO$_2$)	---	11.02	0.81	0.647	5.7	Hamakawa, 5-th EC Phot. Sol. Fn., Athens, 1983
a-Si:H	---					
Ag						

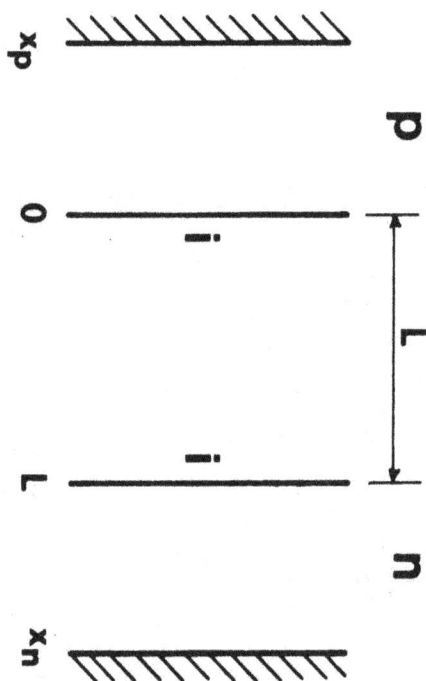

Fig. 1 - Configuration and boundary conditions for p-i-n cell

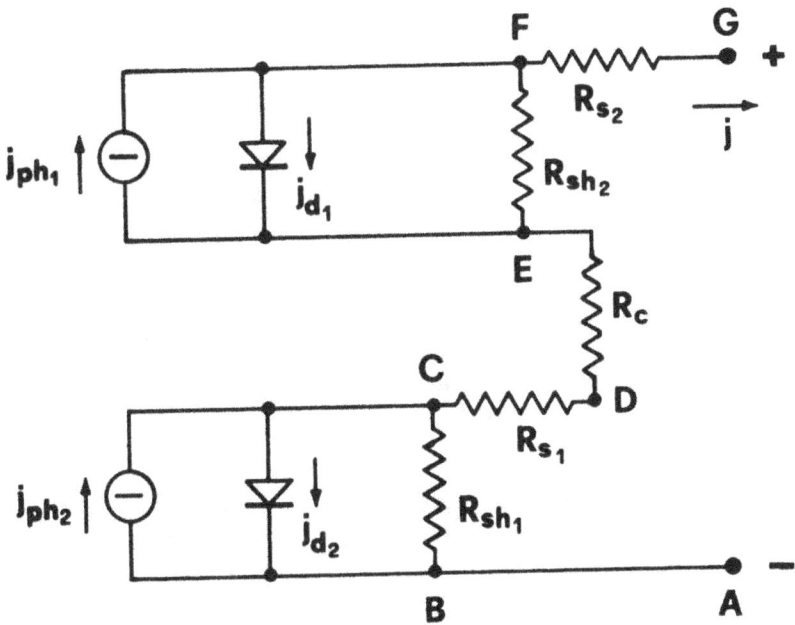

Fig. 2 - Equivalent circuit of amorphous/poly c cell

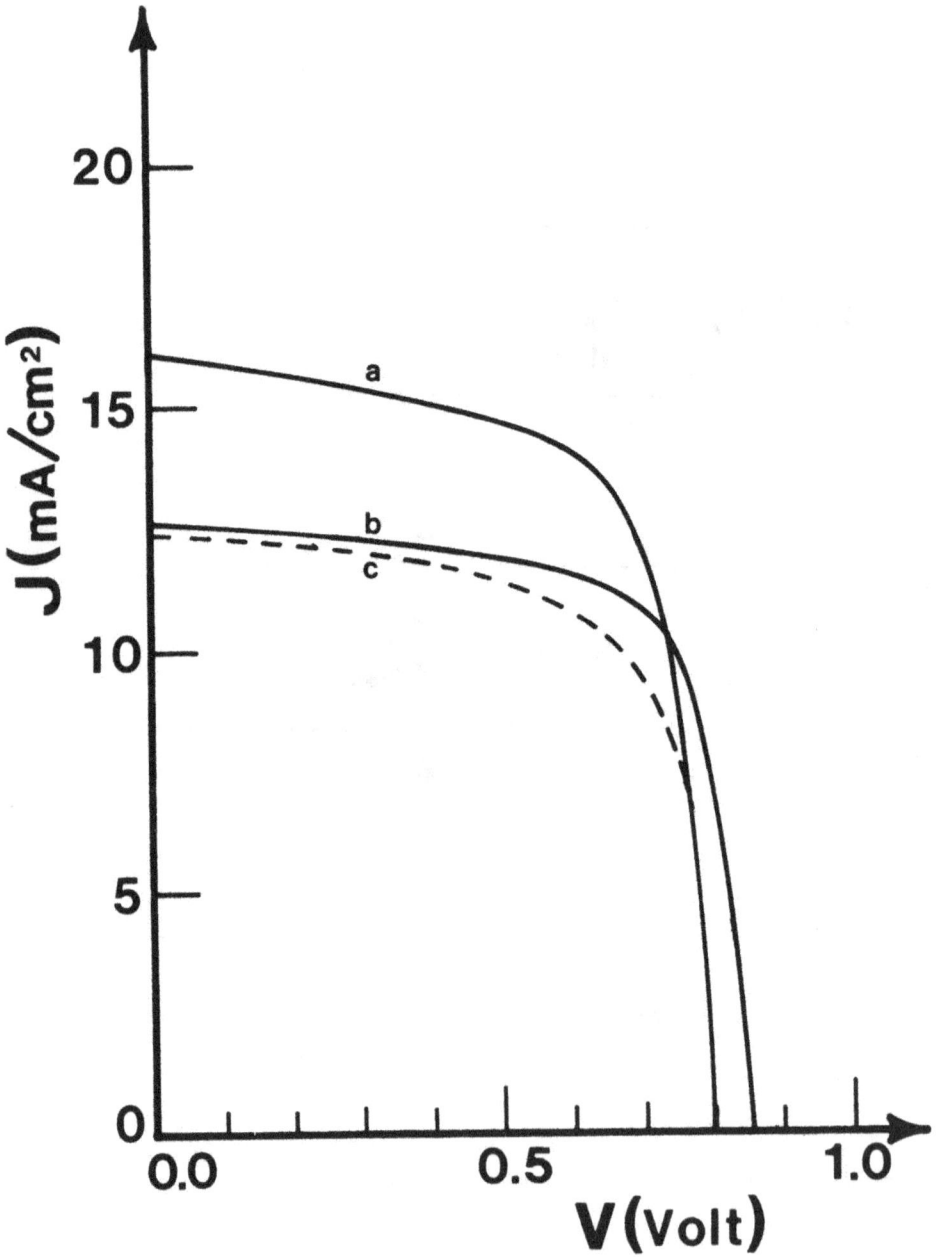

Fig. 3) Calculated J-V characteristic for a a-Si$_{0.48}$Ge$_{0.52}$:H (a) cell, for a p-i-n a-Si:H (b) cell and experimental p-i-n cell (c).

AMORPHOUS SEMICONDUCTOR HETEROSTRUCTURES

F. Evangelisti

Dipartimento di Fisica
Universita' di Roma "La Sapienza"
P.le A.Moro 2, 00185 Roma, Italy

INTRODUCTION

The microscopic study of crystalline-semiconductor heterostructures by photoelectronic techniques has been the subject of much work in the last years.[1] This effort is justified by fundamental reasons and by the widespread applications of semiconductor heterojunctions in solid-state electronics. Recently, very promising heterojunction devices made by amorphous semiconductors have been realized in the field of superlattices[2] and photovoltaic solar cells.[3] Since the behavior of a heterojunction device is strongly dependent on band discontinuities, interface states and diffusion potential, there was an immediate interest in extending to amorphous semiconductors the microscopic investigation of these parameters.[4-6] In the present work we will review some of the results obtained in this field which is still at its first stages of development. It will be shown, moreover, that the technique offers the interesting possibility to investigate fundamental problems like the effect of disorder and/or hydrogenation on the electronic structure of the materials.

EXPERIMENTAL TECHNIQUE

The surface sensitivity of U.V. photoemission spectroscopy makes this technique an ideal probe for interface studies. The physical information carried out by the ejected electrons is inherent to the outermost region of the solid. The thickness of this region varies with the kinetic energy of the electron according to the escape depth curve. Therefore, by using a tunable photon source like synchrotron radiation, it is possible to modulate the surface sensitivity of the technique and to enhance surface effects.

The microscopic study of a heterojunction imply the determination of parameters like the band discontinuities, the position of the Fermi level at the interface and the magnitude of band-bending effects. Photoemission spectroscopy is the most direct way to measure these parameters. The experimental approach consists in growing in situ and step-by-step one semiconductor on top of the other and following the evolution of the valence band and core levels at each step.

The formation of the heterojunction between two semiconductors of different energy gap will produce either a valence band or a conduction band discontinuity or both. Valence band discontinuities larger than few tenths of eV can be measured directly in the spectra by growing on one semiconductor a thin overlayer of the other semiconductor. If the escape depth of the electrons is larger than the thickness of the overlayer, both valence band edges are directly measurable. ΔE_v is obtained by taking the difference between the extrapolation to zero of the two leading edges. This situation is schematically shown in Fig.1.

When the valence band discontinuity is small (0-0.4 eV) the two valence edges are not resolved, however, and ΔE_v is to be derived from the evolution of the leading edge provided that band-bending changes during the interface formation are duly taken into account. This second case requires an analysis of both valence and core structures. First of all, we must remember that a photoemission measurement is always affected by the initial band bending since the probed region is, in general, much smaller than the extention of the band bending itself. When this undergoes a variation all the valence and core structures shift rigidly relative to the Fermi level. Therefore, the determination of ΔE_v proceeds as follows. The valence band maximum E_v and one or more core levels are measured on the clean substrate. Then the measurements are repeated at different overlayer thicknesses. The valence-band top usually shifts and so do the core levels. The shift of E_v is the sum of the band discontinuity plus the change in band bending caused by the variation of the local charge distribution. The band-bending changes at different coverages can be deduced from the shift of the substrate core-level peaks. Therefore, ΔE_v is determined as the difference between the total shift of E_v and the band bending change. It is worth mentioning, however, that often the situation is made more complicated by chemical shifts of the core levels due to the formation of interface chemical bonds. In this last occurrence the analysis is more complex and requires either the measurement of different core lines or a fitting procedure to unfold the different components. It is important to point out that the major source of error in the value of ΔE_v is due to the determination of the valence-band top from the experimental energy distribution curves (EDC). Usually this is done by a linear extrapolation to zero of the leading spectral edge and can be easily affected by 0.1-0.15 eV uncertainty.

For heterojunctions involving amorphous semiconductors it is necessary to clarify which electronic levels are compared by defining the valence band maximum E_v as the linear extrapolation to zero of the valence-band EDC's leading edge. Under usual experimental conditions this edge merges into the noise at a signal level 0.05-0.1 times the intensity of the highest in energy valence-band feature (3p or 4p states for Si or Ge, respectively), i.e. E_v corresponds to $\sim 10^{21}$ states cm^{-3} eV^{-1}. In the crystaline case, this definition does not cause any ambiguity because the top of the valence band is a well defined concept and the density of states $g(E)$ increase rapidly (f.e. in c-Si, $g(E)$ varies from 0 to 10^{21} in ~ 0.05 eV, i.e. over an energy distance smaller than the experimental uncertainty). Things are different in the amorphous case due to the continuum of localized states tailing toward the Fermi level. However, a state density in the 10^{20}-10^{21} cm^{-3} eV^{-1} range is a reasonable estimate[7] for the demarcation level between extended and localized states. Therefore, we can assume that E_v as defined above locates the valence-band mobility-edge, within 0.1-0.2 eV uncertainty.

Finally, let us mention that the conduction-band discontinuity too could be determined experimentally by studying the evolution of the onset of the core-to-conduction-band transitions as measured by partial yield spectroscopy. In practice, however, these transitions are affected by large excitonic effects[8] which make difficult to locate the bottom of conduction band.

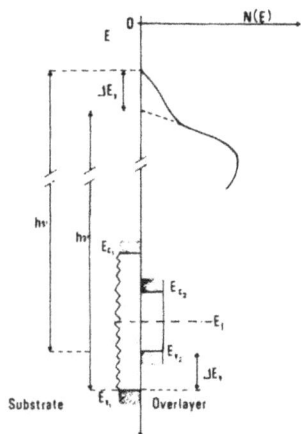

Fig. 1 Transition scheme showing
the low binding-energy side
of the valence-band photoemission spectra in the case
of a thin overlayer. N(E) is
the photoemission intensity.

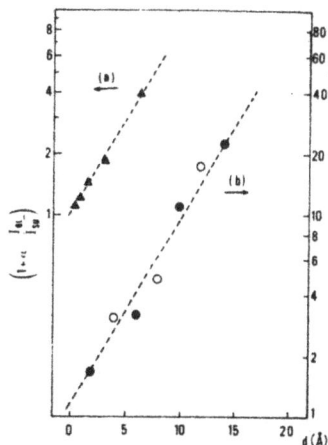

Fig. 2 Ratio of the core-level
intensity as a function
of overlayer thickness.
The expression in abscissa
is discussed in the text.
a) a-Si:H/a-Ge interface;
b) a-Si$_x$C$_{1-x}$:H/a-Si (full
dots) and a-Si$_x$C$_{1-x}$:B, H/a-Si
(open circles) interfaces.

RESULTS AND DISCUSSION

In the following we review and discuss briefly the results obtained
on a number of heterojunctions involving amorphous silicon and germanium
as well as silicon-carbon alloys.

An important characterization of a heterojunction is the quality of
the interface, i.e. its abruptness or the possible interdiffusion of the
different species or the growth of the overlayer through island formation.
This information can be obtained by the intensity dependence of the core
lines as a function of the overlayer thickness. In particular, if the interface is abrupt and the overlayer is uniform we have:

$$(1 + \alpha\, I_{OL}/I_{SU}) = \exp(d/\lambda)$$

where I_{OL} and I_{SU} are the core intensities of the overlayer and substrate,
respectively, α is a factor which accounts for the different photoemission cross-sections, λ and d are the overlayer escape depth and thickness. The experimental values found for the interfaces a-Si$_x$C$_{1-x}$:H/a-Si
and a-Si:H/a-Ge are reported in Fig.2. The exponential behavior is
strictly obeyed showing that the growth at room temperature of amorphous
interfaces between tretravalent elements proceeds without interdiffusion.
An equally abrupt junction has been obtained by Abeles et al.[6] by growing
a-Si:H on a-SiN$_x$:H .

As mentioned previously, the heterojunctions between amorphous semiconductors studied until now exhibit small valence-band discontinuities,
very close to the experimental limit of the technique (a notable exception

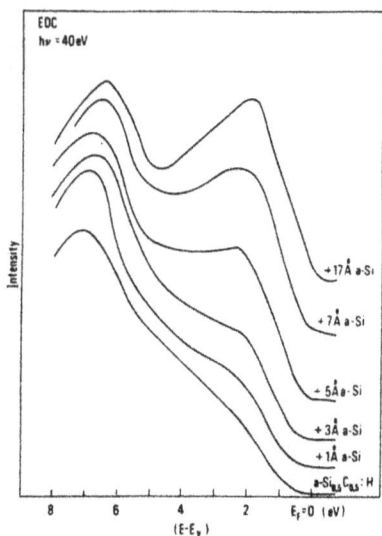

Fig. 3 Photoelectron energy distribution curves (EDC's) of a clean a-Si$_x$C$_{1-x}$:H substrate covered by a-Si overlayer of increasing thickness. Energy scale is referred to the Fermi level E$_F$.

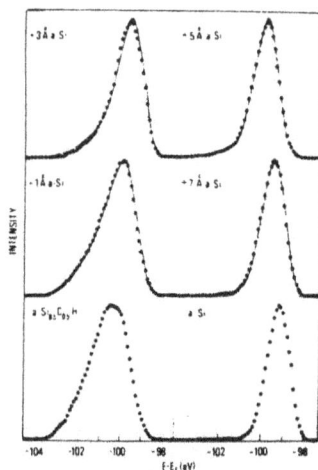

Fig. 4 Si 2p spectra for clean and a-Si covered a-Si$_x$C$_{1-x}$:H taken at hν = 135 eV. The solid lines correspond to the results of the fitting procedure explained in the text.

is the system a-SiN$_x$:H/a-Si:H at large nitrogen concentration[6]). As a consequence, at small overlayer-thicknesses the configuration sketched in Fig.1 is not found experimentally and the band discontinuity cannot be inferred unambiguously from the inspection of the valence band edge. Quite to the contrary, the edge evolves continuously from that of the substrate to that of the overlayer. An example is shown in Fig.3 for the system a-Si$_x$C$_{1-x}$:H/a-Si. The silicon-carbon alloy was grown by glow-discharge in a mixture of 1:3 SiH$_4$ to CH$_4$. The atomic composition was determined by Auger spectroscopy and a relative concentration x=0.4-0.5 was estimated. It is not certain, however, whether this concentration reflects the bulk composition or is the consequence of a preferential removal of one component by Ar sputtering used to clean the surface. Two features are evident in Fig.3 for increasing overlayer thickness: the increase of the first structure (extending in the range 0-5 eV) due to Si 3p bonding states and the movement of the onset toward the Fermi level. However, from the behavior of the valence band top is not possible to establish whether a discontinuity is developing or a simple band bending is occurring. In order to distinguish between these two possibilities it is convenient to analyze the core level spectra. The Si 2p core level EDC are shown in Fig.4. The photon energy (hν = 135 eV) was selected to obtain a photoelectron escape depth similar to that of the valence bands of Fig.3. The two bottom curves refer to the clean substrate and to a thick a-Si overlayer respectively. Notice the larger linewidth and the energy shift of the a-Si$_x$C$_{1-x}$:H peak with respect to the a-Si peak. The lineshape of the a-Si$_x$C$_{1-x}$:H core is due to the simultaneous presence of large amounts of C and H which introduce a large compositional disorder and a variety of

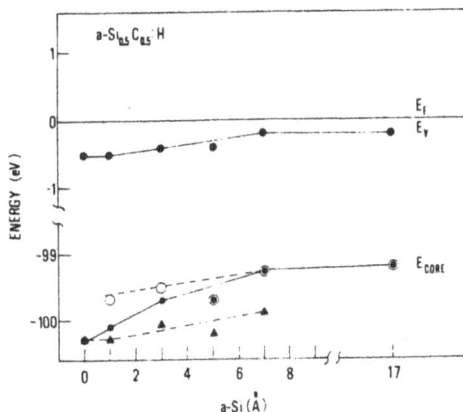

Fig. 5 Position of the valence-band maximum
and of Si 2p core level (dots and solid
lines) as a function of a-Si coverage.
Open circles and triangles refer to the
two components of the core level decon-
volved with the procedure explained in
the text.

bonding configurations for the Si atoms. The resulting lineshape is the
envelope of different Si 2p peaks affected by different chemical shifts.
The middle and top spectra of Fig.4 show the evolution of the linshape
from that of a-Si_xC_{1-x}:H to that of a-Si for progressive a-Si coverages.
These spectra were fitted by a linear superposition of the a-Si and
a-Si_xC_{1-x}:H lineshapes with fixed energy separation, using the peak intensities as fitting parameters. The solid lines show the results of the
fitting procedure and the dotted lines show the two components for each
spectrum.

The results of the analysis of valence-band and core level spectra
can be summarized as shown in Fig.5. The top part shows the shift of the
valence-band maximum as a function of a-Si coverage. The bottom part of
the figure shows the position in energy of the two components of the fit
(dashed lines) together with the overall shift of the Si-2p (solid line).
All energies are reported as a distance from the constant Fermi level.
Notice that the two components of the Si-2p band closely follow the shift
of E_v . This demonstrates that all these shifts are simply due to changes
in the band bending during the a-Si deposition and that there is no discontinuity in the valence band.

Identical results are obtained when the a-Si_xC_{1-x}:H is doped by adding
0.1% of B_2H_6 to the gas mixture. This doped material, with E_F ~0.3 eV
above the top of the valence band, is the one which is actually used as a
p-doped layer in the p-i-n photovoltaic devices.

By combining the photoemission information with the knowledge of the
optical gap and Fermi level position in the bulk, a complete characterization of the heterostructure during its formation is obtained, as sketched
in Fig.6. Notice the p-character of the clean surface, a result found
consistently in the hydrogenated samples investigated and probably due to

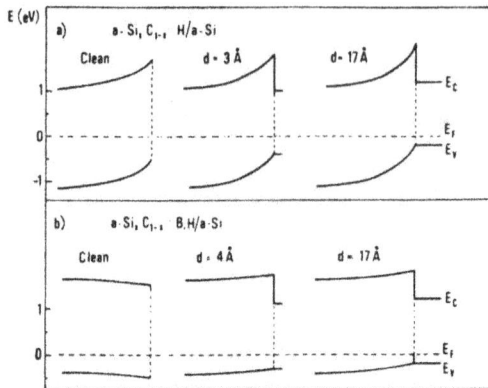

Fig. 6 Scheme of the heterojunction formation
showing the change of surface band
bending and the band discontinuities.

the sputtering procedure used for cleaning the samples which introduces surface states that pin the surface Fermi level close to the valence band maximum. The density of these surface states was estimated by Wagner et al.[9] for a-Si:H and found to be 10^{13} cm^{-3} eV^{-1}. For the boron doped material, however, the distance $E_F - E_V$ at surface is larger than that in the bulk causing a depletion layer for holes. With increasing overlayer thickness the p-character is enchanced and results in a change from depletion to accumulation of holes.

As an example we have discussed in same detail the results obtained for the heterostructure a-Si$_x$C$_{1-x}$:H/a-Si. However, the a-Si$_x$C$_{1-x}$:H/a-Si:H system too has been studied[4,6] and a negligible ΔE_V found. It seems, therefore that alloying the amorphous hydrogenated silicon with a percentage of C not higher than 40-50% does not cause any substantial recession of the valence-band maximum, although an appreciable depletion of states in the Si 3p region is evident. The negligible magnitude of ΔE_V implies that the difference between the two pseudogaps is entirely accomodated by the conduction band discontinuity, as sketched in Fig.6.

It is worth emphasizing that the knowledge of the band discontinuities is fundamental to the understanding of the behavior of a heterostructure device. In particular, the use of a-Si$_x$C$_{1-x}$H as a p-doped layer in the p-i-n photovoltaic devices resulted in a substantial improvement of the efficiency of amorphous silicon solar cells.[3] Besides the beneficial window effect due to the larger optical gap, the band alignment we have shown suggests the presence of a second advantage, i.e. the conduction band discontinuity hinders the back-diffusion of the electrons while the hole collection is not affected at all due to the negligible ΔE_V .

The interfaces between Si and Ge have been extensively investigated in the crystalline case[10] and are one of the test systems[11] for the theories on the heterojunction formation. As a consequence, the study of their amorphous counterpart, besides new information on these disorder systems, can provide new insight on the driving force that determine the

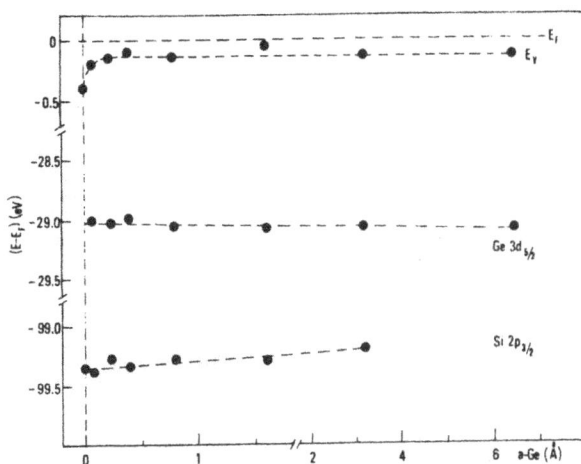

Fig. 7 Position of the valence-band maximum,
Si 2p and Ge 3d core levels as a
function of a—Ge coverage for the
a—Si:H/a—Ge heterojunction.

band alignement, diffusion potential, etc. Moreover, the study of the
amorphous Si/Ge interface has an immediate technological interest, since
amorphous hydrogenated silicon-germanium alloys are used in photovoltaic
devices to obtain an absorbing layer with optical gap smaller than that of
a—Si:H. The band discontinuity of the amorphous Si/Ge system gives an
upper limit to the discontinuities that can occur at the interface between
silicon and the alloys.

The results for the a—Si:H/a—Ge heterojunction[12] can be summarize as
shown in Fig.7, where the energy of the valence band maximum E_v , of the
Ge 3d and the Si 2p core levels are shown as a function of the overlayer
thickness. A shift of E_v by 0.2 eV occurs upon deposition of a fraction
of monolayer, while the Si 2p does not move, indicating that a valence
band discontinuity is setting in. By increasing the overlayer thickness,
a slight decrease of E_v —Si 2p distance is detectable, without any displacement of the valence-band maximum and of the Ge 3d core level. It is
a small effect, close to the sensitivity of the technique, nevertheless,
it points to a complex evolution in the interface formation. It is worth
remarking that the valence band discontinuity is practically established
at the very beginning of the overlayer growth. This surprising and somewhat mysterious behavior is consistently found in the study of crystalline
heterjunctions as well as of the Schottky barrier formation.[1]

Quite to the contrary, no valence band discontinuity is found for
a—Ge grown on a—Si.[12] This result must be compared with the crystalline
counterpart where ΔE_v = 0.17 eV is found.

One particularly interesting aspect of the research on amorphous heterojunctions is the possibility to investigate the effects of disorder
and/or hydrogenation on the valence band edge of the semiconducting material. This is obtained by studying an amorphous overlayer of a given material on the crystalline or hydrogenated amorphous counterpart. This investigation has been performed on c—Si/a—Si and a—Si:H/a—Si systems [5]
The results are summarized in Fig.8.

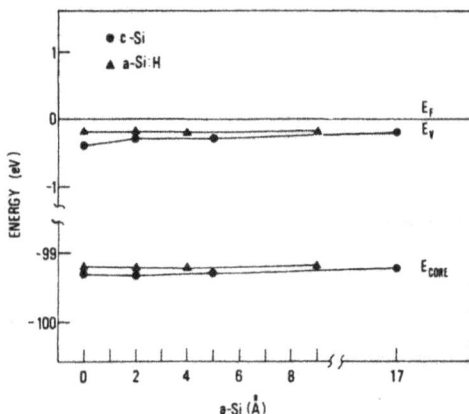

Fig. 8 Position of the valence-band
maximum and of Si 2p core level
for a-Si:H and c-Si as function
of a-Si coverage.

The data on c-Si/a-Si interface suggest a possible upward shift of
0.1 eV which is, however, comparable to the experimental uncertainty.
This result clearly demonstrate that the valence-band mobility-edge coincides within 0.1 - 0.2 eV with the valence-band maximum of the corresponding crystal. Since it is commonly estimated[13] that the mobility gap in
amorphous silicon is larger than the forbidden gap of the crystal, it follows that the conduction-band mobility-edge should undergoes the larger
shift.

A similar situation results from the study of the a-Si:H/a-Si system.
Fig.8 shows that E_v of amorphous hydrogenated silicon is aligned to that
of a-Si. The data refer to samples with an hydrogen content of ~10 %. By
comparing the distance of E_v from the Si 2p core level the same conclusion
can be inferred from data published recently.[14,15] However, from ref.15,
it is evident that at far larger hydrogen concentration (C_H >25%) the valence-band maximum moves closer to the core level, i.e. there is recession toward higher binding energy. On the other hand, at 10 % hydrogen
content the optical gap of a-Si:H is already much larger than that of a-Si
(~1.7-1.8 eV compared to ~1.2 eV). The recent photoemission data show,
therefore, that this widening of the optical gap is primarily due to a
shift in energy of the conduction band edge.

It is instructive to compare some of the experimental results reviewed above with the current theories on the heterojunctions. For crystalline substrates, the problem of the interface formation was the subject
of many theories and experiments over the past 20 years. It cannot be regarded as completely solved, however. For amorphous substrates, the problem is even harder due to the presence of several kind of disorder in the
matrix.

From the theoretical point of view, there are few microscopic band
calculation trying to determine the local electronic structure of the interface in detail, by taking into account the local structure, coordination, etc.. None computes ΔE_v for the crystalline counterpart of the

systems we have discussed. On the other hand, there are several theories trying to calculate the band lineup from bulk crystal properties by suggesting different "driving forces" for the alignement. The first example of these theories is the well-known Anderson's "electron affinity rule",[16] which states that the conduction-band discontinuity is equal to the difference between the electron affinity of the constituent semiconductors. This rule, which proved hard to test experimentally due to the lack of reliable values for the electron affinity within an accuracy of a tenth of eV, was later criticized because it uses free-surface parameters to describe interface parameters. Presently, there are several different suggestions for the "driving force" determining ΔE_v.[17-20] It was recently shown by Katnani and Margaritondo[11] that all the above models predict values for ΔE_v in unsatisfactory agreement with the experimental findings. However, the Harrison's and Frensley-Kroemer's models give the best agreement and the correct chemical trend within 0.2 eV, i.e. very close to the experimental uncertainty itself. More recently a new and interesting suggestion for the "driving force" for band alignement was put forward by Tersoff.[21] These three models, which are not necessarily in contrast, will be discussed briefly in the following.

Harrison used a tight-binding approach to calculate the absolute position of the valence-band maximum in the bulk relative to the vacuum level and argued that in a heterojunction ΔE_v is simply the difference between E_v of the constituent semiconductors. Let us apply this criterion to amorphous interfaces. Moving from crystalline to amorphous semiconductors the velence-band top is blurred by localized states induced by disorder. The width of this tail (which is mainly the result of fluctuation of the dihedral angle) is expected to be of the order of 0.3 eV.[22] The mobility edge should move over a distance less than that or not move at all. As a consequence within few tenth of eV uncertainty, the Harrison criterion is in agreement with the experimental results for c-Si/a-Si and a-Si/a-Ge heterojunctions.

In the Frensley-Kroemer's criterion the valence-band maximum is calculated relative to the average interstitial potential and it is assumed that this average potential coincide in the two semiconductors when the heterostructure is formed. Since in the amorphous case the average local coordination, bond length and bond angle remain unchanged just as the average valence density of states does, we expect that the average interstitial potential too does not change. Therefore, the only difference between amorphous and crystalline case is again a slight shift of the mobility edge. The Frensley-Kroemer model also is in agreement with the experimental results.

Tersoff's criterion states that the interface dipole is the dominant factor determining band lineup and that the actual lineup is that which gives a zero interface dipole. In order to find this "canonical" lineup one has to find the demarcation energy E_B where the gap states shift from dominant valence (i.e. bonding) character to dominant conduction (i.e. antibonding) character. A zero-dipole lineup is obtained by aligning E of the respective semiconductors. This theory, which is remarcably successful in predicting the experimental ΔE_v for a series of crystalline heterojunctions, gives $\Delta E_v = 0.18$ eV for the c-Si/c-Ge system resulting from $E_B = 0.36$ eV above the valence-band maximum in c-Si and 0.18 eV in c-Ge, to be compared to $\Delta E_v = 0.17$ eV found experimentally. Therefore, the zero valence-band discontinuity found for the amorphous Si/Ge interface would mean that E_B is located at equal distance from E_v in the two amorphous semiconductors and that there is a displacement of the demarcation level between bonding and antibonding caracter in one or both semiconductors. On the other hand, the data on the c-Si/a-Si heterojunction are consistent with a displacement of 0.1 eV of E_B in a-Si bringing the expected ΔE_v for a-Si/a-Ge system to 0.08 eV, provided that E_B of Ge does not change. The study of the c-Ge/a-Ge system will be, therefore, an experimental test for the application of the theory to the amorphous semiconductors. It is worth noticing that the determination of E_B is of great importance because it directly impinges on the crucial problem of the extent of donor- and acceptor-like states into the pseudo-gap.

From this short discussion we have seen that the most successful theories on the heterojunction formation are consistent with the results obtained on the amorphous semiconductors. On the other hand, it would be even more interesting to use the above theoretical criteria to infer from the data new properties of the amorphous semiconductors, above all of the more complexes ones like a-Si:H and the hydrogenated alloys. However, to do so, we must be fully confident on the theory and, therefore, we have to wait until the problem of heterojunction formation can be considered completely solved, at least for the less difficult case, i.e. crystalline heterojunctions.

REFERENCES

1. G. Margaritondo, Solid State Electron. 26, 499 (1983), and references therein.
2. B. Abeles and T. Tiedje, Phys. Rev. Lett. 51, 2003 (1983).
3. Y. Tawada, H. Okamoto, and Y. Hamakawa, Appl. Phys. Lett. 39, 237 (1981);   A. Catalano,   R.V. D'Aiello,   J. Dresner,   B. Faughman, A. Firester, J. Kane, H. Schade, Z.E. Smith, G. Swartz and A. Triano, 16th IEEE Photovoltaic Specialists Conference, San Diego 1982.
4. F. Evangelisti, P. Fiorini, C. Giovannella, F. Patella, P. Perfetti, C. Quaresima, and M. Capozi, Appl. Phys. Lett. 44, 764 (1984).
5. F. Patella, F. Evangelisti, P. Fiorini, P. Perfetti, C. Quaresima, M.K. Kelly, R.A. Riedel, and G. Margaritondo, "Optical Effects in Amorphous Semiconductors", AIP Conf. Proc. 120, 402 (1984).
6. B. Abeles, I. Wagner, W. Eberhardt, J. Stohr, H. Stasiewski, and F. Sette, "Optical Effects in Amorphous Semiconductors", AIP Conf. Proc. 120, 394 (1984).
7. D. Adler and F.R. Shapiro, Physica 117B-118B, 932 (1983).
8. F. Evangelisti, F. Patella, R.A. Riedel, G. Margaritondo, P. Fiorini, P. Perfetti, and C. Quaresima, Phys. Rev. Lett. 53, 2504 (1984).
9. I. Wagner, H. Stasiewski, B. Abeles,, and W.A.Lanford, Phys. Rev. B28, 7080 (1983).
10. G. Margaritondo, N.G. Stoffel, A.D. Katnani, and F. Patella, Solid State Commun. 36, 215 (1980); P. Perfetti, N.G. Stoffel, A.D. Katnani, G. Margaritondo, C. Quaresima, F. Patella, A. Savoia, C.M. Bertoni, C. Calandra, and F. Manghi, Phys. Rev. B24, 6174 (1981).
11. A.D. Katnani and G. Margaritondo, Phys. Rev. B28, 1944 (1983).
12. F. Evangelisti, S. Modesti, F. Boscherini, P. Fiorini, P. Perfetti, and C. Quaresima, MRS 1985 Spring Meeting, San Francisco, 1985 and to be published.
13. M.H.Cohen, C.M. Soukoulis, and E.N. Economou, "Optical Effects in Amorphous Semiconductors", AIP Conf. Proc. 120, 371 (1984).
14. D. Wesner and W. Eberhardt, Phys. Rev. B28, 7087 (1983).
15. J. Reichardt, L. Ley, and R.L. Johnson, J. Non-Cryst. Solids 59-60, 329 (1983).
16. R.L. Anderson, Solid-State Electron. 5, 341 (1962).
17. W.R. Frensley and H. Kroemer, Phys. Rev. B16, 2642 (1977)
18. W. Harrison, J. Vac. Sci. Technol. 14, 1016 (1977).
19. M.J. Adams and A. Nussbaum, Solid-State Electron. 22, 783 (19749.
20. O. von Ross, Solid-State Electron. 23, 1069 (1980).
21. J. Tersoff, Phys. Rev. B30, 4874 (1984).

22. F. Yonezawa and M.H. Cohen, "Fundamental Physics of Amorphous Semiconductors", Ed. F. Yonezawa, Springer-Verlag, 1981, p.119.

Polycrystalline Thin-Film Solar Cells

John D. Meakin
Institute of Energy Conversion
University of Delaware
Newark, Delaware 19716

INTRODUCTION

In order to discuss the development and prospects for solar cells in which the active layers are polycrystalline semiconductors it is best to approach the subject in a logical rather than a historic manner. The latter approach tends to make the polycrystalline cells appear as a degraded form of the single crystal Si cells and does not reflect the many advantages of using thin layers of compound semiconductors.

It is relatively straightforward to set out the necessary and sufficient material and device design criteria for a stable, high efficiency photovoltaic device. There are also reasonably clear requirements which must be met if there is to be the potential to produce a given cell in large areas at low cost. These criteria are first discussed in a general way and then applied to a number of material systems that have been developed to at least the laboratory prototype stage.

An essential stage in the transition from concept to pilot scale production, which has often been insufficiently pursued,can be termed chemical process research. This branch of engineering provides the link between laboratory scale research and prototype production.If it is absent, or inadequately carried out,as has been the case on a number of occasions during the short history of commercial solar cells, the result can be total failure to bring a product to market on time, if at all. Examples of this type of engineering research will be described.

We are still at a very early stage in the development and deployment of photovoltaic systems and the history of probably every technology would suggest that extreme caution should be exercised before deciding that we can safely select "winners and losers" in the solar cell arena.

THE GENERIC SOLAR CELL

A generalized solar cell is shown in Figure 1. The essential photovoltaic event is the absorption of a photon with the production of an electron-hole pair. In order to utilize the energy represented by the pair a charge separation junction must exist at which the minority carrier is converted into a stable majority carrier. The absorbing layer must be a semiconductor and a number of authors have analyzed the forbidden energy gap which, under idealized conditions, will yield the highest theoretical conversion efficiency (1,2). Figure 2 is an annotated version of the curve originally published by Prince (2).

The charge separating junction can created between semiconductors, giving a homo- or heterojunction, or between a metal and a semiconductor forming a Schottky junction. The other components of the cell, such as the transparent contact, all have very specific material requirements but these will in general be qualitatively the same for all material systems e.g. high transparency coupled with a system-specific minimum conductivity for the front contact.

THIN-FILM SOLAR CELLS

Taken as a given in theoretical analyses of efficiency is the complete collection of all generated carriers and a barrier to reverse carrier flow that is set by the absorbing semiconductor band gap. The practical realization of these conditions that impose other essential material requirements on the absorbing semiconductor. The absorption of the incident photons follows an exponential law with the characteristic length given by $1/alpha$, where alpha is the absorption coefficient. Alpha is a structure insensitive property of the semiconductor and its magnitude determines the minimum carrier properties which are necessary for good current collection. Figure 3 shows some selected alpha vs photon energy curves which immediately reveal the major property difference between crystalline silicon and various semiconductors used in thin film cells. Silicon, with a band gap of about 1.0eV, has an absorption coefficient which slowly rises from about $10 cm^{-1}$ just above the band gap to about $10^4 cm^{-1}$ at the limit of the useful solar spectrum. In contrast the semiconductors which have been used as polycrystalline thin-films for solar cells show direct absorption edges with alpha in the range of 10^4 to $10^5 cm^{-1}$ immediately above the band gap.

In order to effectively absorb the incident photons the active semiconductor layer must have a thickness of $2-3 \times alpha^{-1}$. Crystalline silicon must therefore be about 100 microns thick whereas the compound semiconductors in Figure 3 need only be of the order of 1 micron thick. The consequent saving in material is much less important than the effect on the "material properties" that are necessary for high efficiency.

To collect electrons or holes over distances of ~100 microns mandates high purity and perfection, e.g. semiconductor grade Si, but much "poorer" material will still give diffusion or drift distances of ~1 micron. This relaxation in required material properties means that a very large number of film preparation techniques can be used with attendant reductions in the cost of making large areas of solar cells. Therein lies the conceptual and practical difference between thin film cells, whether amorphous or polycrystalline, and the traditional Si cells. In the remainder of this paper we will discuss the practical realization of polycrystalline thin-film cells (PTFC) using a number of different compound semiconductors.

COMPOUND SEMICONDUCTORS

There are essentially no sources of "off the shelf" polycrystalline thin films and even the bulk materials are only of limited availability. A cell development effort may therefore have to start with the synthesis of the desired compound. The systematic development of an efficient PTFC cannot rely on a corpus of basic materials knowledge in the way that was possible with Si and as a consequence a logic of the type illustrated in Figure 4 for CuInSe$_2$ must be followed. However the resources available to most programs preclude developing a full basic understanding of the preparation-property relations and the material work may of necessity be highly empirical.

The first PTFC of reasonable efficiency was the CdS/Cu$_2$S heterojunction, unfortunately named the CdS cell. This cell was extensively studied for over two decades and a number of efforts have been made to commercialize it. This cell will be used to demonstrate many of the features of PTFC development which are common to most material systems. However major attention will be devoted to CdTe and CuInSe$_2$ heterojunctions with CdS or (CdZn)S as these are the systems being actively pursued at both the research and pre-production stage.

A material which is not under current development in America at the moment, but which has many attractive features, is Zn$_3$P$_2$ and the status of this and a number of other promising compounds will be briefly described.

THE CdS/Cu$_2$S HETEROJUNCTION

A photovoltaic effect in CdS/Cu$_2$S was discovered by Reynolds and Leies (3) over 30 years ago and since that time there have been innumerable research programs and a number of commercialization efforts on this system. It is fair to say that to a very considerable degree all efforts were heavily empirical and only more recently were many of the underlying phenomena fully understood. With the advantages of hindsight this is explicable in terms of the complexity of the Cu-S system and the range of properties that CdS can show, particularly after interacting with the adjacent Cu$_2$S layer.

A thorough review of the structure and operation of this cell has been given by Rothwarf (4) and only certain features will be dealt with here as exemplary of general effects in PTFC. These will be divided into either material or device related effects in as far as such a division is possible.

Underlying much of the variability in cell performance both within and between different research groups was the unrecognized range of properties that the two component semiconductors can exhibit. Cu$_2$S is particularly troublesome as it is essentially unstable in air and tends towards Cu$_2$O with resultant major changes in electronic properties. Figure 5 shows how the relation between hole density and stoichiometry. The usual Cu$_2$S layer is about 0.5 microns thick, so that the formation of a monolayer of oxide is enough to make orders of magnitude changes in hole density (5). Also changing with stoichiometry is the absorption behavior and electron diffusion length, both of which directly affect the device performance. Until some measure of recognition of these effects was achieved it is understandable that reliable production of high efficiency cells eluded many groups.

CdS is much more stable than Cu_2S but even using a single preparation technique such as vapor deposition, it is possible to make films that span the range from insulating to highly conducting. To add to the complexity, cell production almost invariably involved heating the CdS in intimate contact with the Cu_2S containing layer resulting in a variable degree of Cu compensation. The photoconductive nature of the CdS and hole trapping effects were a fruitful source of misleading analysis when testing was carried out in the dark or under monochromatic illumination. Eventually it was recognized that to explore the normal operation of the cell it was essential that measurements be taken under full white light bias conditions.

There are a number of material lessons to be learnt from the history of this cell. A major focus of the research must be on the materials themselves and it is fatal to forget that, what may in chemical terms be a single material, can show an entire spectrum of semiconductor properties.

An other major class of potential problems arises when an unjustified assumption of uniformity is made. Conventional testing always assumes homogeneity and the output at the terminals is taken to represent the behavior of the junction at all positions in the cell. With a Si wafer this may be well founded but it most certainly is not a forgone conclusion with a 2 micron film deposited from the vapor. In the specific case of the CdS/Cu_2S heterojunction it was found that localized defects were in fact dominating the time dependent behavior of the cell and were a major cause of instability. Laser scanning has proved invaluable in thin-film cell research (6) and the technique is able to both locate and neutralize potentially damaging defects. There is now very good evidence that a stable cell can be made if appropriate steps are taken to prevent or neutralize localized shunt paths (7).

Various aspects of heterojunction behavior more related to device design than materials were exemplified by this PTFC. These included current controlling effects such as front surface reflection and light trapping. Voc is also significantly affected by the junction area which relates to the method of forming the Cu_2S.

The interplay between empirical developments and fundamental understanding is shown by the history of the front surface treatments of the cell. Traditionally the CdS was briefly etched with HCl before the Cu_2S "barriering" step. The beneficial effect of this was shown to be due to an anti-reflection effect but it was not until much later that it was realized that the textured surface also caused total internal reflection and hence total absorption of long wavelength light reflected from the substrate. The texturing also gave a much increased junction area which lowered the achievable Voc and at one stage a cell design with a planar junction seemed a promising higher voltage alternative. The idea was to take care of front surface reflection using a single or double layer anti-reflection coating and to form the Cu_2S using the te Velde (8) process rather than the solution reaction developed by Clevite (9). It was only after considerable effort that it was realized that with a planar surface the second pass light was leaving the cell and hence limiting the achievable current. A potential solution to this problem, appropriately texturing the substrate, has not been fully explored (10).

The controlling effect of the electron affinity mismatch on the barrier height and hence Voc was demonstrated by Palz et al(11).However in all early attempts to raise Voc there was an apparently related drop in current (11,12). The group at IEC was convinced that the modelling of cell behavior was sufficiently reliable to be certain that the loss in Jsc was not inevitable and that with appropriate material properties the gain in voltage could be obtained while maintaining the currents. After a very substantial experimental effort this was confirmed (13).

An other outcome of the CdS/Cu_2S research was a quantitative model of heterojunctions in which the importance of the field distribution in the vicinity of the interface was recognized. It is not unusual to read that good lattice matching is needed to avoid efficiency limiting recombination at interface defects. In fact the match at the CdS/Cu_2S and many other heterojunctions is probably sufficiently poor to ensure a limiting density of interface dislocations. The junctions nevertheless give very usable efficiencies because the interface recombination losses are controlled by a function that involves both the interface recombination rate and the local field namely,

$$IRF = \mu F / (\mu F + S).$$

IRF is the interface recombination factor or the fraction of minority carriers which are successfully converted to majorities and collected (14). μ is the carrier mobility, F the field sweeping the carriers away from the junction and S the interface recombination rate. By ensuring a high field close to the junction the effects of interface recombination can be minimized.

There have been at least two recent and unsuccessful attempts to commercialize the CdS/Cu_2S solar cell for terrestrial use. SES Inc utilized a vapor deposition process for the CdS layer and Photon Power a pyrolitic spray process. Both companies formed the Cu_2S layer using the Clevite process. Many factors contributed to the difficulties faced by each company in their attempts to bring a stable product to the market place but a contributing factor was that there was a dearth of "engineering research" into the production processes themselves. In the case of vapor deposition the laboratory scale research had yielded essentially empirical recipes for CdS deposition. Virtually nothing was known about the mass and heat balances in the CdS sources and in fact subsequent research revealed that they were poorly designed for reliable and reproducible operation (15).

A research program specifically aimed at developing a
continuous and reproducible CdS deposition process on a moving
substrate proved successful and demonstrated the benefits of
applying process chemistry to solar cell production processes
(16). At the present time the only known commercialization effort
is underway at Nukem in Germany using a process based on the
University of Stuttgart research (17).

HETEROJUNCTIONS BASED ON CdTe

CdTe has ideal properties for a photovoltaic semiconductor.
The band gap is direct with a value of 1.45eV at room temperature
and the absorption coefficient is high, Figure 3. Although CdTe
is amphoteric and hence a homojunction can be made this is not an
attractive device design for an efficient solar cell. The very
high absorption coefficient would require a thin n layer and the
cell would be sensitive to surface recombination. In common with
all strongly absorbing semiconductors the most promising design
is a heterojunction with a wide band gap "window" layer. With a
back-wall design the light enters the absorbing semiconductor
through the junction reducing to a minimum the distance that the
minority carrier must diffuse or drift in order to be collected.

Bube(18) and Bube et al (19) have recently reviewed the
status of CdTe based solar cells which include both single
crystal and thin-film devices. For our purposes only
polycrystalline thin-films are of interest and to date there have
been a number of reports of high efficiency cells (20-25) with at
least two groups reporting over 10% efficiency (22,23).

Most wide band gap window materials are n-type and as
electrons are generally more mobile than holes most efforts have
been directed at p-type CdTe. This has been used in combination
with CdS, ZnO and ITO. The major difficulties encountered with p-
type CdTe are controlled doping to achieve high conductivity and
the related problem of making a low resistance stable contact.
The latter is inherently difficult because the work function is
~5.7eV so that no simple metal is expected to give ohmic contact.

The techniques which have given good CdTe films are quite
varied and include electrodeposition (22,24), close spaced vapor
transport (CSVT) (19), chemical vapor deposition (CVD) (25) and
screen printing (23). The salient features of each technique and
the solar cell performance achieved will now be briefly
described.

The CdTe layer in the first thin-film cell reported with a
reasonable efficiency was deposited using a type of hot-wall
system (20). The cell was completed with a vapor deposited
CdS/CdS(In) bilayer. Performance was limited by high contact
resistance but under AMO conditions an efficiency of about 6% was
obtained (20). After ten years of unprotected storage some of the
cells were re-tested and under 100mW/cm^2 they still showed Jsc
over 20mA/cm^2 (26).

Both CdS and CdTe can be cathodically electrodeposited and research at Monosolar (21) yielded a very thin high efficiency CdS/CdTe heterojunction using this growth technique. ITO coated glass was used as a substrate for an initial layer of CdS less than 0.1 micron thick; onto this 1.2-1.5 microns of CdTe was deposited. It was found that the CdTe was initially n-type but a heat treatment for 8-10 minutes at 400°C in the presence of oxygen causes a type conversion. The surface of the CdTe is then etched for the stated purpose of removing oxides and the cell completed with a Au back contact layer. Small area cells ($2mm^2$) with efficiencies of 9.35% have been reported (21); Voc=.73V, Jsc=20mA/cm^2 and FF=64%. Spectral response measurements confirmed that the device operates as a heterojunction.

A high efficiency Schottky barrier solar cell using electrodeposited CdTe has also been reported but in this case the CdTe is n-type (24). Deposition has been carried out onto metallic substrates with a layer of Cd acting as the back ohmic contact. Various metals have been used to form the Schottky barrier including Ni and Au. In spite of the limitations imposed by the transparent metal contact through which the light enters the cell, an efficiency of 8.65% has been reported for a cell of $2mm^2$ area; Voc=0.723V, Jsc=18.7mA/cm^2 and FF=64% (24).

CSVT has been used by researchers at Stanford University (19) and a closely related technique called close spaced sublimation (CSS) at the Kodak Company (22). The latter have reported over 10% efficiency with a cell design very similar to the Monosolar electrodeposited cell. About 0.1 microns of CdS is first deposited in air at 1 Torr onto ITO coated glass. - The junction is completed with about 4 microns of CdTe deposited in oxygen at a pressure of 1.5 Torr. Au is used as the back contact and cells of $10mm^2$ area were delineated by scribing. Under simulated sunlight of 75mW/cm^2 intensity, an efficiency of 10.5% was reported; Voc=.75V, Jsc=17mA/cm^2 (23mA/cm^2 at 100mW/cm^2) and FF=64%. The importance of oxygen was later explored more fully (27) and it was concluded that in an as yet undefined way oxygen serves to promote p-type conductivity in CdTe.

Thin-film CdTe/CdS solar cells have been reported by Bube et al (19) in which the CdTe was deposited onto graphite using either CSVT or hot-wall vacuum evaporation. The graphite contact is not ideal and various efforts to increase the CdTe conductivity are reported. Cells were completed by the physical vapor deposition of a CdS layer with an In grid contact. The CSVT material yielded $14mm^2$ cells with up to 6.4% efficiency when tested under 91mW/cm^2; Voc=.52V, Jsc=17.4mA/cm^2 (19.1mA/cm^2 at 100mW/cm^2) and FF=64%.

The other high efficiency CdTe/CdS heterojunction has been reported by Matsumoto et al (23) who used a screen printing and sintering process. A continuous CdS film is first prepared on glass and onto this is deposited an interdigitated pattern of Ag/In electrodes and a CdTe layer. The back contact to the CdTe is a graphite/Ag structure while the Ag/In stripes contact the CdS. In this way the need for a transparent layer on the glass is avoided. These authors confirm a previous report (20) that a small amount of Cu helps reduce the contact resistance to p-type

CdTe. A 78mm^2 area of electroded CdTe on a continuous layer of CdS gave an efficiency of 12.8% under 100mW/cm^2 simulation; Voc=.754V, Jsc=27.9mA/cm^2 and FF=60.6%. Small modules made up from these cells have been tested under roof top conditions with encouraging results (28).

Direct combination of the elements has been used in a CVD process by Chu et al to grow p-type CdTe films on W coated graphite (25). They report unusually low contact resistances of 2ohm-cm^2 by the use of an Sb interlayer, however the completed cells show some instability which may be contact related. The conductivity of the CdTe films was controlled by either growth under a Te excess or by intentional doping with As introduced into the gas stream as AsH$_3$. Both techniques led to a minimum resistivity of about 200 ohm-cm. Various n-type window layers were tried including ZnO, CdO and CdS but the best results were obtained with sputtered ITO. The top contact on the most efficient cell reported was ITO with an Al grid and a Ta$_2$O$_5$ anti-reflection coating. Tested under 100mW/cm^2 simulation an efficiency of 8.1% was reported for a 1cm^2 cell;Voc=.74V, Jsc=18mA/cm^2 and FF=61%.

There has recently been increased commercial interest in the CdTe/CdS cell and a number of firms have either announced plans to initiate manufacturing efforts or are believed to be engaged in such efforts. Future developments may involve alloys such as (CdZn)Te or (CdHg)Te which would allow adjustment of the band gap for specific device designs including multi-junction tandem systems.

HETEROJUNCTIONS BASED ON CuInSe$_2$

The I-II-VI$_2$ chalcopyrites are an extensive series of semiconductors which often show very useful optical and electronic properties. CuInSe$_2$ has attracted considerable interest for photovoltaic purposes since the first report of high conversion efficiency in a device constructed on a single crystal wafer (29). CuInSe$_2$ has a band gap of about 1eV and the highest absorption coefficient known to this author. In contrast to many II-VI compounds CuInSe$_2$ thin-films cannot be made by vapor deposition from the compound which does not evaporate congruently. Various vapor deposition techniques have been developed as described in a recent review article (30). Cells of more than 10% efficiency have been reported by two groups (31,32) both of which are using elemental Cu, In and Se sources. The structure of the cells comprises a Mo back contact on a glass or Al$_2$O$_3$ substrate. The CuInSe$_2$ film is deposited onto this in two steps, a Cu rich film followed by an In rich film. It seems likely that the initial layer is necessary to ensure ohmic contact to the Mo. The heterojunction is completed with a CdS or (CdZn)S layer which is either In doped throughout (32) or just in the final region (31). There are some differences in the processing of the IEC (32) and Boeing (31) cells but it is evident that in each case a heterojunction is produced with most if not all of the depletion layer in the CuInSe$_2$.

For the same reasons that (CdZn)S was used to improve the Cu_2S/CdS cell it was expected that higher voltages could be obtained in a $CuInSe_2$ cell with (CdZn)S. However the measurements of barrier heights available (33) suggest that the electron affinity mismatch between CdS and $CuInSe_2$ is quite small and much less than for Cu_2S and CdS. To date no significant increase in Voc has been reported although the enhanced transparency of (CdZn)S has yielded somewhat higher Jsc values. The highest efficiency cell reported by the group at Boeing (31) is 11.0% for a $1cm^2$ cell under ELH simulation at $100mW/cm^2$; Voc=.436V, Jsc=38.6mA/cm^2 and FF=65%.

A major reason for the widespread interest in $CuInSe_2$ cells is the high stability that even laboratory prototype cells have shown. Without any specific effort directed to stability it has been found that unencapsulated cells can be maintained under continuous illumination for thousands of hours and show essentially no decline in output (34). This can be ascribed to two main factors, the inherent stability of the $CuInSe_2$ itself and the back-wall cell design in which the CdS layer protects the critical junction region.

There have already been some reports of sub-module assemblies in which a number of cells are connected in series. Although still requiring considerable optimization those made by the Boeing group have shown efficiencies of 9 and 6% for areas of 20 and 91cm^2(35), and Arco Solar (36) have announced a sub-module of100cm^2 with an efficiency of about 5%.

All the published high efficiency cells have been made with $CuInSe_2$ produced by evaporation from the constituent elements which requires a Cu source temperature of well over 1000°C. An encouraging finding at IEC is that composition control need not be very exact and high efficiency cells can be made within the range Cu=24.5±1.5%, In=26.3±0.9% and Se=49.2±1.5%. Under active development at the moment are other large area deposition techniques including reactive sputtering of Cu and In in H_2Se (37), electrodeposition of $CuInSe_2$ directly (38) and selenization of Cu/In layers produced by various techniques (39). Experience with the vapor deposited films has shown that homogenization of $CuInSe_2$ layers occurs rapidly at temperatures of the order of 350°C.

The major parameter limiting efficiency is the open circuit voltage which is low for a band gap of 1eV. In addition to the modification of the n layer there are also efforts underway to increase the $CuInSe_2$ band gap by alloying with $Cu(InGa)Se_2$ (40). For this to be successful it will be necessary to maintain the high currents presently being achieved with $CuInSe_2$. An alternative approach to enhanced efficiency is to use the $CuInSe_2$ cell in a tandem structure as discussed below.

OTHER POTENTIAL MATERIALS

There are many compound semiconductors with band gaps appropriate for solar cells including GaAs, InP, Cu_2O, $CuInS_2$, Zn_3P_2 etc. All of these materials and many others have been made into photovoltaic junctions but are not being actively pursued for PTFC's at the moment. GaAs is under extensive development for many purposes but experience with thin polycrystalline films has been unsuccessful due to grain boundary shunting effects. High efficiencies have been achieved by growing thin single crystal layers (41). Cu_2O was the subject of many research efforts which were not able to overcome the chemical instability of the Cu_2O in contact with other materials (42). Zn_3P_2 remains an inherently attractive material from both the technical and economic viewpoints. As it only exists as a p-type semiconductor the primary requirement was to identify an appropriate n-type partner which has as yet eluded the groups that have studied it (43).

There are undoubtedly innumerable other potential semiconductors for photovoltaic purposes. Very few ternary compounds have been even superficially examined and as proposed by Loferski et al (44) there is an even greater degree of flexibilty in the range of physical properties available with quaternary and higher order compounds.

MULTIJUNCTION SOLAR CELLS

It is impossible to precisely define the efficiency that will ultimately be achieved with a single junction PTFC, nor is there a fixed minimum efficiency that is required for large scale economic power generation. A commonly accepted value for the latter is 15% (45) which is probably rather close to the efficiency that is achievable with a single junction thin-film solar cell. The concept of stacking two junctions is now 30 years old (46) and although most attention seems to have been devoted to applying this concept to single crystal Si or GaAs, and recently to amorphous cells, it is equally adaptable to PTFC. The basic requirements for an optically and electrically coupled tandem cell are a high band gap cell with transparent contacts, a low resistance interconnect and a low band gap cell. There are appropriate PTFC candidate cells namely the CdTe and $CuInSe_2$ heterojunctions with CdS. These do not have ideal band gaps but are sufficiently close to warrant investigation particularly as each has demonstrated over 10% efficiency. It can readily be shown that if the published individual performances can be maintained in a tandem, and there are not excessive optical or resistive losses in the interconnect, the tandem cell should be about 15% efficient (47).

A program is under way at IEC to develop such a tandem with
support from the U.S. Department of Energy through the Solar
Energy Research Institute. The cell design is shown in Figure 6.
The primary research need is to develop a CdTe cell which can be
grown on an existing CuInSe$_2$ cell. Most published CdTe/CdS cells
are produced by depositing the CdTe onto an existing CdS film
which is not applicable to the present tandem design. Live
devices have been produced as illustrated in Figure 7. As the
performance of the top CdTe cell improves there will be a
corresponding gain in the tandem cell. There are extensive
opportunities for future modifications to the basic tandem design
such as band-gap adjustments using (CdZn)Te, (CdZn)S and
Cu(InGa)Se$_2$.

SUMMARY

The present focus of both research and commercialization in
photovoltaics is on the amorphous materials based on Si:H which
may well be consuming over 75% of the efforts worldwide. To some
degree this emphasis reflects the fact that there are seen to be
uses for the amorphous materials other than low cost solar cells
thus increasing the potential pay-off from research expenditures.
Why then pursue polycrystalline materials? The question requires
no answer if the work is viewed as part of the overall search for
fundamental knowledge but it is a valid question if more
immediate and practically useful results are offered as
justification for industrial or governmental expenditures. Such
results are indeed projected and are based on the following
observations. Firstly there is an almost unlimited supply of
crystalline compounds which can be expected to contain materials
with any physically possible combination of specific properties.
Secondly, in spite of the relatively minor investment of funds, a
number of PTFC's have been developed with efficiencies well over
10% and these include the only high efficiency, stable cell with
a band-gap as low as 1eV. Such a cell is essential for a high
efficiency tandem cell.

It is this authors opinion that continued development will
yield a wide range of new candidate materials for solar cells and
other purposes and that multijunction PTFC's will take their
place in the photovoltaic market.

ACKNOWLEDGMENTS

The research at IEC has been supported by many industrial
organizations, the federal government and the state of Delaware.
Currently the U. S. Department of Energy is supporting the
research on tandem cells through the Solar Energy Research
Institute under sub-contract XL-4-04024-1. The author
acknowledges his indebtedness to the Principal Investigators on
that program, R. W. Birkmire and J. E. Phillips and to all
members of the IEC staff who have contributed so much to the
development of polycrystalline solar cells.

REFERENCES

1. J. J. Loferski, J. Appl. Phys., 27 (1956) 777
2. M. B. Prince, J. Appl. Phys., 26 (1955) 534
3. D. C. Reynolds and G. M. Leies, Electr. Eng., 73 (1954) 734
4. A. Rothwarf, Solar Cells, 2 (1980) 115-140
5. B. N. Baron, A. W. Catalano and E. A. Fagen, 13th IEEE
   Photovoltaic Specialists Conf., (1978) 406-410.
6. B. C. Plunkett and P. G. Lasswell, SPIE 248 (1980) 142-146.
7. J. E. Phillips, R. W. Birkmire and P. G. Lasswell, 16th IEEE
   Photovoltaic Specialists Conf.,(1982) 719-722.
8. T. S. te Velde, Energy Convers., 15 (1975) 111
9. l. R. Shiozawa, F. Augustine, G. A. Sullivan, J. M. Smith and
   W. R. Cook Jr., Final Report, Clevite Corp., AF-33(615)-5224,
   1969.
10. J. A. Bragagnolo, R. W. Birkmire and J. E. Phillips, 14th
    IEEE Photovoltaic Specialists Conf., (1980) 1400-1401.
11. W. Paltz, J. Besson, T. Nguyen Duy and J. Vedel, Proc. 10th
    IEEE Photovoltaic Specialists Conf., (1973) p. 69.
12. L.C.Burton and T. L. Hench, Appl. Phys. Lett., 29 (1976) 612
13. R. B. Hall, R. W. Birkmire, J. E. Phillips and J. D. Meakin,
    Appl. Phys. Lett., 38 (1981) 925
14. A. Rothwarf, L. C. Burton, H. C. Hadley Jr. and G. M. Storti,
    Proc.11th IEEE Photovoltaic Specialists Conf., (1975) p.476
15. R. E. Rocheleau, B. N. Baron and T. W. F. Russell, AIChE
    Journal 28 (1982) 656-662.
16. T. W. F. Russell, B. N. Baron and R. E. Rocheleau, AIChE
    Symposium Series 77 (1981) 70-77.
17. H. Huschka, B. Schurich and J. Worner, 4th EC Photovoltaic
    Solar Energy Conf.,(1982) p.399.
18. R. H. Bube, in Proc. Symp. on Materials and New Proc. Technol.
    for Photovoltaics, Eds. J. A. Amick, V. K. Kapur and J. Dietl,
    Electrochemical Soc.,83-11 (1983) p.359
19. R. H. Bube, A. L. Fahrenbruch, R. Sinclair, T. C. Anthony, C.
    Fortmann, W. Huber, C-T. Lee, T. Thorpe and T. Yamashita, IEEE
    Trans. Electron Devices, ED-31, (1984) 528-538.
20. D. Bonnet and H. Rabenhorst, 9th IEEE Photovoltaic Specialists
    Conf., (1972) p.129.
21. B. M. Basol, J. Appl. Phys.,55 (1984) 601-603.
22. Y-S. Tyan and E. A. Perez-Albuerne, 16th IEEE Photovoltaic
    Specialists Conf., (1982) 794-800.
23. H. Matsumoto, K. Kuribayashi, H. Uda, Y. Komatsu, A. Nakano
    and S. Ikegami, Solar Cells, 11 (1984) 367-373.
24. G. Fulop, M. Doty, P. Meyers, J. Betz and C. H. Liu, Appl.
    Phys. Lett.,40 (1982) 327-328.
25. T. L. Chu, S. S. Chu, F. Firszt, H. A. Nassem and R. Stawski,
    17th IEEE Photovoltaic Specialists Conf., (1984) 835-839.
26. D. Bonnet, 5th EC Photovoltaic Solar Energy Conf.,(1983) 897-
    900.
27. Y-S. Tyan, F. Vazan and T. S. Barge ,17th IEEE Photovoltaic
    Specialists Conf.,(1984) 840-845.
28. A. Nakano, S. Ikegami, H. Matsumoto, H. Uda and Y. Komatsu,
    Solar Cells, To be published.
29. J. L. Shay, S. Wagner and H. M. Kasper, Appl. Phys. Lett.,27
    (1975) 89-90.
30. J. D. Meakin, SPIE 543 (1985) 108-118

134

31. R. A. Mickelsen, W. S. Chen, Y. R. Hsiao and V. E. Lowe, IEEE Trans. Electron Devices, ED-31 (1984) 542-546.
32. R. W. Birkmire, L. C. DiNetta, J. D. Meakin and J. E. Phillips, Solar Cells, To be published.
33. M. Turowski, M. K. Kelly, G. Margaritondo and R. D. Tomlinson Appl. Phys. Lett., 44 (1984) 768-770.
34. R. A. Mickelsen and W. S. Chen, 16th IEEE Photovoltaic Specialists Conf.,(1982) 781-785.
35. B. J. Stanbery, W. S. Chen and R. A. Mickelsen, J. Electroch. Soc.,To be published.
36. R. Potter and C. Eberspacher, SERI Polycrystalline Thin-Film Review Meeting, SERI/CP-211-2548 (1984).
37. J. A. Thornton, D. G. Cornog, R. B. Hall, S. P. Shea and J. D. Meakin, 17th IEEE Photovoltaic Specialists Conf., (1984) 781-785
38. R. N. Bhattacharya, J. Electrochem. Soc.,130 (1983) 2040-2042.
39. T. L. Chu, S. S. Chu, S. C. Lin and J. Y. Yue, J. Electrochem. Soc.,131 (1984) 2182-2185.
40. W. Arndt, H. Dittrich, F. Pfisterer and H. W. Schock, 6th EC Photovoltaic Solar Energy Conf., (1985). To be published.
41. C. O. Bozler, R. W. McClelland and J. C. C. Fan, IEEE Electron Dev. Lett., EDL-2 (1981) 203-205.
42. L. C. Olson, F. W. Addis and W. Miller, Solar Cells, 7 (1983) 247-279.
43. M. Bhushan and J. D. Meakin, Final Report SERI/STR-211-2515 (1985)
44. J. J. Loferski, J. Shewchun, R. Boessler, R. Beaulieu, J. Piekoszewski, M. Gorska and G. Chapman, 13th IEEE Photovoltaic Specialists Conf., (1978) 190-194.
45. R. Taylor,in Proc. Symp. on Materials and New Proc. Technol. for Photovoltaics, Eds. J. A. Amick, V. K. Kapur and J. Dietl, Electrochemical Soc., 83-11 (1983) p.56-59.
46. E. D. Jackson, Trans. Int. Conf. on Uses of Solar Energy-Scientific Basis, 5 (1955) 122-126.
47. J. E. Phillips, R. W. Birkmire, L. C. DiNetta and J. D. Meakin, J. Electroch. Soc., To be published.

FIGURE CAPTIONS

1.A generalized solar cell showing the absorbing and collector/convertor semiconducting layers, the front transparent contact and the back opaque contact.

2.An annotated version of the efficiency versus band gap curve published by Prince (2).

3.Absorbtion coefficients for various semiconductors.

4.A logical sequence for the development of efficient solar cells using a relatively unknown semiconductor.

5.The relation between Cu_2S stoichiometry and hole density.

6.The structure of a monolithic tandem cell based on $CuInSe_2$.

7.Initial results obtained with a monolithic tandem cell based on $CuInSe_2$.

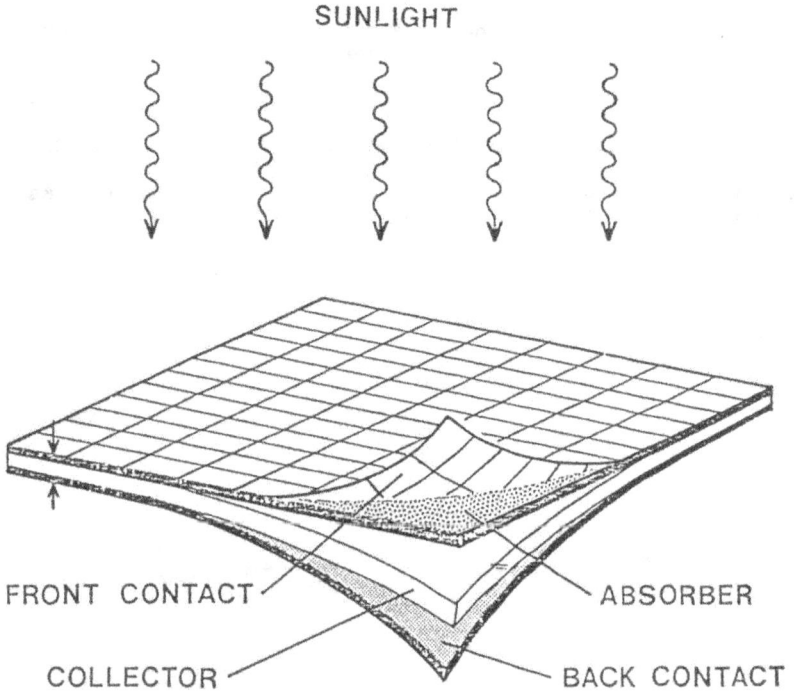

SUNLIGHT

FRONT CONTACT

COLLECTOR

ABSORBER

BACK CONTACT

Fig. 1

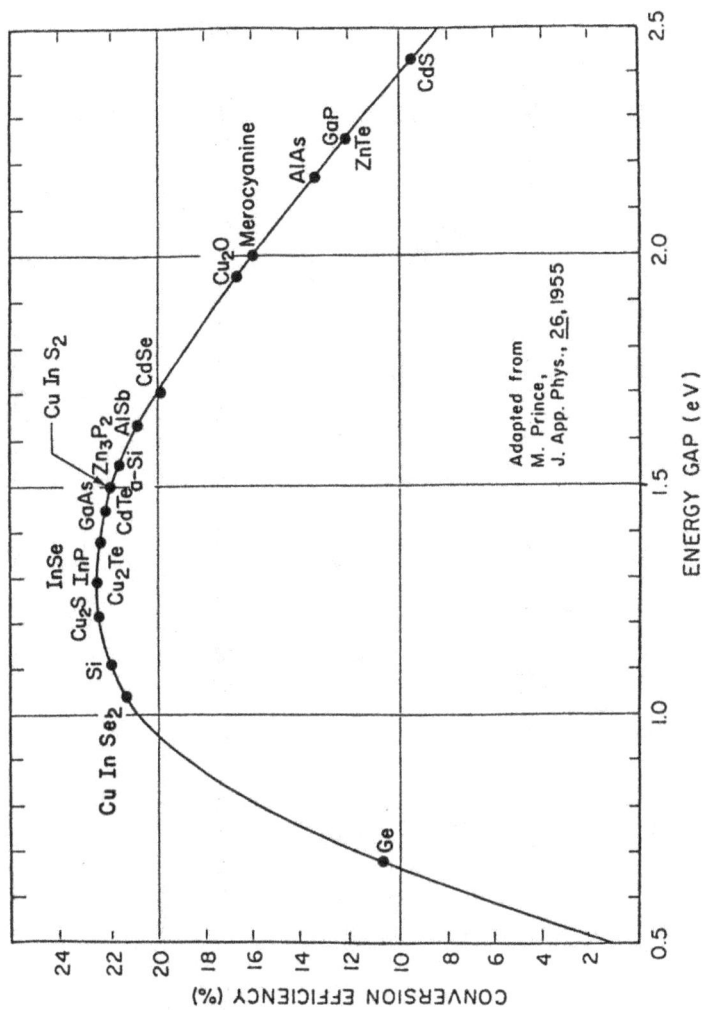

Fig. 2

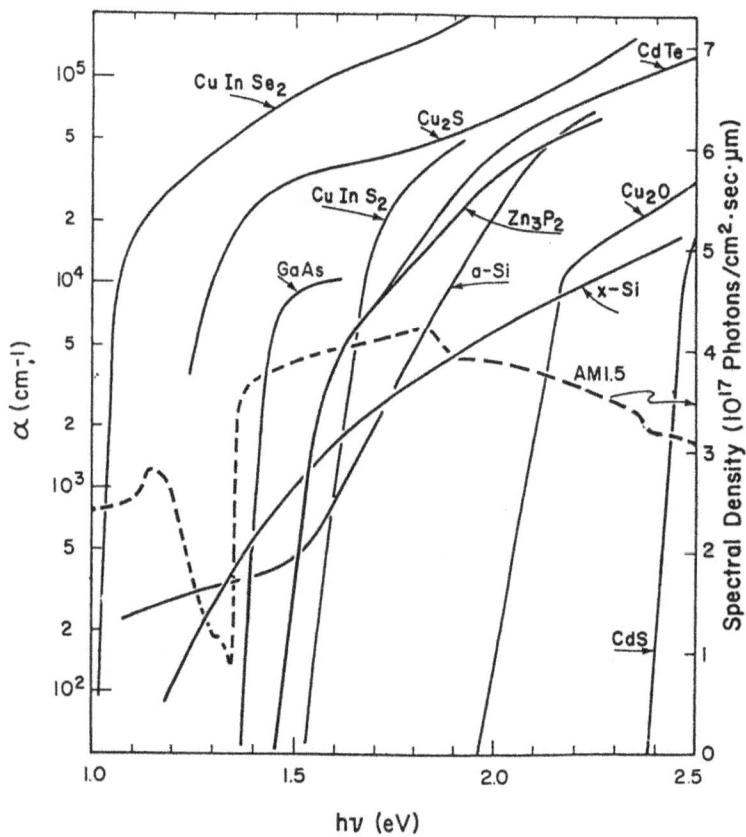

Fig. 3

Cu In Se$_2$
Cell Development and Optimization

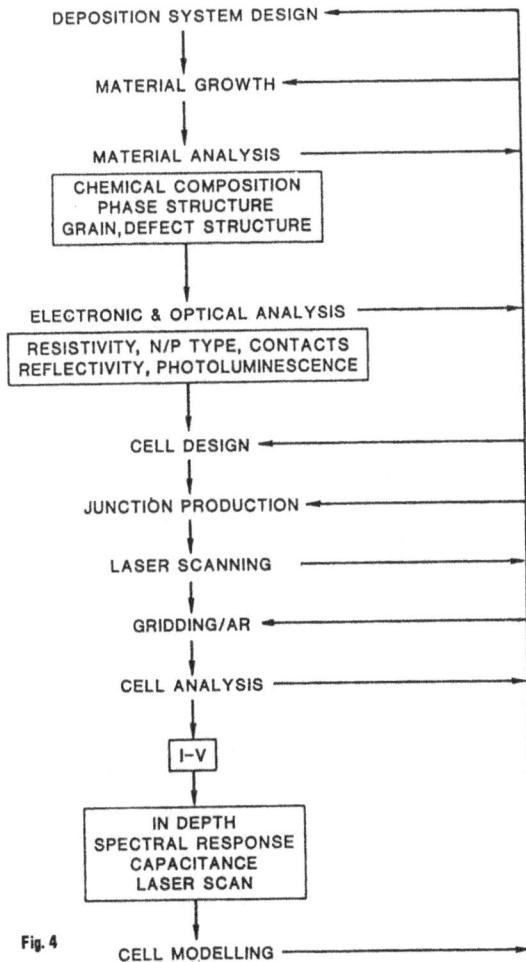

DEPOSITION SYSTEM DESIGN

MATERIAL GROWTH

MATERIAL ANALYSIS

CHEMICAL COMPOSITION
PHASE STRUCTURE
GRAIN, DEFECT STRUCTURE

ELECTRONIC & OPTICAL ANALYSIS

RESISTIVITY, N/P TYPE, CONTACTS
REFLECTIVITY, PHOTOLUMINESCENCE

CELL DESIGN

JUNCTION PRODUCTION

LASER SCANNING

GRIDDING/AR

CELL ANALYSIS

I-V

IN DEPTH
SPECTRAL RESPONSE
CAPACITANCE
LASER SCAN

Fig. 4

CELL MODELLING

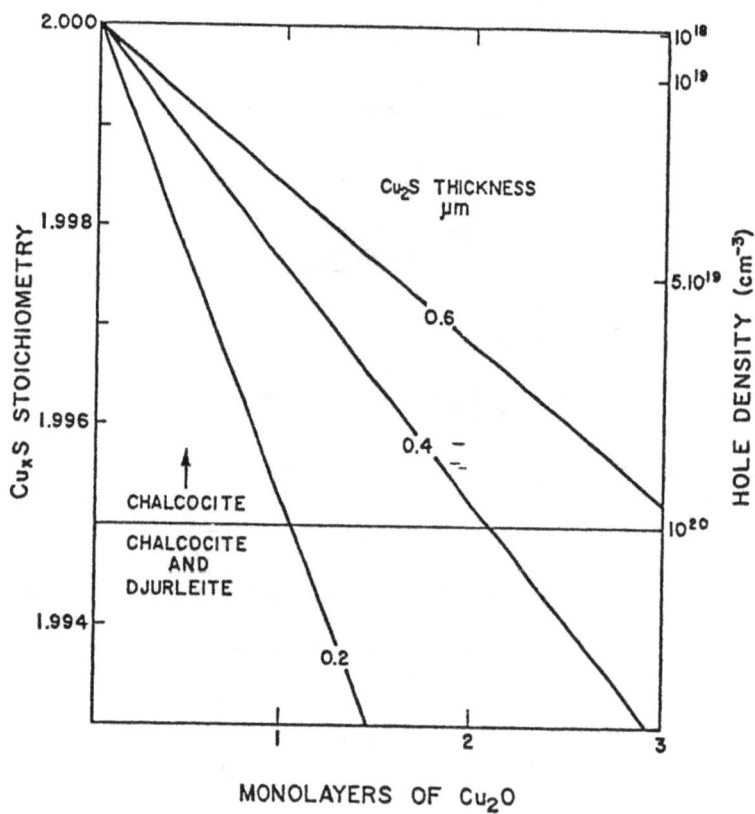

Fig. 5

CdS/(CdHg)Te : CdS/CuInSe$_2$

TANDEM CELL

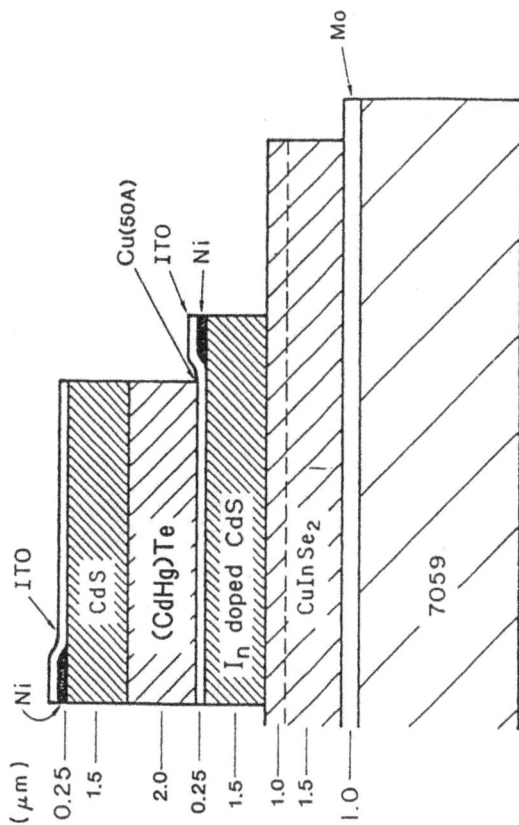

Fig. 6

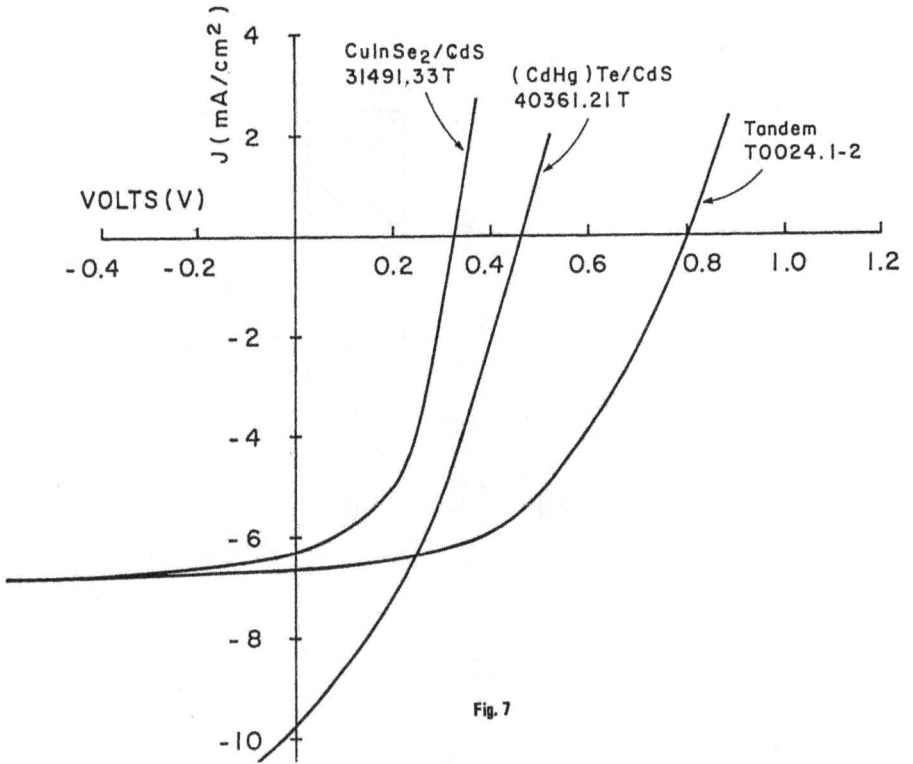

Fig. 7

ADVANCED OPTICAL MATERIALS FOR ENERGY EFFICIENCY
AND SOLAR CONVERSION

Carl M. Lampert

Materials and Molecular Research Division
and
Energy and Environment Division
Lawrence Berkeley Laboratory
University of California
Berkeley, CA 94720

August 1986

This work was supported by the Assistant Secretary for Conservation
and Renewable Energy, Office of Buildings and Community Systems,
Buildings Division of the U.S. Department of Energy under
Contract No. DE-AC03-76SF00098.

ABSTRACT

Materials science properties of optical materials and coatings
are discussed for a broad range of solar conversion, architectural
glazings and greenhouse energy efficient uses. Transparent low
emittance coatings for glazings are discussed for radiative heat
transfer reduction. Both interference multilayer and doped semiconductor low emittance coatings are covered. The use of Drude
theory to model coatings is discussed. Discussion of various types
of selective absorbers include interference multilayer, composite
tandem absorbers and selective paints. Effective medium theories
are used to describe composite absorbers. The basic properties of
radiative cooling materials and antireflection coatings are detailed.
Also, the properties of reflector materials on glass and plastics
are covered. Application of fluorescent concentrators, spectral
splitting and cold mirror films are outlined. Research on transparent aerogel insulation and optical switching films for windows
are introduced.

INTRODUCTION

Optical materials and coatings play an important role in determining the efficiency of solar conversion processes. At present the best known coatings are heat mirrors, selective absorbers, and reflector materials. There are, however, many less known coatings and materials, some of which are under current research and development. Since they are of significant consequence to solar conversion and energy conservation, they will be included in this work. These films and materials are known as antireflective coatings or treatments, fluorescent concentrator materials, holographic films, cold mirrors, radiative cooling surfaces, optical switching films, and transparent insulation materials. The use of such films and materials improves the efficiency and stimulates innovation in passive and active solar energy conversion, photovoltaics, energy-efficient windows, and many hybrid designs. The present study is intended to instruct readers on the properties of solar energy materials, their function, as well as advantages and limitations. It is also intended to expand the horizons of solar invention by considering new materials, techniques, and concepts that can manipulate solar energy into the forms of heat, light, and electrical power. The material stability requirements for effectively collecting and transmitting solar energy are extremely demanding. This combined with the need for inexpensive production methods creates a very broad area for innovative scientific research. The solutions to materials science and design problems are the responsibilities of a number of scientists, engineers, and educators working in many countries and in numerous fields. It is their strong commitment to solar energy that will further advance this energy source.

BASIC THEORY

   The optical engineering properties that are of primary interest for
most solar materials are transmittance (T), reflectance (R), absorptance
(A), and emittance (E). These properties characterize how a particular
coating or a material interacts with incident energy. Furthermore,
these qualities can be related to intrinsic materials properties such as
the index of refraction (n) and the extinction coefficient (k). Also all
these properties have a spatial, wavelength, or temperature dependence.
Formally, the spectral transmittance (T_w) is the ratio of incident radiation transmitted through a medium at wavelength w , to that of the
total incident radiation at w.

   The spectral reflectance (R_w) is the ratio of incident radiation
reflected from the medium at w to the total incident radiation at w.
The spectral absorptance (A_w) is the ratio of incident radiation
absorbed in the medium at w to the total incident radiation at w. A
relationship between these quantities allows for conservation of energy:

$$A_w + R_w + T_w = 1 .$$

   The thermal emittance (E(T)) is defined as the ratio of emitted
radiation from a surface at Temperature T to the corresponding blackbody
radiation at T. The spectral quantities of A_w, and E_w are equivalent provided the measurements are taken at the same temperature and under the
same conditions:

$$A_w(T) = E_w(T).$$

   In practice these properties are best represented as integrated
values. Total solar absorptance can be defined as :

$$A_s = \frac{\int_{0.3}^{2.5} A_w I_w dw}{\int_{0.3}^{2.5} I_w dw} , w \text{ in microns},$$

where (I_w) is defined as the solar irradiance at w at some air mass.
For a selective absorber, T_w = 0 and A_w can be defined from reflectance
measurements as:

$$A_w = 1 - R_w.$$

Similar integrations can be made for R_s and T_s. The total thermal emittance $E_T(T)$ can be defined as:

$$E_T(T) = \frac{\int_1^{100} E_w(T)B_w(T)\ dw}{\int_1^{100} E_w(T)\ dw}, \quad w \text{ in microns},$$

where $B_w(T)$ is the radiation intensity (given by the Planck function) for a blackbody at temperature T, the same temperature of the object being measured. For heat mirrors and other transparent coatings the visible transmittance is important. Although much reported data are cited as visible transmittance, these are usually only averaged data and not integrated, and T_V is defined as follows:

$$T_v = \frac{\int_{0.39}^{0.77} T_w\ P_w\ dw}{\int_{0.39}^{0.77} P_w\ dw}, \quad \text{where } w \text{ is in microns},$$

In this case the distribution function is the photopic or specular eye response distribution, (P_w). This distribution is Gaussian in nature and peaks in the green wavelengths (0.55 microns).

TRANSPARENT HEAT MIRRORS

Heat mirror coatings play a significant role in solar thermal conversion,[1] transparent insulation for architectural windows[2] and photovoltaic conversion.[3] In this work a heat mirror is defined as a coating that is predominately transparent over the visible wavelengths (0.3 - 0.77 microns) and reflective in the infrared (2.0 - 100 microns). Over the near infrared (0.77 - 2.0 microns) the coating may exhibit combined properties depending upon design or application requirements. Figure 1 shows an example of an idealized heat mirror response superimposed on solar and blackbody radiation spectra. Heat mirrors for windows derive their usefulness from their low emittance (or high reflectance) in the infrared. The lower the emittance, the less the magnitude of radiative transfer by the window. The emittance of glass is $E_T = 0.84$; many plastics also have high emittance values. Two scenarios can be invoked to demonstrate the usefulness of single glazed heat mirrors in buildings.[2]   The first scenario, depicted in Figure 2, represents winter heating where solar gain is important to reduce the heating load in a building. The optimum heat mirror for this function would transmit both solar visible and near infrared to about 2-3 microns, the cut off wavelength for glass. The thermal infrared would be reflected back into the building. In this process the majority of the solar energy could be utilized for passive solar gain and daylighting.

A second scenario can be treated as a cooling-load-reducing heat mirror, where all infrared energy is reflected to reduce air conditioning loads. This type of heat mirror is depicted in Figure 3. The coating allows only transmission of visible energy through the window, with the remainder of the wavelengths reflected. A heat mirror alters only the radiative character of a window; the effects of convective and conductive heat losses must be taken into account, too. Window orientation, climate, and building type are important factors in choosing the exact combination of optical properties the heat mirror is to have. A film with adjustable properties might be the best solution.

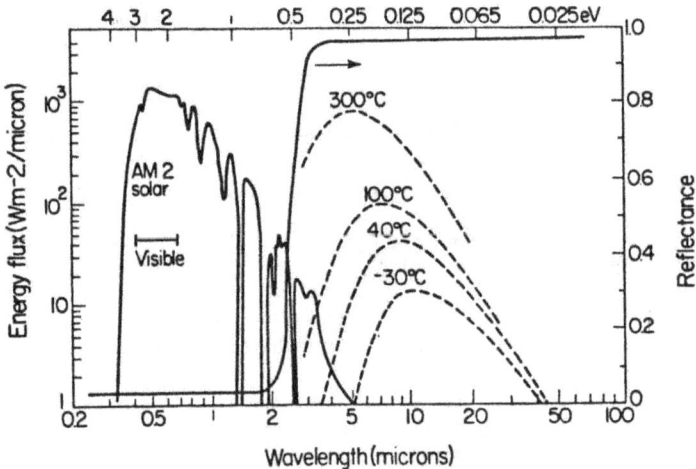

Figure 1.   Solar spectrum (air mass 2) with four blackbody spectra (-30°C-300°C).   Superimposed is the idealized reflectance of a heat mirror coating.

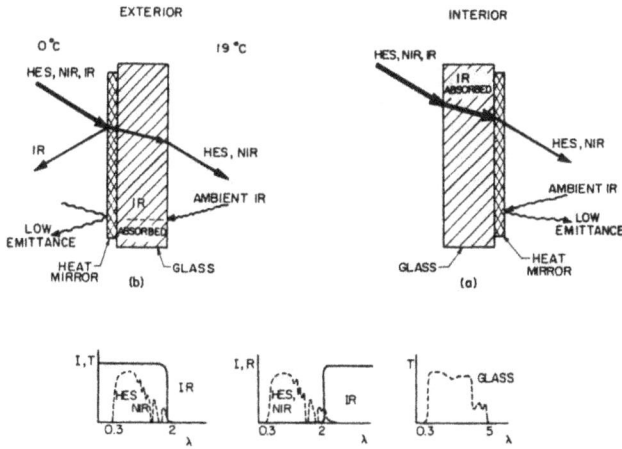

Figure 2. Schematic of heating-load heat mirror properties. Lower
drawings depict optical properties of coating and substrate.

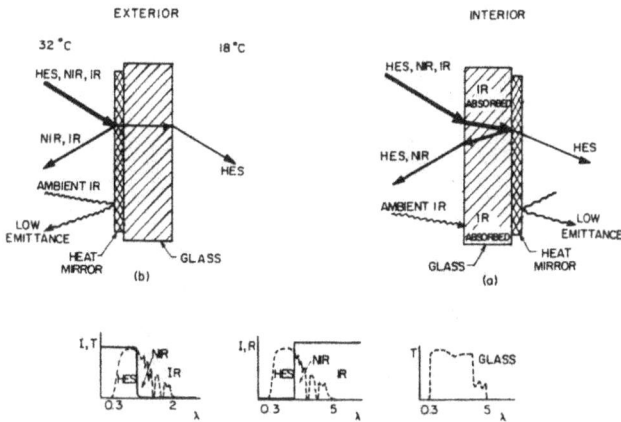

Figure 3. Schematic of cooling-load heat mirror properties. The
exterior placement is subject to convection losses that
can lessen the benefit.

Examples of multiple glazed heat mirrors are shown in Figure 4. Here both a double and triple glazed window are shown. One design de-picts a heat mirror coating on plastic. Deposition of heat mirrors on glass and plastic substrates are of significant importance. One fits well into conventional glass coating processes, and the other is a refined extension of the metallizing process for plastics. By use of a computer model,[4] thermal conductance (U) values have been derived for various types of glazings. These data are shown in Figure 5. A nonglass insu-lated wall has 0.3 W/m^{2}K (R-19) to U = 0.6 W/m^{2}K (R-11).
thermal conductance values of u =

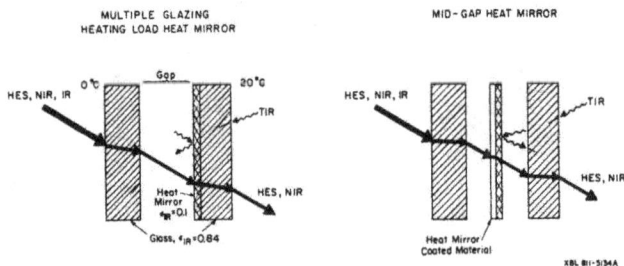

Figure 4. Double-and triple-glazed windows incorporating heat mirrors.

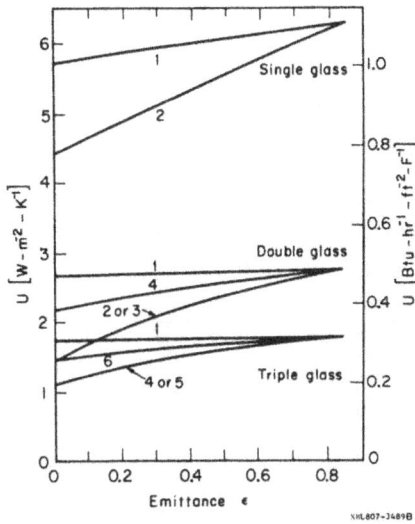

Figure 5. Computer-modeled thermal conductance (U) for various window designs[4] using ASHRAE standard winter conditions (T_{out} = -18°C, wind speed = 24 km/hr). The effect of lowering the emittance of a single surface by the addition of a heat mirror coating is shown. The surfaces on which the coating appears are given as consecutive numbers, starting from the outside surface, labeled 1. Airgap is 1.27 cm.

A heating-load heat mirror could also be used for solar thermal collectors as an alternative to selective absorbers. But, in this case the heat mirror could be used with a nonselective absorber. One modification of this heat mirror is to shorten the transition or cut off reflectance wavelength, which would be more suitable for higher operating temperatures of solar collectors. The results of using a heat mirror coating on a double glazed flat plate collector are shown in Figure 6. Also shown for comparison is the effect of using a selective absorber and antireflective coatings.[5] These coatings will be discussed in subse-

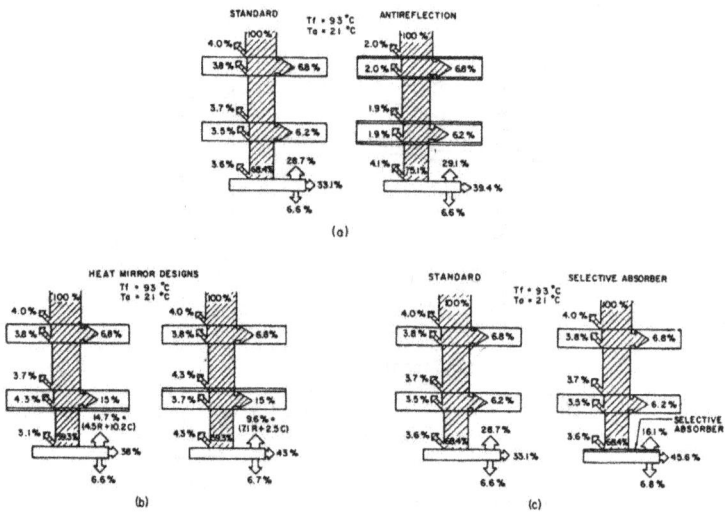

XBL781-6856B

Figure 6. The effect of various coatings on collector efficiency[5] for a flat plate collection operating at 93°C under ambient of 21°C. The notation R stands for reflection losses, and C stands for convection losses.

quent sections. Transparent conductors as heat mirrors can be physically deposited (PVD) on glass and plastic by vacuum evaporative and sputtering techniques. Methods, including chemical vapor deposition (CVD) and polycondensation of organometallics, have been limited mostly to glass. The thermal and chemical stability of the substrate are significant in determination of the proper deposition technique and conditions. With CVD the secondary and competing reactions must be suppressed by control over reaction kinetics and knowledge of system thermodynamics. Also, thermal durability and property stability of substrate and film is significant for long-life designs. Heat mirror films can be classified into two categories: multilayer dielectric/metal-based films such as Al_2O_3/Ag, $ZnS/Cu/ZnS$, and $TiO_2/Ag/TiO_2$, and single-layer semiconductors (highly doped) such as In_2O_3:Sn and SnO_2:F. There is considerable growing interest in both multilayer and semiconductor heat mirrors in Europe, North America, USSR, East Germany, and Japan. Further information on heat mirrors is contained in a number of excellent works.[2,3]

Multilayer Heat Mirror Films

Metal films less than approximately 100 $\overset{o}{A}$ thick exhibit partial visible and solar transparency. Dielectric overlayers serve to both protect and to partly antireflect the metal film in the visible region, thereby increasing transmission. Generally, though, further protection is required of the metal/dielectric film as it is quite thin and vulnerable to corrosion and abrasion. The design of the appropriate dielectric type and thickness is detailed elsewhere.[6] The dielectric film, when used to overcoat a metal, must exhibit high infrared transmittance in order to preserve the infrared reflectance of the metal. Example systems[2] are SiO_2/M, polymer/M, Al_2O_3/M, Bi_2O_3/M, $Bi_2O_3/M/Bi_2O_3$, $ZnO/M/ZnO$, $TiO_2/M/TiO_2$, and $ZnS/M/ZnS$, where M is a metal of Ag, Al, Au, Cu, Cr, Ni Ti. Additional designs include X/M and X/M/X where X is an appropriate semiconductor or polymer. Infrared transparent polymers of polyethlene, polyvinylidene chloride, polyacrylonitrile, polypropylene, and polyvinyl fluoride might be used for this application. Multilayer films have an advantage over the doped semiconductors of broad wavelength tunability. Some selected multilayer films on glass and plastic substrates [7-10] are shown in Figures 7 and 8. Detailed property data on these films is shown in Table 1. Extensive data on D/M/D coatings is covered elsewhere.[2] Durability improvement of multilayer films still remains an important research area for the materials scientist. A recent review details the properties of various types of commercial heat mirror coatings, including both multilayer and single layer types.[11]

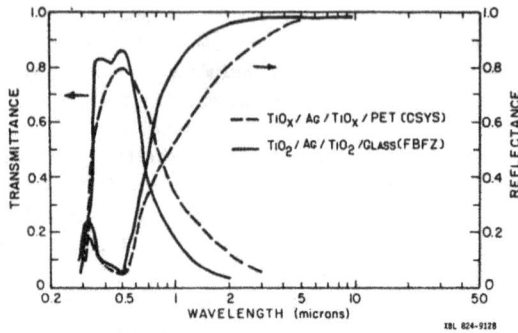

Figure 7. $TiO_2/Ag/TiO_2$ coating on polyethylene terephthalate[7] (PET) and glass.[8]

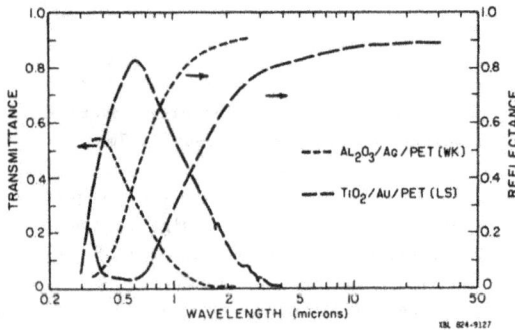

Figure 8. Dielectric/metal coatings on PET.[9,10]

TABLE 1

Multilayer Heat Mirror Films

Material	Al_2O_3/Ag	$TiO_2/Au/PET$	$TiO_2/Ag/TiO_2$	$TiO_x/Ag/TiO_x$	$ZnS/Ag/ZnS$
Deposition tech.	Ion Beam Sputter	e-Beam Evap. & Chemical Dep.	RF Sputter	Chemical Dep. & Vac. Evap.	Vac. Evap.
Sheet Resist. (ohm/sq)	–	10	–	–	10
Thickness (angstroms)	–	–	180/180/180	270/150/270	520/100/770
T_{vis}(ave) or T_s	0.47, 5µm	0.80	0.84	~0.75	0.68
R_{ir} or (E_{ir})	0.93, 2.5µm	0.87	0.99, 10µm	0.98, 5µm	(0.06)
Reference	9	10	8	7	23

Doped Semiconductor Films

Certain doped semiconductors can exhibit high infrared reflectance due to the proper combination of high mobility (>10 cm^2/V sec) carrier concentration (10^{20}-$10^{23} cm^{-3}$) effective mass, and lattice relaxation frequency. Materials science details are outlined elsewhere.[12] The best known transparent semiconductors are SnO_2:F, SnO_2:Sb, In_2O_3:Sn, and Cd_2SnO_4. Characteristic spectral transmission and reflectance is shown in Figure 9 for research-grade films on glass.[13-15] The transmission of these films can be increased by etching microgrids in the coatings.[16] Obtaining a reproducible high-quality commercial coating is considerably more difficult than making research-grade films. Present techniques need to be improved to deposit these coatings on polymeric substrates.[17,18] Specific examples of these films are shown in Figure 10. Table 2 lists the physical properties of selected doped semiconductor films. Detailed information on deposition processes given elsewhere.[2,12,19-22] There are other promising heat mirror materials. They include some of the rare earth oxides, borides, transition metal nitrides, carbides, and selected ternary systems. Although little knowledge has been obtained optically about these materials, they are known to exhibit Drude-like electrical conduction. Also, graded index and surface textured heat mirror coatings remain to be developed. They show promise of improved solar transmission characteristics.

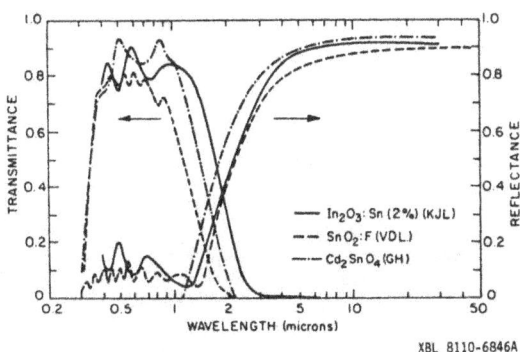

XBL 8110-6846A

Figure 9. Spectral normal transmittance and reflectance of heat mirror research-grade film on glass based on In_2O_3:Sn, SnO_2:F, and Cd_2SnO_4.[13-15]

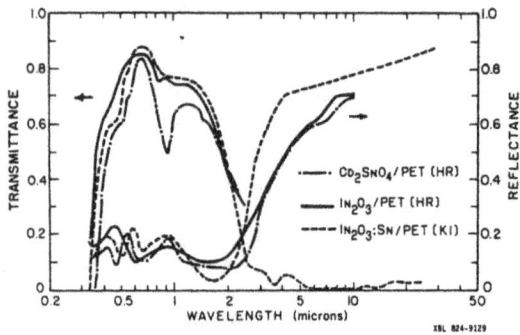

Figure 10. Spectral normal transmittance and reflectance of single-layer heat mirror coatings deposited on polyethylene terephthalate.[17-18]

TABLE 2

Selected Heat Mirror Films

Material	SnO_2:F	Cd_2SnO_4	In_2O_3:Sn	In_2O_3:Sn
Deposition tech.	Spray Hydro. 500-570°C	RF Sputt. Anneal 420°C	Spray Hydro.	RF Sputt. etched microgrid
Sheet Resist. (ohm/sq)	4	26-43	15-20	3
Thickness (microns)	1.0	< 0.3	-	0.35
Mobility (cm^2/V-sec)	37	-	-	-
Carrier Density (cm^{-3} x 10^{20})	4.4	-	15	-
T_{vis}(ave) or (T_s)	(0.75)	(0.86)	~0.9	(0.90,AM2)
R_{ir} or (E_{ir})	(0.15)	(0.12, 77°C)	0.85	0.83, 10um
Reference	15	14	13	16

TABLE 2 (continued)

Material	Cd_2SnO_4/PET	In_2O_3/PET	In_2O_3:Sn/PET
Deposition tech.	REACT DC Sputt. & RF BIAS	REACT DC Sputt. & RF BIAS	REACT RF Sputt.
Sheet Resist. (ohm/sq)	24	30	<30
Thickness (microns)	0.28	0.34	0.5
Mobility (cm^2/V-sec)	30	29	-
Carrier Density (cm^{-3} x 10^{20})	3.4	2.1	-
T_{vis}(ave)	~0.65	~0.78	0.8
R_{ir} or (E_{ir})	0.7, 10um	0.7, 10um	0.8, 10um
Reference	17	17	18

Theory of Drude-Like Coatings

Both highly doped semiconductors and thin metal films are linked by the conductive metal-like properties described by classical Drude theory.[24] In this theory, a well-defined plasma edge characterizes the material. It occurs due to excitation of free carriers by incident electromagnetic radiation. Since the charge carriers are actually moving in a potential field of the crystal, an effective mass (m^*) must be used. The effective mass is defined as:

$$m^* = m_r m_e ,$$

where m_e is the rest mass and m_r is the relative mass of the charge carriers in the field. The complex dielectric constant is related to the optical constants by:

$$e = e_1 - e_2 = (n - ik)^2 ,$$

where $e_1 = n^2 - k^2$ and $e_2 = 2nk$.

For a Drude-like reflector, the dielectric constant can be expressed in terms of (V/v_p) and (Y/v_p) as follows:

$$e_1 = e_b [1 - (1 + (\frac{Y}{v_p})^2) / ((\frac{V}{v_p})^2 + (\frac{Y}{v_p})^2)] ,$$

$$e_2 = e_b [\frac{Y}{v_p}(1 + (\frac{Y}{v_p})^2) / \frac{V}{v_p}((\frac{V}{v_p})^2 + (\frac{Y}{v_p})^2)] ,$$

where e_b is the dielectric constant associated with bound carriers (at very high frequency) and Y is the relaxation frequency, $Y = e/um_e$, u is the carrier mobility, and e is the electron charge. The damped plasma frequency (v_p) is derived as:

$$v_p = (Ne^2/e_b e_o m^*)^{1/2} - Y^2 ,$$

where N is the carrier density and e_o is the dielectric constant of air. From this theory, n and k can be extracted in terms of frequency (W). Reflectance in air (normal-specular) can be derived by:

$$R = ((n-1)^2 + k^2)/((n+1)^2 + k^2).$$

The theoretical reflectance for a highly doped transparent conductor is shown in Figure 1. The Drude-modeled reflectance agrees well with that obtained by experimental doped semiconductor films. One should note that in the Drude equation as $\dot{V} \gg v_p > Y$, n becomes constant and $k \rightarrow 0$, indicating that the material is transparent at short wavelengths and for $V \ll Y < v_p$, n = k are about equal and large in magnitude, indicating a high reflectance.

SOLAR SELECTIVE ABSORBERS

Absorbers for collectors have been one of the most active materials science research fields over the last several years. There are two categories of absorbers, selective and nonselective. The selective absorber has optical properties that vary greatly from one spectral region to another. The selective absorber or selective surface efficiently captures solar energy in the high-intensity visible and the near-infrared spectral regions while exhibiting poor infrared radiating properties. This characteristic is depicted in Figure 11. In contrast a nonselective absorber, such as ordinary black paint, has a flat spectral response and loses much of its absorbed energy by reradiation. The optimum characteristics of a solar absorber are high solar absorptance and a minimum emittance (or maximum reflectance) in the infrared wavelengths. The exact transition wavelength is determined by application, solar concentration, and operating temperatures. Examples of concentrating collector designs are shown in Figure 12. Solar selectivity can be obtained by a variety of methods. These techniques consist of intrinsic solar selective materials, optical trapping surfaces, semiconductor/reflector tandems, composite coatings, multilayer thin films, and quantum size effects. Significant reviews have been written about solar absorbers, giving insight and detail beyond the scope of this work.[1, 25-29] A selected group of solar selective absorbers is outlined in Table 3 after Herzenberg and Silberglitt.[26]

Intrinsic Absorbers

No known natural compound or material exhibits ideal solar selectivity. However, there are some materials that show wavelength selectivity to some degree. Possibly, synthetic materials could be fabricated once enough knowledge is obtained to design and predict fairly complex electronic structures and their relationships to the necessary optical properties. Both transition metals and heavily doped semiconductors exhibit at least one desireable absorber characteristic. Unfortunately, metals exhibt a plasma edge too early in the solar spectrum to be good absorbers. Semiconductors tend to be good absorbers but exhibit transmission in the infrared. These differences have resulted in the

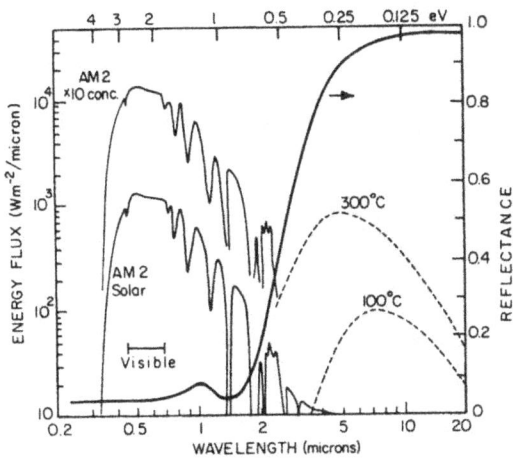

XBL799-7000

Figure 11. Wavelength relationship between a characteristic black
chrome solar selective surface in terms of reflectance,
to that of solar energy and blackbody spectra.
Both concentrated (x10) and regular solar spectra for
air mass 2 are given.

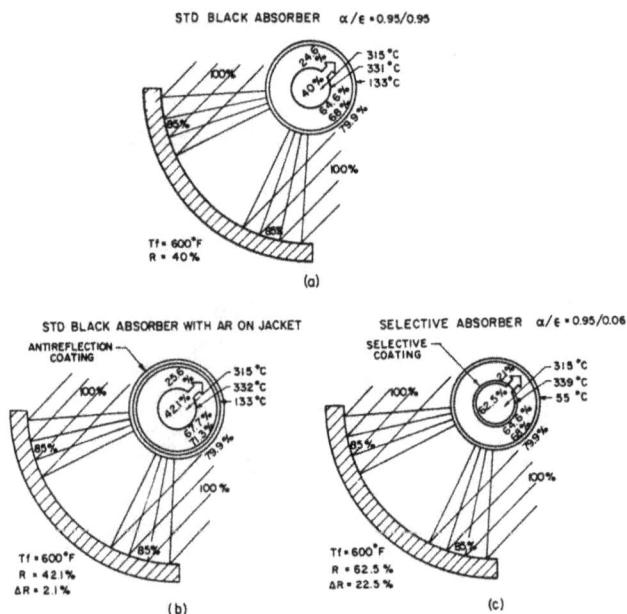

Figure 12. Fixed conditions for a parabolic concentrating evacuated collector.[80] The collector is operating at 315°C with ambient of 21°C. Notation: R = net collection efficiency. ΔR = change in collecting efficiency over standard black absorber.

use of tandem absorbers. Materials such as Cu_2S, HfC, ZrB_2 Mo-MoO_3, Na_xWo_3 Eu_2O_3, V_2O_5, ReO_3, and LaB_6 (as shown in Figure 13) have reasonably intrinsic selective reflectance. Here, too, the dominant transition takes place too early in the solar spectrum for a good absorber. Also, the absorptance and transmittance properties are not well characterized. Lanthanum hexaboride may be a good heat mirror, since it has fair visible transmittance.

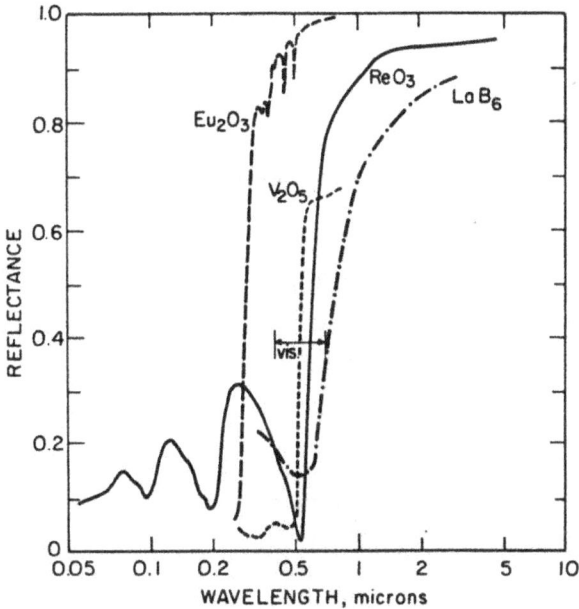

XBL 783-4706B

Figure 13. Spectral reflectance of various oxides showing intrinsic wavelength sensitivity.[12]

Optical Trapping Surfaces

It is possible to roughen surfaces so they will enhance absorption geometrically in one wavelength region but appear smooth in another. This technique is possible since the high-energy solar spectrum is distributed far enough in wavelength away from the thermal infrared spectra. Materials can be grown as dendrites, rough crystallites, or roughened surfaces to form optical trapping surfaces as in Figure 14. These dendritic materials do not require a high intrinsic absorption as they rely upon multiple reflection and partial absorption to give a large effective absorption. Materials such as $NiAl_x$, W, Mo, Ni, Cu, Fe, Co, Mn, Sb, and stainless steel have been grown as dendrites or textured by sputter etching.[1,31,58] Texturing of surfaces is also an excelllent method to antireflect other types of absorbers. This has been done with great sucess with the $Mo-Al_2O_3$ composite absorber.[54]

Solar Radiation

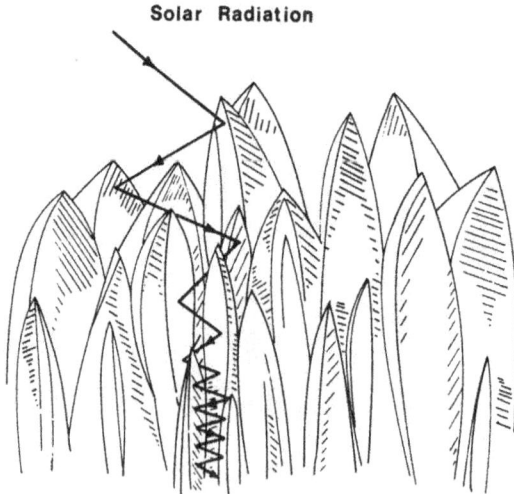

XBL 7712-11146

Figure 14. Dendritic selective surface.

Semiconductor/Metal Tandems

Semiconductor coated metals provide complementary functions as a spectrally selective surface. The semiconductor provides high absorptance in the high-energy solar region and becomes transparent beyond its absorption edge in the infrared. In the infrared the underlying metal layer gives the tandem or metallic reflectance as low emittance. A simple tandem is an oxidized metal. Usually the natural thickness of oxide is not optimum or is not stable in operation. Many chemical conversion processes are used to oxidize metals.[1] Metals like stainless steel, copper, and titanium have been oxidized. Other tandems can be made with deposits of oxides on a variety of metals. Examples are shown in Figure 15. High temperature absorbers can also be designed, such as the Si/Ag absorber which is stable to 500°C.[27] Tandem absorbers can be made by simple chemical conversion, electrochemical, deposition chemical vapor deposition (CVD) and physical vapor deposition processes (PVD) (sputtering and evaporation). Composite coatings can also be used as the absorber portion of tandem absorbers.

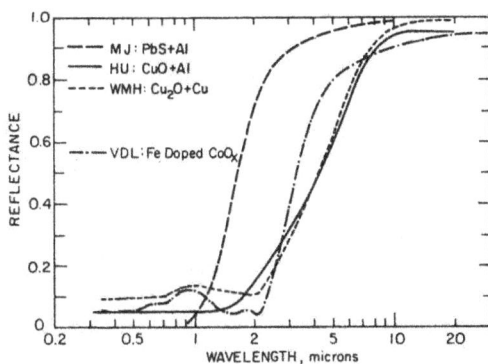

X BL 792-57888

Figure 15. Reflectance of various tandem absorbers: PbS/Al,[59] CuO/Al,[61] Cu$_2$O/Cu,[60] CoO$_x$:Fe/Ni.[15]

Metal-Dielectric Composite and Graded Coatings

Composite coatings consist of two or more components graded and interdispersed in a film. These components are usually phase separated. Reflective scattering can take place purely by the geometry of imbedded surface particles. Resonant scattering is dependent upon particle size, shape, and the effective index of refraction with respect to the surrounding media. Both Mie and Maxwell-Garnett theories describe the role of sub-wavelength particles as resonant scatters.[62] An example of a complex composite material is the well-known black chrome absorber depicted in Figure 16. This coating consists of a predominantly oxide region that grades to a Cr_2O_3-Cr region that is responsible for solar absorption. Infrared properties are derived from a deeper metallic chromium region coupled with a metallic substrate.[63-66] Further research on black chrome has been reviewed elsewhere.[26] Other types of composites range from $Ni-Al_2O_3$, $Zn-ZnO$, $Cr-Cr_2O_3$, $Pt-Al_2O_3$, $Mo-Al_2O_3$, and metal carbides. Many of their properties are detailed in Table 3. Composite coatings, black chrome, $NiCrO_x$, and $Cr-CrO_x$ have been deposited on thin metal films for stick-on applications.[39, 55,57] The spectral characteristics of black chrome and $Cr-Cr_2O_3$ cermet coatings are shown in Figure 17.

Electrodeposited black nickel has a very interesting structure. By changing plating parameters during deposition, layers of ZnS and NiS can be formed on top of a metallic substrate. This absorber shows a combination of tandem and multilayer interference effect. The reflectance for the two varieties is shown in Figure 18 . Multilayer effects will be discussed in a subsequent section. It is important to note that in a number of cases combined effects in a single absorber may make a superior design, although a more complicated one.

Of theoretical interest regarding tandem absorbers is the theory of Quantum Size Effects (QSE).[79] QSE occurs in thin film for metals, perhaps less than 10-20 Å for metals and 500 Å for degenerate semiconductors. QSE relates the influence of the geometrical dimensions of the sample to that of distribution of electron states. This distribution can be optimized by size effects to interact strongly with incident electromagnetic radiation. A combination of a QSE material and a reflector metal layer can result in a tandem absorber. QSE has been experimentally verified for InSb/Al, and InSb/Ag.[79]

Two effective-medium theories that can be used to describe composite solar absorbers are the Maxwell-Garnett theory[69,70] and the Bruggerman theory[71]

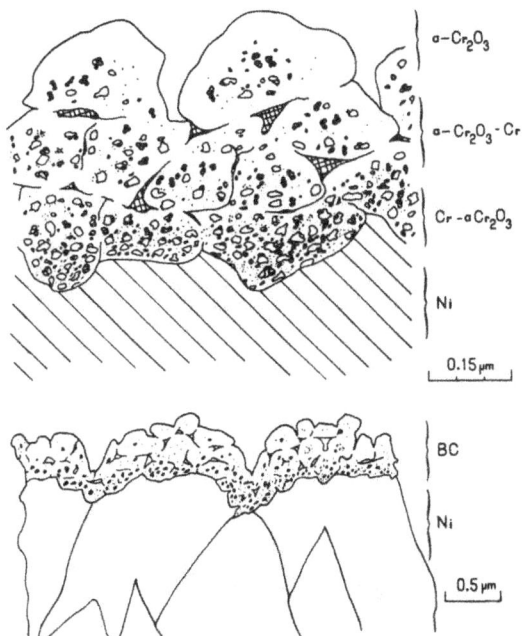

XBL 799-11767

Figure 16. Schematic cross section of black chrome.[64] Three distinct
regions are shown in the as-plated structure:
1. Top layer of amorphous or fine crystalline Cr_2O_3.
2. Intermediate area of metallic Cr in Cr_2O_3.
3. Bottom layers, consisting principally of Cr and Ni
   substrate.

TABLE 3: PROPERTIES OF SELECTIVE ABSORBERS[26]

Absorber[1]	Type	Deposition Technique	Maturity[2]	A_s	$E(^\circ C)$	Stability[3]$^\circ C$	Ref.
Copper	Textured	Sputter etch	1	0.9-.95	0.08-.11	300(air)	31
SS	Textured	Sputter etch	1	0.9-.96	0.22-.26	350(air)	31
Ni	Textured	Sputter etch	1	0.9-.95	0.08-.11 (27)	250(air)	31
ZrB2/Si3N4	AR,intrinsic	CVD	1	0.93	0.08-.09 (102)	500(air)	30
Cr-Cr2O3 (black chrome)	Graded Composite	Electroplating	5	0.92-0.97	0.04-.06 (100)	400(air)	32-34
Cr-Mo-Cr2O3	Graded Composite	Co-electroplating	1	0.96-0.97		400(air)	35
Ni-Al2O3/Al2O3	Graded Composite	Anodic Oxidation	5	0.92-0.97	0.1-.26 (65)	300(air)	36
Zn-ZnO	Graded Composite	Anodic Oxidation	1	0.98	0.18(100)	<300(air)	37
Cu2O-CuO-Cu	Composite/Tandem	Anodic Oxidation	1	0.95	0.34(100)	130(air)	38
NiO-Ni-Cr/NiCrOx	Graded Composite	Chemical Conversion	5	0.97-0.99	0.07-.1 (100)	250(air)	39,57
SS-C	Graded Composite	Reactive Magnetron Sputtering	5	0.94	0.03-.1 (100)	300(vacuum) 200(air)	40
SS-C (on rough sputtered copper)	Textured Composite	Reactive Mag.Sput.	3	0.9	0.04(67)	450 (vacuum)	41
SS-SSOx/SSOx	AR Composite	Reactive Mag. Sput.	1	0.89-0.93	0.08(20)	150(air)	42
Cr-Al2O3	Graded Composite	Dual Source Mag. Sput.	1	0.92	0.09(20)		42
Mo-MoO2/Si3N4	Composite/Intrinsic	CVD	1	0.91	0.11(500)	500(vacuum) 300 (air)	43
Tellurium	Textured Tandem	Angled Vapor Deposition	1	0.92	0.03		47
a-Si/Si3N4	AR Tandem	CVD	1	0.75	0.08(500)	500(air)	48

TABLE 3: PROPERTIES OF SELECTIVE ABSORBERS[26]

Absorber[1]	Type	Deposition Technique	Maturity[2]	A_s	E(°C)	Stability[3],[°C]	Ref.
Cu2S	Tandem	Chemical Spray Dep	1	0.89	0.25(100)	130(air)	49
CoO	Textured Graded Tandem	Electropl.+ Heat Oxidation	1	0.98	0.2(100)	425(air)	50
CoO-FeO3-Co3O4	Textured Tandem	Electropl.+ Heat Oxidation	1	0.9	0.07(100)	300(air)	51
Ge-CaF2	Composite	Sputtering	1	0.65-0.72	0.01-.1 (100)		52
Ge	Textured Tandem	RF Sputt. H2O2 etching	1	0.98-0.99	0.58		53
AMA (M-Cr)	3-layer	Re.Mag.Sput.	3	0.95	0.12(20)	300(air)	45
AMA (M-Ni)	3-layer	Re.Mag.Sput.	3	0.91	0.08(20)	350(air)	45
AMA (M-Ta)	3-layer	Re.Mag.Sput.	3	0.89	0.12(20)	300(air)	45
Proprietary Al2O3/Pt-Al2O3/Al2O3	3-layer	Elect.B. Evap.	3	0.92-.96	0.05-.08	250(air)	26
	3-layer	RF Mag. Sput.	3	0.91-0.93	0.08-.1 (20)	>600(air)	45
Al2O3/Mo-Al2O3 Mo	Composite	Co-evap.	1	0.99	0.2(500) 0.08(200)	750(vac)	54
ZrC/Zr Cr-CrOx	Tandem Composite on Al foil	RF Sput. Re. Evap.	3 3	0.93 0.9	0.25 0.05	625(vac) 175	55 55
Ni-C	Composite	Sputtering	1	0.8	0.028-0.035 (150)		56
NiNx	Tandem	Sputtering	1	0.84	0.09 (150)		56

Notes: 1 Absorber layers are separated by a /. Constituents of composite layers are separated by a -.
2 Maturity of absorber coatings: 5 commercial, 3 development, 1 research
3,4 Temperature stability
for most absorbers is not well known.

Abbreviations: AR - Antireflected
CVD - Chemical vapor deposition
Mag.Sput. - magnetron sputtering
Electrpl. - Electroplating
RF - Radio Frequency
Re. - Reactive
Elect. - Electron
Evap. - Evaporation

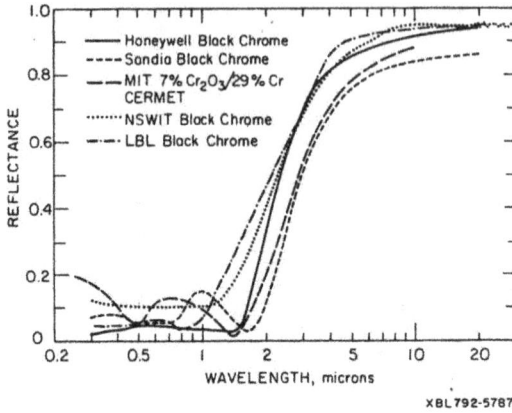

Figure 17. Reflectance for different types of selective black chrome absorbers.[1, 67, 68]

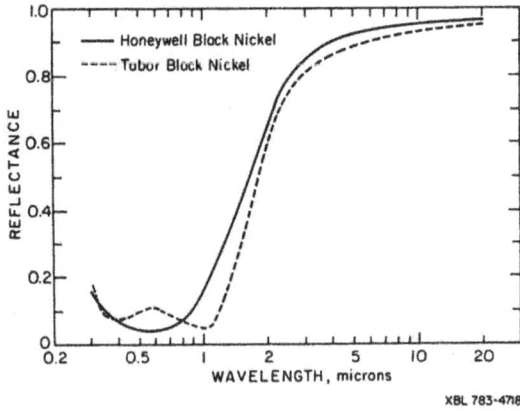

Figure 18   Spectral reflectance for black nickel.[77,78]

A novel graded index coating has been devised by D.C. anodization of aluminum.[36] The coating is made by phosphoric acid anodization followed by an A.C. electrolysis of a nickel pigmenting bath. The structure of this coating consists of porous Al_2O_3 with the lower portion of the pores filled with bundles of needle-shaped nickel particles. A thin Al_2O_3 barrier layer protects the aluminum substrate. A spectral plot of this absorber is shown in Figure 19 .

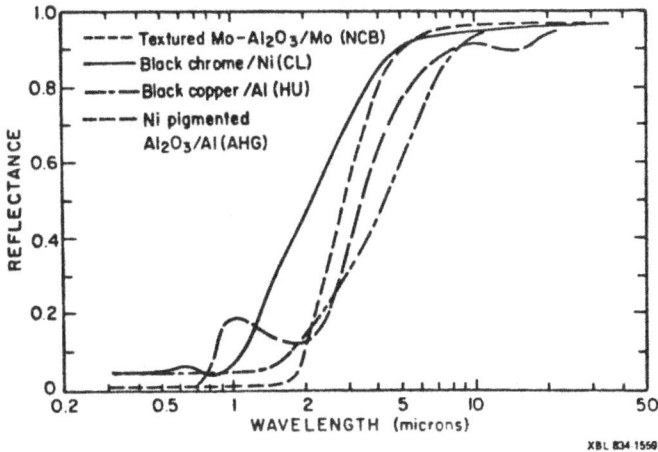

XBL 834 1558

Figure 19 . Spectral reflectance of nickel-pigmented Al_2O_3 absorber[36] shown with textured $Mo-Al_2O_3/Mo$ surface,[54] black chrome,[64] and copper.[61]

Maxwell-Garnett Theory

The Maxwell-Garnett theory (MG) is employed to help elucidate the absorption properties of extremely fine particles in a solid matrix. The MG theory was developed from Mie Scattering theory, which describes the scattering properties of spherical particles larger than those that could be handled by Rayleigh theory. The original MG theory treated isolated spherical metal particles in a dielectric matrix. Also, it was assumed that the particles exhibited identical properties to the bulk metal, which is oversimplified. The MG theory was further developed to take a random distribution of particles characterized by a metal filling factor (f) and simple particle shapes.[62,72] The diameter (d) of the particles is small compared to the film dimensions and wavelengths (2), so d < 0.1w.

Historically, the MG theory has been used to described the color obtained from colloidal suspension of metal in glass, and it is currently used to predict the properties of uniform ultrafine particulate solar absorbers.[72]

In general, a homogenous isotropic material is optically character-ized when the refractive index and the extinction coefficient are known as a function of all wavelengths of interest.

Consider a medium composed of metal spheres suspended in a vacuum. There are N spheres per unit volume. Each sphere behaves like an elec-tric dipole with dipole moment $\overline{P}$ in an applied external field $\overline{E}$. The external field is the sum of the field due to the incident light or energy $\overline{E}_1$ plus contributions from other dipoles, as expressed in the following relationship:

$$\overline{E} = \overline{E}_1 + \frac{4}{3}\overline{P},$$

where $\overline{P}$ due to $\overline{N}$ individual dipole moments per unit volume is given as:

$$\overline{P} = \overline{N}p = \overline{N}a^3\left[\frac{(e-1)}{(e+2)}\right].$$

The complex dielectric constant is given as e and is defined as:

$$e = (n - ink)^2 = e_1 + ie_2.$$

The wavelengths of interest for solar energy applications are orders of magnitude larger than the particle sizes, showing that the optical prop-erties should be discussed in terms of a spatially averaged dielectric constant:

$$\bar{e} = \bar{e}_1 + i\bar{e}_2 .$$

To use the Maxwell Garnett theory, the fill factors of metal to matrix ratio must be ≤ 0.2. Otherwise, corrections have to be made for retardation effects. The Maxwell-Garnett theory gives:

$$\bar{e} = e_d \frac{1 + \frac{2}{3} fa}{1 - \frac{1}{3} fa} ,$$

where (a) is the polarizability factor; for metal spherical particles in a dielectric matrix:

$$a = \frac{(e_m - e_d)}{e_d + Q(e_m - e_d)} , \qquad Q = \frac{1}{3},$$

where,
e_m = the complex dielectric constant for the metal,
e_d = the dielectric constant for the oxide, and
Q = the total depolarization factor.

Also $\bar{e}$ can be expressed as:

$$\bar{e} = \frac{e_d \left[2e_d(1-f) + e_m(2f+1) \right]}{e_d(f+2) + e_m(1-f)} ,$$

The above equation can be expressed in its symmetric form:

$$\left[\frac{\bar{e} - e_d}{\bar{e} + 2e_d} \right] = f \left[\frac{e_m - e_d}{e_m - 2e_d} \right] .$$

Bruggeman Effective-Medium Theory

In the Bruggeman theory (BR) an inhomogeneous two-phase material is treated as a system of spherical particles. These particles are composed separately of pure phase A and pure phase B. The theory solves for the local electric field around a typical two-phase element, embedded in an effective medium. In the BR theory first-order scattering shall vanish on the average; that is, all field fluctuations will average to zero. In this fashion, a self-consistent local field is equivalent to the choice of an active medium such that the average single site scattering is zero. Considering a metal-oxide system, the effective-medium permeability results as:

$$\overline{e} = e_{ox} \frac{(1 - f + \frac{1}{3}fa)}{(1 - f - \frac{2}{3}fa)} \quad ,$$

where (a) for spheres is:

$$a = \frac{(e_m - \overline{e})}{\overline{e} + \frac{1}{3}(e_m - \overline{e})} ,$$

and e_{ox} = dielectric permeability of oxide,
e_m = dielectric permeability of metal,
$\overline{e}$ = effective dielectric permeability, and
$\overline{f}$ = chromium fill factor.

Also, the above expression can be seen in its symmetric form:

$$f \left[\frac{e_m - \overline{e}}{e_m + 2\overline{e}} \right] = (f - 1) \left[\frac{e_{ox} - \overline{e}}{e_{ox} + 2\overline{e}} \right] .$$

Effective-Medium Bound Theory

A range of effective-medium theories, including various microstructured shape effects in cermet systems, can be bounded by mathematical theory. An effective dielectric function within bounds can be calculated with only knowledge of separate dielectric functions for phase A and B volume fractions. This can be done regardless of their geometrical configuration.[74-76] The bounds can be developed using the function F(s), which is defined as:[75]

$$F(s) \equiv (e_d - \overline{e})/e_d , \text{ and}$$

$$s \equiv e_d/(e_d - e_m) .$$

The bounds can be derived using the parametric representation, as:[74-75]

$$F_e(s) = \frac{f(s - s_o)}{[s - s_o(s - \frac{1}{3}(1-f))]} \quad \text{for } \frac{2}{3} < s_o < 1$$

and

$$F_f(s) = \frac{f(s - s_0)}{(s-s_0)(s - \frac{1}{3}(1-f)) - \frac{2}{3}(1-f)(1-s_0)} \quad \text{for } \frac{2}{3} < s_0 < 1.$$

Both $F_e(s)$ and $F_f(s)$ represent arcs that intersect on the complex F plane. The area bounded represents allowed values of $\bar{e}$. Their inter-section points are exactly the MG theory result. If percolation is con-sidered for the metal (or phase A), then $F_f(s)$ becomes:[76]

$$F_{fa}(s) = \frac{1}{s}\frac{f(s-s_0)(s- \frac{2}{3}) - fs_0(1-s_0)/ 3(1 - 2s_0)}{(s-s_0)(s- 1 + f/3) - fs_0(1-s_0)/ 3(1 - 2s_0)}$$

for $0 < s_0 < \frac{1}{3}$.

If the dielectric (or phase B) percolates, then $F_f(s)$ becomes:

$$F_{fb}(s) = \frac{f(s-s_0)}{(s-s_0)[s- \frac{1}{3} - (1-f)s_0(1 - s_0)]/ 3(2s_0 - 1)}$$

for $\frac{2}{3} < s_0 < 1$.

Graphical representations showing the relationship between these bounds and various effective-medium theories are detailed elsewhere.[76]

Multilayer Absorbers

Multilayer thin films can make excellent solar absorbers. Dielectric/metal/dielectric combinations known as interference stacks behave like selective filters for energy absorption. The desired effect of an interference stack is to capture energy between metal-dielectric alternations. Specific solar wavelengths are absorbed by multiple reflection in the layers. Other wavelengths not of the absorption or tuning frequency of the multilayer films are reflected. For solar energy absorption a broad band filter is required. Thin films are gen-erally produced by CVD or PVD processes. Examples film responses are shown in Figure 20. The disadvantage with most multilayer coatings is that they are fairly expensive to fabricate compared to a single-layer process. These coatings can also suffer from interdiffusion and corro-sion at elevated temperatures and humidity. In spite of all this, stable multilayer coatings do exist, like that of the $Al_2O_3/M/Al_2O_3$ design.[73] The metal layer in these coatings is typically 50-100Å thick to appear semi-transparent to incoming radiation. The dielectric layers

need not be intrinsically absorbing to solar radiation since this three-
layer structure behaves as a resonant cavity tuned to a band of solar
wavelengths. A novel multilayer coating has been devised by D.C. anodization of aluminum.[36] The coating is made by phosphoric acid anodization followed by an A.C. electrolysis of a nickel pigmenting bath. The
structure of this coating consists of porous Al_2O_3 with the lower portion of the pores filled with bundles of needle-shaped nickel particles.
A thin Al_2O_3 barrier layer protects the aluminum substrate.

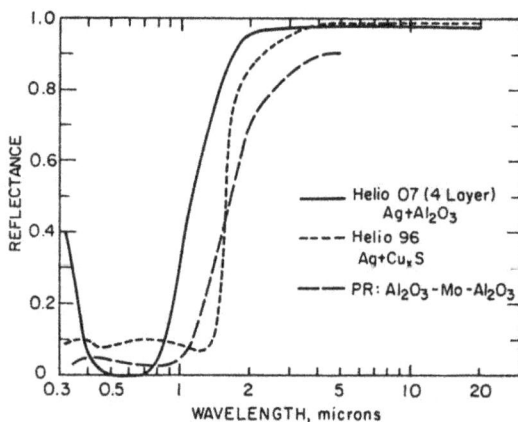

XBL783-4746

Figure 20. Reflectance for sample multilayer absorbers of the di-
electric/metal type. Also included is the Al_2O_3-Mo-Al_2O_3
high-temperature absorber.[73]

Absorber Paints

If a thermally resistant, property stable paint could be developed that offered the solar selectivity of   black chrome for a low cost, the solar coating industry would be revolutionized. Up to this time a highly selective paint does not exist. There are a few that exhibit partial solar selectivity and are commercially available. Others, however, are still in the research stage.   Some properties of selective paints are shown in Table 4.   Three approaches can be taken in designing a selective paint. One makes the paint coating very thin so the low-infrared emittance of the metal substrate dominates the infrared. Notice that a metal substrate is specified. For passive applications such as Trombe walls the substrate would probably be concrete or brick, and this type of paint would not be selective. The major problem with thickness-sensitive paints is their binder material. This is usually a polymer, which exhibits dominant infrared - absorption bands. This absorption increases the infrared emittances of the paint. To remedy this either a special polymer must be devised or an inorganic system must be used. Another approach to the paint is to use a coated metal flake in the composition. In this way the selective effect is created as a distributed tandem, with each particle serving as an individual tandem absorber; recent research has documented this effect.[44] A third approach is to use an intrinsic absorber material   such as ZrB_2 and disperse it in a low-absorbing binder. The final concept relies upon the development of both the intrinsic absorber and binder.

RADIATIVE COOLING MATERIALS

The earth naturally cools itself by radiative transfer through high-transmission windows in the atmosphere to the cold troposphere. This effect is most noticeable on clear nights. A significant atmospheric window occurs from 8-13 microns wavelength, as shown in Figure 21. One could conceivably design an upward-facing surface that   would emit over this wavelength range.[81-84] To model radiative cooling one must first consider a surface radiating towards the sky. A simplified radiation balance can be used to determined idealized radiative cooling power in the absence of convective and conductive heat-transfer effects.[81] The radiative cooling power (P) is given by:

$$P = bE_s(T_s^4 - T_a^4) + E_{sc}(1 - E_{ac})\int_8^{13} B_w(T_a)dw \ ,$$

TABLE 4 SELECTIVE ABSORBER PAINTS

All data taken from manufacturers' product literature or research reports, accuracy or completeness has not been verified.

Type	Highest Operating Temp. in Air.	$A_s(T), /E_{ir}(T)$	Ref.
Soot/polyurethane alkyd	>70-100°C	0.90, 0.30 (100°C)	46
Thurmalox Silicone based	537°C	0.96, 0.52 (84°C)	Dampney Co. Everett, MA
Solkote-Hi/Sorb	880°C	0.95, 0.44	Solec Princeton, NJ
Fe,Mn,Cu oxides + silicone	200°C	0.9 , 0.31 (100°C)	44
Fe, Mn,Cu oxides and silicone epoxy	200°C	0.9 , 0.31 (100°C)	44
Fe, Mn,Cu oxides + * /Al flake and silicone	>600°C	0.91-0.93, 0.8-0.1 (20°C)	44

*Thickness insensitive

where

w = wavelength in microns,

E_s = hemispherical emittance of the cooling surface,

b = Stefan Bolzmann constant,

T_s = Temperature of the cooling surface,

T_a = Temperature of the atmosphere,

E_{ac} = Average hemispherical emittance of atmosphere from 8-13 microns,

E_{sc} = Average hemispherical emittance of the surface from 8-13 microns.

E_{ac} is given by:

$$E_{sc} = \frac{\int_8^{13} B_w(T_a) E_s \, dw}{\int_8^{13} B_w(T_a) \, dw}$$

The term E_{ac} can be defined in a similar manner.

As a figure of merit the largest cooling-temperature differential requires that E_{sc}/E_s be maximum and that the cooling power at near ambient is governed by E_{sc}.

This implies that a material would have to have high reflectance for 0.3-50 microns, excluding the 8-13 micron region. In the 8-13 micron region the material would have to have a very low reflectance or high emittance. It is theoretically possible for such a surface to reach 50 °C below ambient, with typical temperatures about 15 °C below ambient. Temperatures below the dew point should be avoided. The cooling power is about < 100 W/m^2 at near ambient.

Solid-state materials used for radiative cooling include SiO/Al, Si_3N_4/Al (see Figure 22), and polymer-coated metals.[81] Polymers such as polyvinylchloride (PVC), polyvinylfluoride (PVF, Tedlar), and poly-4-methylpentene(TPX) can be used as covers. A radiative-cooling device can also consist of two separate materials, a selective cover and an emitter. Infrared emitters are easier to find, but the selective cover is a challenge. Materials like polyethylene with coatings of Tellurium or dispersions of TiO_2 have been experimented with. Materials need to be designed that not only satisfy the optical requirements but are also resistant to weathering and solar degradation. For the materials investigated thus far, the emittance of the coatings needs to be optimized to take full advantage of the 8-13 micron window. Finally, methods of coupling these surfaces with heat-transfer media need to be devised. By using a selectively emitting gas as the emitter, the gas can serve as its own heat transfer medium. Gases such as ammonia, ethylene and ethylene oxide have been investigated.[84]

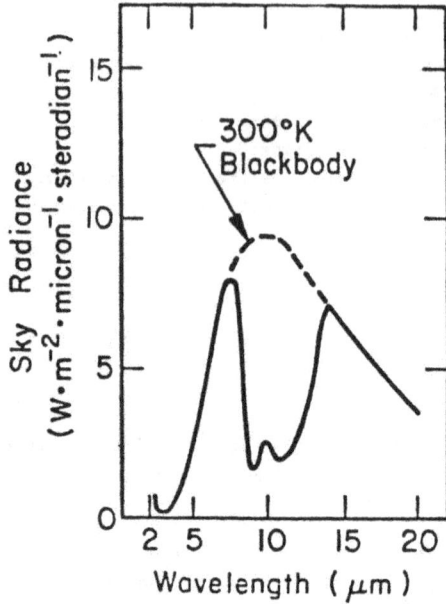

Figure 21.. Infared spectral sky radiance (solid curve)
for clear-sky conditions depicting the atmospheric
transparency window at 8 to 13 microns wavelength.[82]

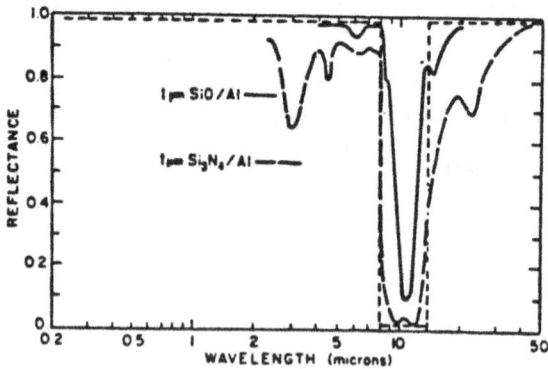

Figure 22 Spectral properties of radiative-cooling tandems of
SiO/Al and Si$_3$N$_4$/Al.[83] Ideal properties are shown by the short
dashed line.

ANTIREFLECTION FILMS

Transparent covers and films used in solar energy conversion systems and architectural windows can have their reflection losses reduced by antireflection treatments. Reflection losses are caused by optical interference from the boundary formed between different media. As a propagating electromagnetic wave in one medium enters another, there is a change in phase velocity, wavelength, and direction. Because of interference between the incident wave and the atoms of the new medium, a reflected wave is radiated. This effect is depicted in Figure 23 .

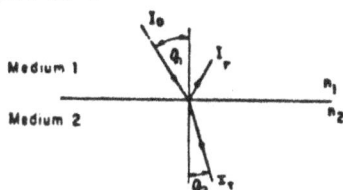

Figure 23 . Angles of incidence and refraction in media having refractive indices of n_1 and n_2.

When the incident medium is vacuum, the ratio of phase velocities (vacuum/medium) is the index of refraction, n. If the medium is energy absorbing, n can be a complex quantity $n = n - ik$. In absorbing media the extinction coefficient k is related to the absorption coefficient $a = 4\pi k/w$, where w is wavelength. Examples of refractive indices (589 nm) of typical media are: air, n = 1.000393; water, n = 1.3336; crown glass, n = 1.523; and diamond, n = 2.42. Materials such as window glass (n = 1.51) and polymers like polymethyl methacrylate (PMMA), polyvinyl fluoride (PVF), polyethylene terephthalate (PET), polyethylene, and polycarbonate (n = 1.46 – 1.54) are used for solar apertures (see Table 5).

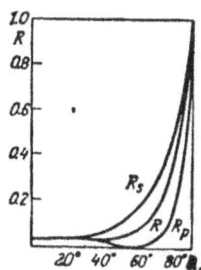

Figure 24 . Dependence of reflectance R, R_p, and R_s upon incident angle for air/glass interface.

Table 5

Optical Properties of Transparent Cover Materials[85]

Material	Index of Refraction	Normal Incident Short-wave Transmittance ($\lambda=0.4-2.5\mu$)	Normal Incident Long-wave Transmittance ($\lambda=2.5-40\mu$)	Thickness (m)
Glass	1.518	0.840	0.020	3.175×10^{-3}
Polymethyl Methacrylate (Acrylic, Plexiglas)	1.490	0.900	0.020	3.175×10^{-3}
Polycarbonate (Lexan)	1.586	0.840	0.020	3.175×10^{-3}
Polytetrafluoroethylene (Teflon)	1.343	0.960	0.256	5.080×10^{-5}
Polyvinyl Fluoride (Tedlar)	1.460	0.920	0.207	1.016×10^{-4}
Polyethylene Taraphthalate (Polyester, Mylar)	1.640	0.870	0.178	1.270×10^{-4}
Polyvinylidene Fluoride (Kynar)	1.413	0.930	0.230	1.016×10^{-4}
Polyethylene (Marlex)	1.500	0.920	0.810	1.016×10^{-4}
Fiberglass Reinforced Polyester (Sunlite)	1.540	0.870	0.076	6.350×10^{-4}

In Figure 24 the dip in the R_p curve corresponds to the Brewster
angle (a_b), which is the maximum polarization in reflection. For glass the
Brewster angle is 57° given by:

$$a_b = \tan^{-1} \frac{n_2}{n_1} .$$

At this angle (a_b) p-polarized light is completely transmitted into
the medium without reflection. The reason for this effect is that
glass partially polarizes (favoring s-polarization) unpolarized
incident light by reflection. Also the refracted (transmitted)
portion of the beam is partially polarized (favoring p-polarization).
At the Brewster angle the electromagnetic vectors for p-polarized
light are such that they pass into the glass without interference.
Common glare is caused by polarization by reflection and can be
supressed by polarizing filters. Several glass sheets or many layers
of optical thin films can also produce polarization.   One glass
plate removes about 8% of light polarized in one direction, and four more
plates remove about 67% of polarized light .

The reflectance, R at an interface between two media has been
derived by Fresnel as:

$$R = \frac{I_r}{I_o} = \frac{1}{2}\left[\frac{\sin^2(a_2 - a_1)}{\sin^2(a_2 + a_1)} + \frac{\tan^2(a_2 - a_1)}{\tan^2(a_2 + a_1)}\right] = \frac{1}{2}\left(R_s + R_p\right) .$$

Non polarized reflectance in this expression is the average of two
reflected componets of light polarization, R_s perpendicular (s-wave)
and R_p, parallel (p-wave). The s-wave electric field oscillates in
a plane spatially perpendicular to the plane of incidence. P-waves

oscillate in a plane spatially parallel to the plane of incidence. The angles a_1 and a_2 are related to the index of refraction of the non-absorbing media by Snell's law:

$$\frac{n_1}{n_2} = \frac{\sin a_2}{\sin a_1}, \quad \text{for absorbing media:} \quad \frac{n_1 - ik_1}{n_2 - ik_2} = \frac{\sin a_2}{\sin a_1}.$$

Also, reflectance can be written for nonabsorbing media as:

$$R = \frac{1}{2}\left[\frac{\left(n_2 \cos a_1 - n_1 \cos a_2\right)^2}{\left(n_2 \cos a_1 + n_1 \cos a_2\right)^2} + \frac{\left(n_2 \cos a_2 - n_1 \cos a_1\right)^2}{\left(n_2 \cos a_2 + n_1 \cos a_1\right)^2}\right].$$

For two absorbing media $\bar{n}_1 = n_1 - ik_1$, $\bar{n}_2 = n_2 - ik_2$, then the two reflectance polarizations become:

$$R_s = \left[\frac{\left(n_2 \cos a_1 - n_1 \cos a_2\right)^2 + \left(k_2 \cos a_1 - k_1 \cos a_2\right)^2}{\left(n_2 \cos a_1 + n_1 \cos a_2\right)^2 + \left(k_2 \cos a_1 + k_1 \cos a_2\right)^2}\right],$$

$$R_p = \left[\frac{\left(n_2 \cos a_2 - n_1 \cos a_1\right)^2 + \left(k_2 \cos a_2 - k_1 \cos a_1\right)^2}{\left(n_2 \cos a_2 - n_1 \cos a_1\right)^2 + \left(k_2 \cos a_2 + k_1 \cos a_1\right)^2}\right],$$

and

$$\frac{\sin a_2}{\sin a_1} = \frac{n_1 - ik_1}{n_2 - ik_2}.$$

If reflectance is plotted as a function of incident angle a_1 for a hypothetical air-to-glass (n = 1.52) interface, the relationships between R, R_s, and R_p can be noted, as shown in Figure 24 .

For radiation at normal incidence $a_1 = a_2 = 0$, so the reflectance equation becomes normal reflectance, R_n:

$$R_n = \frac{\left(n_2 - n_1\right)^2}{\left(n_2 + n_1\right)^2} \, ,$$

and for two isotropic absorbing media the normal reflectance is:

$$R_n = \left[\frac{\left(n_2 - n_1\right)^2 + \left(k_2 - k_1\right)^2}{\left(n_2 + n_1\right)^2 + \left(k_2 - k_1\right)^2} \right] \, .$$

So for most polymers and glasses for solar use, other than low-index halocarbons the normal incidence reflectance losses are about 3.5 − 4.5% per surface. The net increase in performance for each aperture cover would be about 7 − 9 % maximum. This gain in transmittance can be very significant when multiple glazings are used, as demonstrated in Figure 25 .

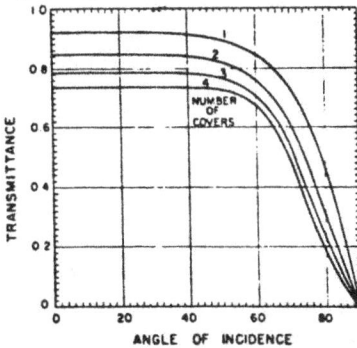

Figure 25 . The idealized transmittance of different numbers of covers showing the effect of reflectance only (no absorption) for a medium with n = 1.52.

Now consider the addition of an antireflection layer to the cover surface. In order that two beams annul each other, two conditions must be satisfied; (1) the amplitudes must be equal; (2) the phase difference must be 180°. Consider the situation for isotropic nonabsorbing media at normal incidence. Let n_0 be the refractive index of air, n_1 be the refractive index of the coating, and n_2 the refractive index of the substrate. The amplitudes will be equal if the reflectance from the air/coating interface is equal to the reflectance from the coating/substrate interface:

$$\frac{\left(\dfrac{n_1}{n_0}\right) - 1}{\left(\dfrac{n_1}{n_0}\right) + 1} = \frac{\left(\dfrac{n_2}{n_1}\right) - 1}{\left(\dfrac{n_2}{n_1}\right) + 1} \; ,$$

and when $n_1 = (n_2)^{1/2}$ this condition is satisfied. The phase change takes place in the coating for both waves. To produce a phase difference of $180°$ it is necessary that the phase length be equal to multiples $^{(N)}$ of $w/4$ or $2n_1 t = (2N+1)1/2w$. Generally no suitable solid material exists with $n_1 = 1.225$ for antireflection of a $n_2 = 1.5$ substrate.

At normal incidence the reflectance is given as :

$$R_n = \left[\frac{n_0 n_2 - n_1^2}{n_0 n_2 + n_1^2}\right]^2 .$$

Formulas for multilayer films are given elsewhere.[6]

Antireflective coatings, if designed properly, can also serve as durable overcoating materials. For photovoltaics, some polymeric and elastomeric protective coatings can be effective antireflective materials if the coating is thin enough, although protective coatings are generally used in thick-film form. Popular protective materials are silicones, fluorocarbons, halocarbons, and acrylic resins. One major need is to develop a coating that serves both protective and antireflective functions. Some polymers having a low refreactive index (n) can antireflect glass (n = 1.5) and other high-index plastics. Dispersions of fluorinated ethylene propylene (n = 1.34) can be used for this purpose. Polyvinyl fluoride (n = 1.46) can be antireflected by dipping in acetophenon. Graded-index films present a versatile range of coatings having refractive indices that are not

readily found. Fluorosilicic acid can give a graded-index,
antireflective coating to glass (see Figure 26). It primarily
roughens the surface by etching out small pores, in non-silica
regions.[86,87] Silica films deposited from sodium silicate or
collodial silica can be used for acrylic, polycarbonate, and several
glasses. A treatment for polyethylene terephthalate (polyester) and
glass materials has been devised.[88-89] The coating is made from a
steam-oxidized aluminum film; this processng causes a needle-like
structure of aluminum hydroxide [AlO(OH)] to form. A polyester film
treated in this fashion can serve in glazing applications where solar
transmission must be optimum[90] (see Figure 27). Inorganic thin
films have been used for a wide range of single and multiple
interference-coating applications Compounds such as MgF_2, CeO_2,
SiO, SiO_2, and TiO_2 in various combinations have been used for
antireflection applications. Other than the traditional PVD
techniques, a number of oxides can be dip-coated onto optical
substrates. Coatings of hot hydrolyzed metal alkoxides can be
polycondensed, forming oxides of transition metals, refractory metals,
and some rare earths.[91] A similar technique known as the sol-gel
process has formed mixed TiO_2-SiO_2 antireflective films on
silicon[92] and black chrome. Diamond-like (i - carbon) transparent
coatings have been used for antireflective films. They are formed
from plasma decomposition by hydrocarbons and ion beam deposition.[93]
Coatings of about n = 1.9 can be made that are suited to
photovoltaics. However, the absorption properties of i-carbon films
must be reduced before they can be utilized for optical applications.

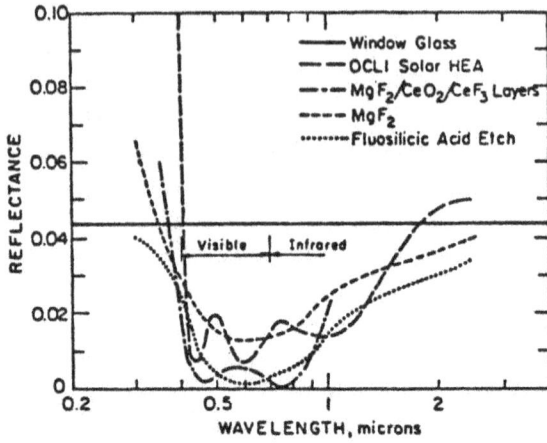

Figure 26. Specular reflectance for a number of antireflection treatments on glass.[1]

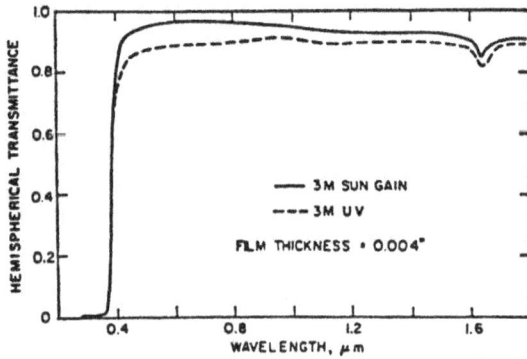

Figure 27. Hemispherical transmission of antireflected 3M Sungain polyester film compared to the uncoated substrate. [84]

REFLECTOR MATERIALS

Reflector materials consist of a metal reflector on a substrate either in a front-surface or back-surface configuration. They can differ according to the method of metallic deposition, be it either silver, aluminum, or an alloy reflector. The substrate material can be flexible or rigid.

Front-surface mirrors offer the best initial optical reflectance but suffer from abrasion and atmospheric corrosion and delamination. Application of a durable overcoating material is required. Second-surface mirrors are conventionally produced by a wet chemical process. For mirrors made with this process, there is considerable lack of understanding of various interfacial reactions and degradation mechanisms that can occur with time. A recent workshop addressed many of these investigated.[94] Also, more durable reflector layers for second-surface mirrors have been devised by sputtering, evaporation, and decomposition of organometallic resinates.[95] For both types of reflectors an understanding of the stability between metal/polymer and metal/glass mirrors is a significant issue. Dirt and dust can be responsible for considerable decline of efficiency of reflector surfaces. Techniques to limit dusting and washing of surfaces need to be devised.

Due to the variety of solar applications for reflector material, the optical requirements for solar mirrors vary greatly, but all are sensitive to the integrated solar reflectance. The spatial distribution of reflected light from a mirror surface is another important parameter since a mirror can range from highly diffuse to highly specular. The amount of this spatial variation is known as specularity. Specularity is dependent upon the exact roughness and contour variations of the mirror surface. Ranges of specularity are shown in Figure 28 .

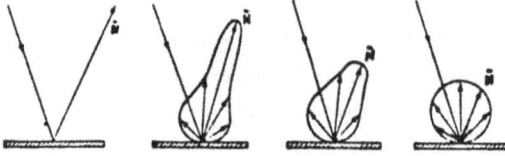

Figure 28 . Angular light distribution in ideal specular reflecton
(left) and total diffuse reflection (right).

The normal clear-day distribution of direct solar beam radiation
has a normal distribution and dispersion of d_g = 3.5 (mrad). milli radians The
effective dispersion d_e after reflection from a mirror with
dispersion d_m is given by:

$$d_e = \left[d_s^2 + d_m^2 \right]^{1/2}.$$

The dispersion for the mirror surface is measured on a bidirectional
reflectometer with a variable collection slit.[96]   An example beam
profile is shown in Figure 29 .

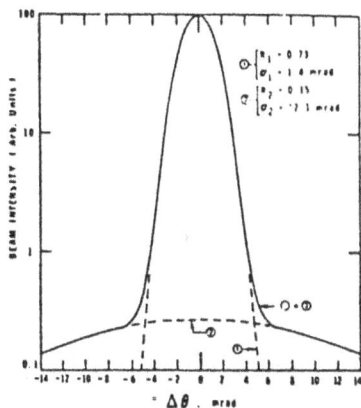

Figure 29 . Reflected beam profile for aluminized teflon film at
400 nm as a functon of angular aperture (after 96).

The specular reflectance properties for several mirror materials are
shown in Figure 30 and Table 6.

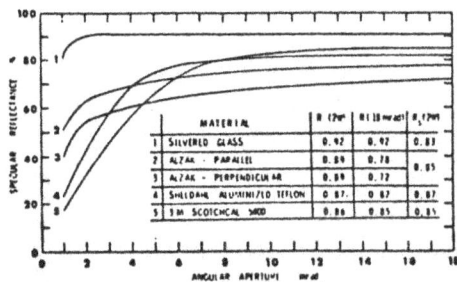

Figure 30   Specular reflectance properties for several solar mirror
materials. The inset table lists hemispherical
reflectance R_w, integrated specular solar reflectance
R_s, and specular reflectance at 18 milliradians. 96

Table 6.

Specular Reflectance Properties of Mirror
Materials[97]

Material	Supplier	Estimates of Solar Weighted Reflectance[b] at Receiver Acceptance Angle τ			
		τ=4mr	10mr	18mr	$R_s(2\pi)$
I. Second-Surface Glass					
(a) Laminated Float Glass - 2.7mm thick -silvered	Carolina Mirror Co.	0.83	0.83	0.83	0.83
(b) Laminated Low-Iron Sheet Glass - 3.35mm thick - silvered	Gardner Mirror Co.	0.90	0.90	0.90	0.90
(c) Corning Silvered Microsheet Co.-0.114 mm thick- Mounted on optically flat plate	Corning Glass	0.76	0.87	0.92	0.95
(d) Corning 0317 Glass - 1.5 mm thick - Evaporated silver	Corning Glass	0.95	0.95	0.95	0.95
II. Metallized Plastic Films					
(a) 3M Scotchcal 5400 Laminated to backing sheet	3M Company	0.60	0.84	0.85	0.85
(b) 3M FEK-163 Laminated to backing sheet	3M Company	0.83	0.85	0.85	0.85
(c) Aluminized 2 mil FEP Teflon (640560J) Laminated to backing sheet	Sheldahl	0.70	0.81	0.82	0.87
(d) Silvered 2 mil FEP Teflon (6400300) Mounted on Optically Flat Plate	Sheldahl[a]	0.73	0.82	0.90	0.96
(e) Silvered 5 mil FEP Teflon (6401500) Mounted on Optically Flat Plate	Sheldahl[a]	0.77	0.83	0.89	0.95
(f) Front Surface Aluminized Mylar (200XM648A) stretched membrane	Boeing	0.88	0.88	0.88	0.88
III. Polished, Bulk Aluminum					
(a) Alzak Type I Specular	Alcoa				0.85
Perpendicular to rolling marks		0.61	0.68	0.76	
Parallel to rolling marks		0.68	0.76	0.83	
(b) Kinglux No. C4	Kingston Ind.				0.85
Perpendicular to rolling marks		0.67	0.71	0.75	
Parallel to rolling marks		0.69	0.71	0.75	
(c) Type 3002 High Purity Al - Buffed and Bright Anodized	Metal Fabrications, Inc.	0.44	0.60	0.71	0.84

a) Experimental materials not produced in high production.

b) Estimated from 500 nm specularity data (Ref 96) and solar weighted total hemispherical reflectance data. Standard deviation of the estimates is about 2%.
($1mr = 0.0573°$)

FLUORESCENT CONCENTRATORS

Fluorescent materials can be used to down-convert or alter spectrally incident radiation, and to concentrate and guide it along the plane of the material. This process is shown schematically in Figure 31. A fluorescent concentrator consists of a transparent plate (of polymer or glass) that has been doped with fluorescent dye molecules. Depending upon the dyes used, various spectral compositions can be obtained. A combination of coupled dyes is depicted in Figure 32. Incident light corresponding to the fluorescent absorption of the dye will be captured and emitted isotropically. Because of the index of refraction difference between the plate and surrounding medium, a large portion of light will be trapped and guided to the edges of the plate by total internal reflection. One edge can be favored by silvering the other three edges (see Figure 31). At this favored edge a photovoltaic or photothermal collector can be placed.[98-102] One of the unique advantages of this system is that it will collect diffuse, low-insolation radiation without solar tracking. Other advantages are that there is less heat dissipation in

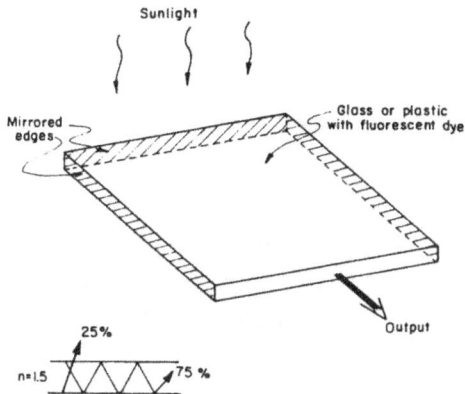

XBL821-5089

Figure 31. Schematic representation of a fluorescent concentrator showing how light can be guided along the plane of the material.

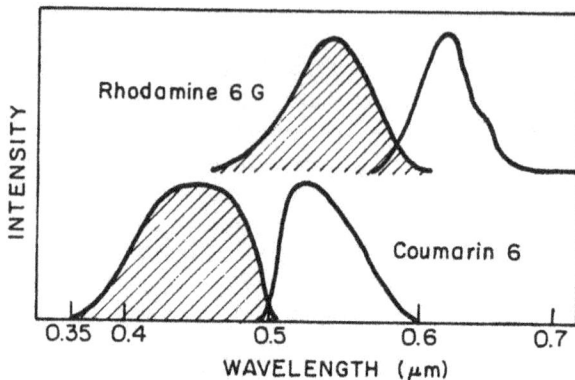

XBL821-5085

Figure 32. Spectra of coupled fluorescent dyes. Absorption wavelengths are crosshatched. Emission spectra are shown
spectrally downshifted.

photovoltaics and high efficiency at low insolation levels. Concentration ratios for these systems can be fairly high (10-100). The collection efficiency of this system is dependent upon a number of factors.
The loss due to light leaving the collector through the boundary planes
by fluorescence for a single plate is given by:

$$L = \frac{1-(n^2 - 1)^{1/2}}{n} \quad ,$$

where n is the index of refraction of the collector plate. If n = 1.5,
then L = 0.255; if n = 2, then L = 0.134. This equation implies that n
should be as large as possible to minimize loss. But another process
is important, that of external reflection loss, which implies a low n.
This loss is given by the Fresnel equation. As a result the total
optical loss of the system is given by:

$$L_T = \frac{1 - 4(n^2 - 1)^{1/2}}{(n + 1)^2} \quad .$$

A minimum occurs at n = 2.0, L_T = 0.23. If we consider the role of the dye and its associated properties, then an overall concentrator efficiency (n_c) can be derived:

$$n_c = n_q n_f n_g n_e n_s \quad ,$$

where n_q is defined as the quantum efficiency of the fluorescent dye; n_f is the efficiency due to fluorescent emission beyond the limiting angle of total internal reflection; n_g is the efficiency due to collector geometry (absorption); n_e is the energy conversion efficiency due to the Stokes wavelength shift associated with fluorescence, and, finally, n_s is the loss of the long-and short-wavelength tails of the solar spectrum. For thermal collectors using fluorescent concentrators, operational efficiencies of 42-59% have been estimated. For photovoltaic systems, operating efficiencies of 32% for a four-glazing plate system have been estimated.[98] By using multiple plates various portions of the solar spectrum can be collected. Each level of collector plate down has a higher absorption energy, so that the innermost level absorbs the highest solar energy. A backup mirror is used on the lowest level to reflect unused energy to the upper lower-energy fluorescent levels. This action is seen in Figure 33.

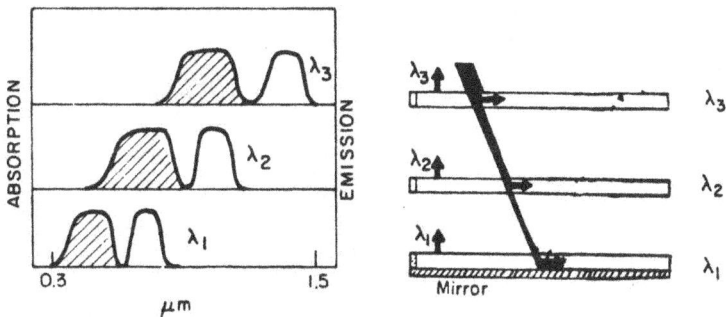

XBL 821-5088A

Figure 33. Three-stage fluorescent concentrator depicting peak wavelengths for each stage.

Current research on fluorescent concentrators has favored polymethyl methacralate (PMMA) and rare-earth-doped glasses as host materials for the dye. Their high optical-transmission property is good to about 1 micron. For a variety of glazing uses there is a need to devise durable glazing materials that can maintain transparency further into the infrared. Dyes that are used for experimentation have not been specifically tailored for solar uses. Experimental collectors generally use laser dyes. Dyes need to be developed with high quantum efficiency, with separated emission spectra with low self-absorption. One of the most severe requirements for the dyes is they be UV stable in the PMMA matrix. It is possible that new materials such as ligands containing rare-earth ions and non-radiative coupled organic systems may offer greater stability. Furthermore, it may be possible to make a polymer-based fluorescent thin film that by index matching could couple energy into a substrate material. This design could minimize reabsorption by the dye. Emission – absorption coupled dyes may offer a wide variety of solar collection and energy redistribution possiblities.

SPECTRAL SPLITTING AND COLD MIRROR FILMS

Spectral splitting coatings are used to divide the solar spectrum into various broadband regions. In this fashion various regions can be tailored to particular photovoltaic or photothermal needs.[104-106] A simple design utilizing a heat mirror might be used to separate heat and light from the solar spectrum. The infrared energy could be used for photo – thermal uses and the high energy visible could be used for a photovoltaic. Photovoltaics will operate more efficiently if infrared heating is eliminated. Expanding this idea further, a system depicted in Figure 34 could result, for an all-photovoltaic system. If heat mirrors with different spectral characteristics were used, the solar spectrum could be partitioned from low to high energy as the heat mirror transition wavelength becomes shorter. Broadband Lippmann Holograms can also be used for spectral splitting.[106]

A coating known as a cold mirror could also be employed for spectral splitting. A system might consist of a series of cold mirrors, where the transition from reflecting to transmitting moves to longer wavelengths for each successive cell, as shown in Figure 34. The cold mirror has the opposite spectral response to that of the heat mirror. It exhibits high reflectance in the visible region and transmits highly in the infrared. Cold mirrors are generally all dielectric interference films. Material systems such as ZnS/M_gF_2 and TiO_2/SiO_2 have been devised.[107-108] Spectral characteristics of a commerical film are shown in Figure 36. An application for this film is for greenhouses.[108] Plants require only a range of wavelengths(0.3 - 0.75 microns);the remainder of the solar spectrum is unused. This remainder can be utilized as heat to warm the greenhouse indirectly. A baffle-type greenhouse is depicted in Figure 36. Heat mirrors can also be used in this fashion.[109]

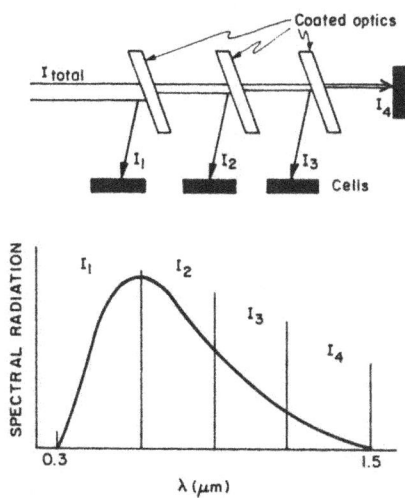

XBL 821-5090

Figure 34. Spectral splitting scheme for photovoltaics.

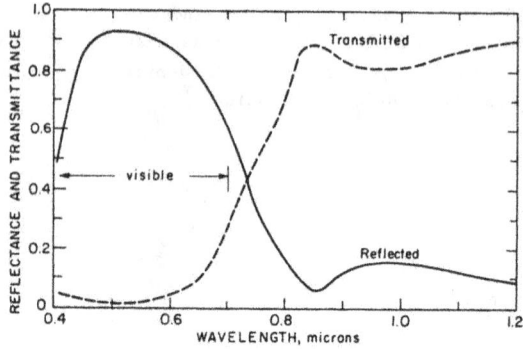

Figure 35. Cold mirror coating for greenhouse baffle collector.[108]

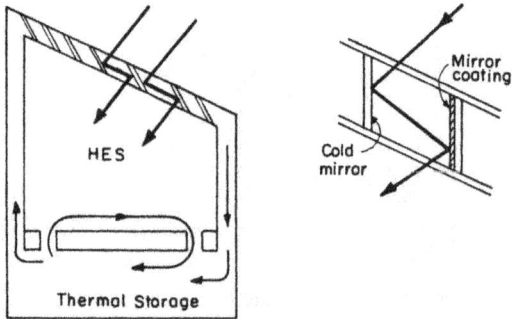

XBL 821-5091

Figure 36. Selective baffle greenhouse using cold mirror coatings to separate heat from visible energy. The roof is made of air cells that circulate heated air to the thermal storage. In this fashion the solar spectrum is divided to suit both needs without an added collector area.

TRANSPARENT INSULATION MATERIAL

One of the major drawbacks of conventional windows is their high thermal loss characteristics compared to other building elements. As discussed previously, surface treatments and modified window design can do much to solve this shortcoming. Another approach is to develop a highly transparent material that by virtue of its bulk properties has low thermal conductivity. Research on one such material, known as silica aerogel, has begun.[110-112] This material has a microstructure of bonded fine silica particles surrounded with high porosity volume. Since the particles are smaller than a wavelength of visible light, they are not visibly scattering. The thermal conductivity of such a material is better than still air with optical properties similar to silica glass except for $n \approx 1.05$ This material has been modeled optically by Rayleigh Theory.[110] Silica aerogel is made by producing a colloidal silica gel from hydrolysis and polycondensation reaction. This gel is solidified in place by supercritical drying. The disadvantage of aerogel is that it must be protected from shock and moisture. It is possible to form aerogel between two sheets of glass to make a window. For a window of aerogel (20 mm thick), the thermal conductance (U) is calculated[110] to be about $1 W/m^2 K$ (R-5). For a double glazing without the aerogel (20mm spacing), $U = 2.8 W/m^2 K$ (R-2). The solar hemispherical transmission properties for aerogel are $T_s = 0.67$ (20mm thick) and $T_s = 0.9$ (5mm thick). Example optical properties are shown in Figure 37.

OPTICAL SWITCHING MATERIALS AND DEVICES

There are various physical processes that can be used for the regulation of incident solar energy and glare in buildings. Optical switching materials or devices can be used for energy-efficient windows or other passive solar uses. The basic property of an optical shutter is that it offers a radical change in optical properties upon a change in light intensity, spectral composition, heat, electrical field, or injected charge. This optical change can be manifested in a transformation from highly transmitting to reflecting either totally or partly over the solar spectrum. The purpose of such a device would be to control the flow of light and/or heat in to and out of a building window, according to an energy management scheme. This device could also control lighting and heating levels for energy load functions. In general, the idea of an optical shutter is a scientific possibility based on a future research and design idea. Phenomena of interest as optical switching processes can be classified as either discrete mass movement or collective particle movement. Discrete mass movement includes ion and localized electron motion (photorefractive, chromogenic, and redox reactions),

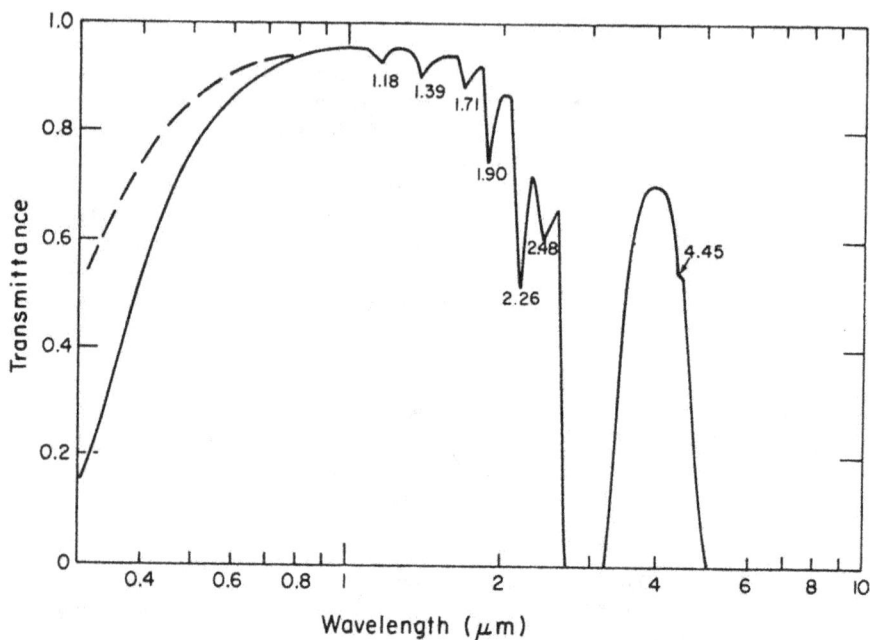

XBL 826-829

Figure 37 .   Example. Transmission properties of
                   4mm thick Silicon Aerogel[110]. Spectral
                   Normal-Normal Transmittance (solid line)
                   and Spectral Normal Hemispherical
                   Transmittance (dashed line).

and ion and delocalized electron population changes (reversible electrodeposition). Collective particle movement includes gas or vacuum
deformable membranes and adjustable diffraction gratings as one
category, and liquid crystals and electrophoresis processes as another.

Only selected physical processes will be covered here. Chromogenic
reactions known as photochromism, thermochromism, and electrochromism
will be discussed along with holographic and liquid crystals.

Photochromic Materials

Photochromic materials alter their optical properties with light
intensity. Generally, photochromic materials are energy-absorptive.
Basically, the phenonemon is the reversible change of a single chemical
species between two energy states having different absorption spectra.
This change in states can be induced by electromagnetic radiation. Photochromic materials have been reviewed.[113]-[115] Probably the best known
photochromic material is photochromic glass for eyeglasses and goggles.
Photochromic materials are classified as organics, inorganics, and
glasses. Within the organics are stereoismers, dyes, and polynuclear
aromatic hydrocarbons. The inorganics include ZnS, TiO_2, Li_3N, HgI_2,
HgCNS, and alkaline earth sulfides and titanates, with many of these
compounds requiring traces of heavy metal or a halogen to be photochromic. Glasses that exhibit photochromism are Hackmanite, Ce, and Eu
doped glasses (which are ultraviolet sensitive), and silver halide
glasses (which include other metal oxides). Of all the photochromic
materials the bulk of the information is on the alkali-halide glasses.
The silver halide glasses transform by color-center formation from an
AgCl crystalline phase. The typical response for a photochromic glass
is depicted in Figure 38. A photochromic plastic has been formulated

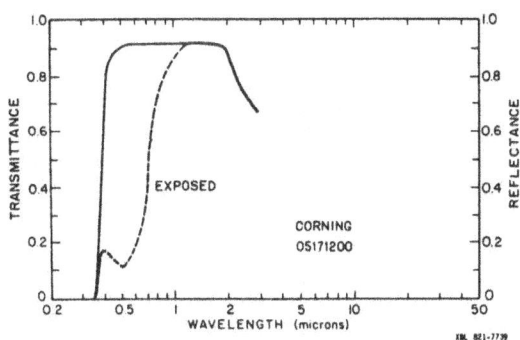

Figure 38. Sample light and dark transmission spectra for Corning photochromic glass.

using derviatives of spiroindolinonaphthoxazine in a matrix of cellulose acetate butyrate and other plastics.[117] This material may have future use for energy regulating glazings for solar energy control.

Thermochromic Materials

Many thermochromic materials are used as non-reversible temperature indicators by displaying acolor change with temperature variation. For an optical shutter one must consider only the reversible materials, although their actual cyclic lifetime may be dependent on non-reversible secondary reactions. Certain organic compounds in the anil, spiropyrans, polyvinyl acetal resins, and hydrozide groups exhibit thermochromism. Inorganic thermochromic compounds include AgI, HgI_2, Ag_2HgI_4, HgI, $SrTiO_3$, Cd_3P_3Cl, along with various copper, cobalt and tin complexes.[118] Another broad group of materials that exhibit thermochromism are the semiconductor to insulator transition compounds. Examples of these compounds are VO_2, V_2O_3, Ti_2O_3, Ti_4O_7, Ti_5O_9, NbO_2, Fe_3O_4, $FeSi_2$ and NiS.[119]
Current research for solar energy application has been on the $V_{1-x}M_xO_2$, (where M is a transition metal) compounds. Such materials, if perfected, can be used to control both the solar transmittance and infrared emittance of a glazing or surface.[120]

Electrochromic Materials and Devices

Electrochromism is exhibited by a large number of materials both inorganic and organic. The electrochromism effect is of current research interest mainly because of its application to large-scale electronic information display devices, optical switching windows and reflectors for architectural and automotive uses. The electrochromic effect, in essence, is a material that exhibits intense color change due to the formation a colored compound. The reaction might follow: $MO_x + yA^+ + ye \leftrightarrow A_yMO_x$ for cathodic materials.

There are three categories of electrochromic materials: transition metal oxides, organic compounds, and intercalated materials. The materials that have gained the most research interest are WO_3, MoO_3, and IrO_x films. These compounds, among other transition metal oxides, are the subject of a few research reviews.[121-123] Organic electrochromics are based on the liquid viologens, anthraquinones, diphthalocyanines, and tetrathiafulvalenes. With organics, coloration of a liquid is achieved by an oxidation-reduction reaction, which may be coupled with a chemical reaction. Intercalated electrochromics are mostly based on graphite and so are not useful for window applications.

A solid-state window device can be fabricated from five (or less) layers consisting of two transparent conductors (TC), an electrolyte or fast-ion conductor (FIC), counter electrode (CE), and electrochromic layer (EC), as shown in Figure 39. Much research is needed to develop a usable panel, better electrochromic materials with high cycle lifetimes, and short response times. Certainly fast-ion conductors and solid electrolytes also require study.

Several research groups are investigating electrochromic materials and devices. The most researched materials are amorphous WO_3, crystalline WO_3 and $Ni(OH)_2$.[124-130] Crystalline WO_3 offers near-infrared modulation, which has the potential to control the infrared portion of the solar spectrum.[127-128] An example of the optical response of electrochromic $Ni(OH)_2$ is shown in Figure 40.[129-130]

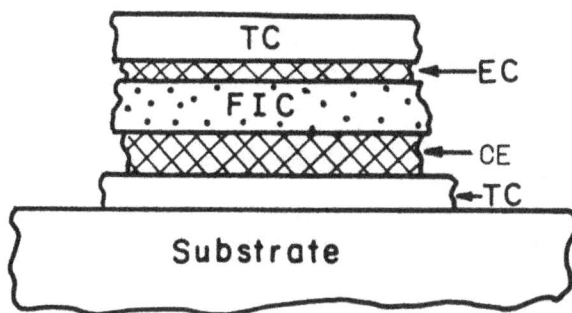

XBL 805-9713

Figure 39. Schematic structure of a solid-state electrochromic device.

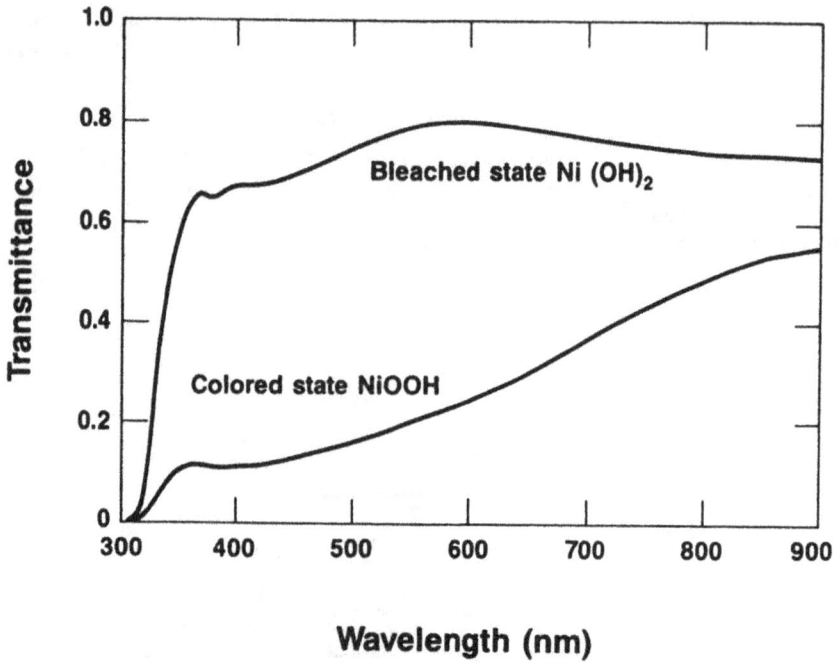

Figure 40.   Spectral Transmission of Electrochromic
Nickel Oxide Film on Tin-Oxide coated
glass substrate. T_v(off) = 0.77
T_v (on) = 0.21. Also Solar Transmission
is T_s (off) = 0.73 and T_s (on) = 0.35.
The film was colored by the application
of 1V pulse at 0.1 Hz.

LIQUID CRYSTALS

Thermotropic liquid crystals are actively used for electronic and temperature indicating displays. Liquid crystals exist in one of the three structural mesophases: smectic, nematic and twisted nematic (cholesteric). Chemically, they are based on azo-azoxy esters, biphenyls and Schiff bases.[131] The most widely used type for electronic displays is the twisted nematic type. [132] For optical shutters, the twisted nematic type is not a good choice since it requires polarizers which reduce transmission. The dynamic scattering nematic liquid crystals offer promise as optical shutter materials. In the activated state, the material becomes translucent white. An example of its optical properties are shown in Figure 41. The switching effect of this device spans the entire solar spectrum, up to the absorption edge of glass. The scattering properties of nematic liquid crystals can be utilized by encapsulating the materials in a polymer matrix of matched optical index. In the off-state, the materials appears translucent white. When activated, the liquid crystal droplets align and the material becomes transparent. [133] Pleochroic dyes can be added to darken the device in the off-state. Liquid crystals can be used as optical filters if they are aligned and solidified by polymerization. This processing can give preset optical properties. In general, the drawbacks of liquid crystals as optical switching materials are cost, limited grey scale and problems in fabricating large area devices (though this is being overcome). Compared to electrochromics, the power consumption is higher due to the need for continuous power in the activated state.

204

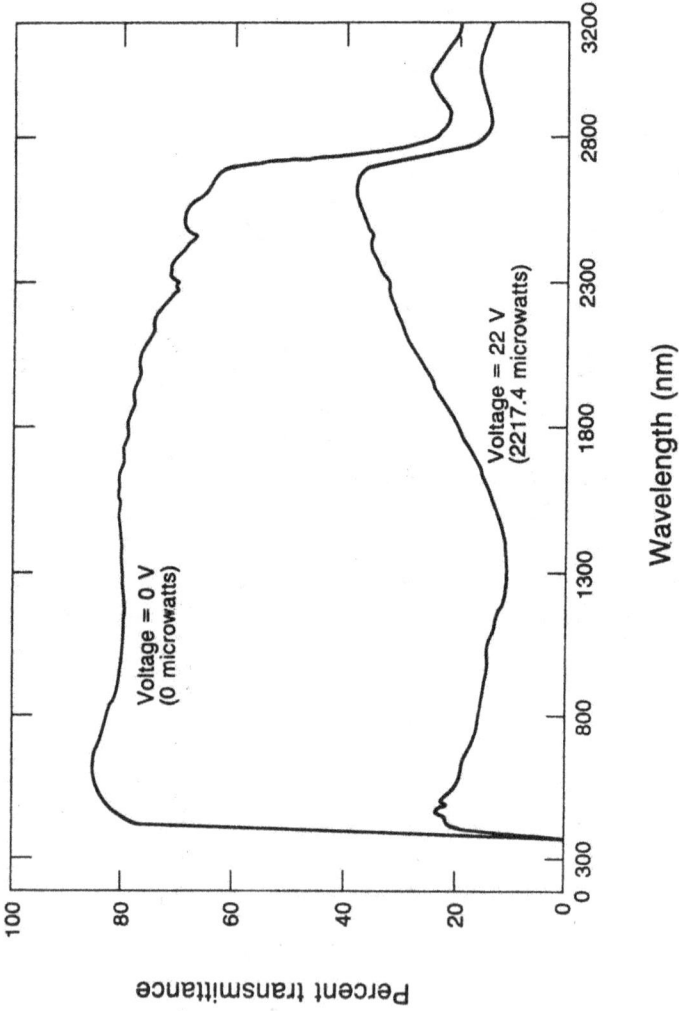

Figure 41. Spectral Transmittance of a Dynamic
Scattering Liquid Crystal Device.

XBL 852-7033

CONCLUSIONS

A very wide range of categories of optical materials have been covered
for solar energy and related applications. Attention has been paid to
current advances and modern optical materials. Also, the potential for
advanced materials (such as optical switching of glazings) has been discussed
from a materials science viewpoint. There are still categories of materials
that have not been covered here. For example, photovoltaic and photoelectrochemical materials have been excluded. Also new categories, such as
large area light guide materials and holographic films, need to be developed
for solar energy applications. In this work, it is hoped that by better
understanding these optical materials will result in new and improved solar
energy conversion devices and systems.

ACKNOWLEDGEMENTS

This paper was created as the result of educational lectures given at four scientific institutions over the course of several years: Gansu Natural Energy Research Institute, Lanzhou, China, April 1986; The International Center for Theoretical Physics, Trieste, Italy, September 1985; University of Split, Yugoslavia, September 1983; and the International Institute for Advanced Studies, Caracas, Venezuela, June 1982.

This work was supported by the Assistant Secretary for Conservation and Renewable Energy, Office of Building Energy Research and Development, Building Systems Division of the U. S. Department of Energy under Contract No. DE-AC03-76SF00098.

REFERENCES

1. C.M. Lampert, Coatings for Enhanced Phtothermal Energy Collection, I & II, Solar Energy Materials 1:319 (1979); 2:1 (1979).

2. C. M. Lampert, Heat-Mirror Coatings for Energy Conserving Windows, Solar Energy Materials 6:1 (1981).

3. O.P. Agnihotri and B.K. Gupta, "Solar Selective Surfaces," J. Wiley, New York (1981).

4. M. Rubin, R. Creswick, and S. Selkowitz, Transparent Heat Mirrors for Windows - Thermal Performance, Proc. 5th National Passive Solar Conf., pp. 990-994, Pergamon NY (1980).

5. R. M. Winegarner, Heat Mirror - A Practical Alternative to the Selective Absorber, Proc. of ISES American Section and Solar Energy Soc. of Canada, Pergamon NY 6:339 (1976).

6. Z. Knitle, "Optics of Thin Films,"J. Wiley , London (1976).

7. K. Chiba, S. Sobajima, U. Yonemura, and N. Suzuki, Transparent Heat Insulating Coatings on   Polyester Film Using Chemically prepared Dielectrics, Solar Energy Materials 8:371 (1983).

8. J.C. Fan, F. Bachner, G. Foley, and P. Zavracky, Transparent Heat Mirror Films of $TiO_2/Ag/TiO_2$ for Solar Energy Collection and Radiation Insulation, Appl. Phys. Lett. 25:693 (1974).

9. W. J. King, High Performance Solar Control Windows, Lawrence Berkeley Laboratory Report, Berkeley, CA, LBL-12119 (1981).

10. B. P. Levin and P.E. Schumacher, A Discussion of Heat Mirror Film: Performance, Production Process and Cost Estimates, Lawrence Berkeley Laboratory Report, Berkeley CA, LBL-7812 (1977).

11. J. Dolenga, Low-E: Piecing Together The Puzzle, Glass Magazine, March 1986, p. 116.

12. C.M. Lampert, Materials Chemistry and Optical Properties of Transparent Conductive Thin Films for Solar Energy Utilization, Ind. Engin. Chem. Prod. R&D21:, (1982).

13. H. Kostlin, R. Jost, W. Lems, Optical and Electrical Properties of Doped In_2O_3 Films, Phys. Status Sol. A29: 87 (1975).

14. G. Haacke, Evaluation of Cadmium Stannate Films for Solar Heat Collectors, Appl. Phys. Lett. 30:380 (1977).

15. M. Van der Leij, "Spectral Selective Surfaces for Thermal Conversion of Solar Energy," Delft University Press, Delft, The Netherlands (1979).

16. J.C.C.Fan, Wavelength-Selective Surfaces for Solar Energy Utilization, Proc of SPIE 85:39 (1977).

17. R.P. Howson and M. I. Ridge, Heat Mirrors on Plastic Sheet Using Transparent Oxide Conducting Coatings, Proc. of SPIE 324:16 (1982).

18. K. Itoyma, Properties of Sn-doped Indium Oxide Coatings Deposited on Polyester Film By High Rate Sputterings, J. Electrochem. Soc. 126:691 (1979).

19. C.M. Lampert, ed. "Optical Coatings for Energy Efficiency and Solar Applications" Proc. of SPIE Vol. 324, The Intern. Soc. for Optical Engin, Bellingham WA (1982).

20. V.A. Baum and A.V. Sheklein, Choice of Materials for Selective Transparent Insulation, Gelioteknika 4:50 (1968).

21. G. Haacke, Transparent Conducting Oxides, Ann. Rev. Mat. Sci 7:73 (1977).

22. J. L. Vossen, Transparent Conducting Films, in "Phys. of Thin Films", G. Hass, M. Francombe and R. Hoffman, eds., Academic Press, New York (1977).

23. M.M. Koltun and S. A. Faiziev, Geliotekhnika 10:58 (1974).

24. P.Drude, Phys. Z. 1: 161 (1900).

25. W. F. Bogaerts and C.M. Lampert, Materials for Photothermal Solar Energy Conversion, J. Material Sci. 18:2847 (1983).

26. S.A. Herzenberg and R. Silberglitt, Low Temperature Selective Absorber Research, Proc. of SPIE 324:92 (1982).

27. B.O. Seraphin, ed., "Solar Energy Conversion-Solid State Aspects," Springer-Verlag, Berlin, FRG (1979).

28. M. M. Koltun, "Selective Optical Surfaces for Solar Energy Converters," Allerton Press, New York (1981).

29. A.B. Meinel and M. P. Meinel, "Applied Solar Energy An Introduction," Addision Wesley, Reading, MA (1976).

30. E. Randich and R.B. Pettit, Solar Selective Properties and High Temperature Stability of CVD ZrB_2, Solar Energy Materials 5:425 (1981).

31. G.L. Harding and M.R. Lake, Sputter Etched Metal Solar Selective Absorbing Surfaces for High Temperature Thermal Collectors, Solar Energy Materials 5:445 (1981).

32. C.M. Lampert and J. Washburn, Microstructure of a Black Chrome Solar Selective Absorber, Solar Energy Materials 1:81 (1979).

33. P.H. Holloway, et al. Oxidation of Electrodeposited Black Chrome Selective Thin Solid Films 72:121 (1980).

34. P.M. Driver, An Electrochemical Approach to the Characterization of Black Chrome Selective Surfaces, Solar Energy Materials 4:179 (1981).

35. G. B. Smith, A. Ignatiev, Black chromium – Molybdenum a New Stable Solar Absorber, Solar Energy Material 4:119 (1981).

36. A. Andersson, O. Hunderi, G. Granqvist, Nickel Pigmented Anodic Aluminum Oxide For Selective Absorption of Solar Energy, J. Appl. Phys. 51: 754 (1980).

37. C. Homhual, O.T. Inal, L.E. Murr, A.E. Torma, I. Gündiler, Microstructural and Mechanical Property Evaluation of Zinc Oxide Coated Solar Collectors, Solar Energy Materials 4:309. (1981).

38. H. Potdar, N. Pavaskar, A. Mitra and A.P.B. Sinha, Solar Selective copper Black Layers by an Anodic Oxidation Process, Solar Energy Materials 4:291 (1981)

39. C.M. Lampert, Selective Absorber Coated Foils For Solar Collectors, Plating and Surface Finishing 67:52 (1980).

40. G.L. Harding, B. Window, Graded Metal Carbide Solar Selective Surfaces Coated Onto Glass Tubes by a Magetron Sputtering System, J. Vac. Sci & Techn. 16: 2101 (1979).

41. S. Craig and E.L. Harding, Solar Selective properties of rough sputtered Copper Films, Solar Energy Materials 4:245 (1981).

42. J. A. Thornton, A.S. Penfold, and J.L. Lamb, Development of Selective Surfaces, US DOE Report DE-ACO4-78CS35306. (August 1980).

43. E.E. Chain, K. Seshan, and B.O. Seraphin, Optical and Structural Properties of Black Mo Photothermal Converter Layers Deposited by the Pyrolysis of Mo(CO)$_6$, J. Appl. Phys. 52:1356 (1981).

44. S.W. Moore, Solar Absorber Selective Paint Research, Solar Energy Materials 12:435,449 (1985).

45. J.A. Thornton and J.L. Lamb, Development of Selective Surfaces, DOE Report. DE-ACO4-80AL13116 (Dec. 1981).

46. S. Lofving, A Paint for Selective Solar Absorbers, Solar Energy Materials 5:103 (1981).

47. F. H. Cocks, M. J. Peterson, and P.L. Jones, The Dependence of the Texture of Tellurium Films on Vacuum Deposition Angle, Thin Solid Films 70:297 (1980).

48. D.C. Booth, O. Allred, and B.O. Seraphin, Stabilized CVD Amorphous Silicon for High Temperature Photothermal Solar Energy Conversion Solar Energy Materials 2:107 (1979).

49. S.B. Gadgil, R. Thangaraj, J.V. Iyer, A.K. Sharma, B.K. Gupta, and O.P. Agnihotri, Spectrally Selective Copper Sulfide Coatings, Solar Energy Materials 5:129 (1981).

50. G.B. Smith, A.Ignatiev, and G. Zajac, Solar Selective Black Cobalt: Preparation, Structure and Thermal Stability J. Appl. Phys. 51:4186 (1980).

51. W. Kruidhof and M. Van der Leij, Cobalt Oxide as a Spectrally Selective Material For Use in Solar Collectors, Solar Energy Materials, 2:69 (1979).

52. J.I. Gittleman, E.K. Sichel, and Y. Arie, Composite Semiconductors: Selective Absorbers of Solar Energy, Solar Energy Materials, 1:93 (1979).

53. P. Swab, S. V. Krishnaswamy, and R. Messier, Characterization of Black Ge Selective Absorbers, J. Vac. Sci. & Technol. 17:1 (1980).

54. G.A. Nyberg, H.G. Crighead, R.A. Buhrman, Roughned, Graded Index, Cermet Photothermal Absorbers With Very High Absorptivities, Proc. of SPIE 324:117 (1982).

55. Y. Noguchi et al., Fabrication of ZrCx/Zr and Cr-CrO$_x$ Films for Practical Solar Selective Absorption Systems, Proc. of SPIE, 324:124 (1982).

56. M. Sikkens, Spectrally selective properties of reactively sputtered NiC and NiN$_x$ Films Proc. of SPIE, 324:131 (1982).

57. J. Mason, and T. Brendel, Maxorb - A New Selective Surface on Nickel, Proc. of SPIE 324:139 (1982).

58. E. Erben, A. Muehlratzer, B. Tihanyi and B. Cornils, Development of Selective Absorber Coatings for High Temperature Application, Solar Energy Materials 9:281 (1983).

59. T.J. McMahon and S.N. Jasperson, PbS-Al Selective Solar Absorbers, Appl. Optics 13:2750 (1974).

60. C.N. Watson-Munro and C.M. Horwitz (1975) cited by Meniel (29).

61. H.C. Hottel and T.A. Unger, The Properties of a Copper Oxide - Aluminum Selective Black Absorber of Solar Energy, Solar Energy 3:10 (1959).

62. H.C. Van deHulst "Light Scattering by Small Particles", Dover New York (1983).

63. I.T. Ritchie, S.K. Sharma, J. Valignat, and J. Spitz, Thermal Degradation of Chromium Black Solar Selective Surfaces, Solar Energy Materials 2:167 (1980).

64. C.M. Lampert, Chemical, Structural and Optical Characterization of a Black Chrome Solar Selective Absorber, Lawrence Berkeley Laboratory Report LBL-9123 (1979).

65. J. Spitz, T. V. Danh, and A. Aubert, Chromium Black Coatings For Photothermal Conversion of Solar Energy, Part 1: Preparation and Structural Characterization, Solar Energy Materials 1:189 (1979).

66. C.M. Lampert, and J. Washburn, Metallurgical Analysis and High Temperature Degradation of the Black Chrome Solar Selective Absorber, Thin Solid Films 72:75 (1980).

67. P. Driver and P.G. Mc Cormick, Black Chrome Selective Surfaces I, II, Solar Energy Materials 6:159, 381 (1982).

68. J.C.C. Fan and S. A. Spura, Selective Black Absorbers Using RF-Sputtered Cr_2O_3/Cr Cermet Films, Appl. Phys. Let. 30:511 (1977).

69. J.C.M. Garnett, Colours in Metal Glasses, in Metallic Films, Phil. Trans. Roy Soc., Lond 203A:385 (1904).

70. J.C.M. Garnett, Colours in Metal Glasses, in Metallic Films and in Metallic Solutions, Phil. Trans Roy. Soc Lond. 205A:237 (1906).

71. D.A.G. Bruggeman, Berechnung Verschiedener Physikalischer Konstanten Van Heterogenen Substanzen, Ann. der Phys. , 24:636 (1935).

72. C.G. Granqvist, Optical Properties of Cermet Materials, J. De. Physics 42:C1 (1981).

73. R.E. Peterson and J. Ramsey, Solar Absorber Study, U.S. Air Force Materials Lab, Wright Patterson AFB, Ohio, AFML-TR-73-80 (1973).

74. G.W. Milton, Bounds on the Complex Dielectric Constant of a Composite Material, App. Phys. Lett. 37:300 (1980).

75. D.J. Bergman, Exactly Solvable Microscopic Geometrics and Rigorous Bounds for the Complex Dielectric Constant of a Two Component Composite Material, Phys. Rev. Lett. 44:1285 (1980).

76. G.A. Niklasson and C. G. Granqvist, Photothermal Conversion with Cermet Films: Implications of the Bounds on the Effective Dielectric Function, Solar Energy Materials, 5:173 (1981).

77. J.H. Lin and R. E. Peterson, Improved Black Nickel Coatings for Flat Plate Solar Collectors, Proc. SPIE 85:62 (1976).

78. H. Tabor, Selective Radiation, Bull. Res. Counc. Isreal 5A: 119 (1956).

79. G. Burrafato et al., Thin Film Solar Acceptors, in Heliotechnique and Development, M. Kettani and J. Sossou, eds., Devel. Analysis Assoc. Inc., NY (1977).

80. R.M. Winegarner, Coating Costs and Project Independence, Photonics Spectra 9:20 (1975).

81. S. Catalanotti, V. Cuomo, G. Piro, D.Ruggi, V. Silvestrini and G. Troise, The Radiative Cooling of Selective Surfaces, Solar Energy, 17:83 (1975).

82. F. Sakkal, M. Martin and P. Berdahl, Experimental Test Facility for Selective Radiative Cooling Surfaces, Proc. of the 4th National Passive Solar Conf. 4:483 (1979).

83. T.S. Eriksson, E.M. Lushihu and C.G. Granqvist, Materials for Radiative Cooling to Low Temperatures, Solar Energy Materials 11:141 (1984).

84. E.M. Lushihu and C. G. Granqvist, Radiative Cooling with selectively Infrared Emitting Gases, Applied Optics, 23:1835 (1984).

85. G. L. Jorgenson, Long Term Glazing Performance, SERI Report TP-31-193. June 1979.

86. J. Jurison, R.E. Peterson, and H.Y.B. Mar, Principles and Applications of Selective Solar Coatings, J. Vac. Sci. and Technology, 12:101 (1975).

87. E.M. Pastirik and M.C. Keeling, A Low Cost, Durable Antireflective Film for Solar Collectors, Proc. of the IEEE 13th Photovoltaic Specialists Conference, Washington, DC, 620 (June 5-8, 1978).

88. C.M. Lampert, Optical Films for Solar Energy Applications, Proc. of SPIE 387:36 (1983).

89. P.K. Lee, and M.K. Debe, Measurement and Modeling of the Reflectance-Reducing Properties of Graded Index Microstructured Surfaces, Photo. Sci. and Engin., 24: 211. (1980).

90. M. Rubin, and S. Selkowitz, Thermal Performance of Windows Having High Solar Transmittance, Proc. of the Sixth National Passive Solar Conference, p. 141-145 (1981).

91. H. Dislich, and E., Hussman, Amorphous and Crystalline Dip Coatings Obtained from Organometallic Solutions: Procedures, Chemical Processes and Products, Thin Solid Films, 77:129 (1981).

92. C.J. Brinker and M.S. Harringston, Sol-gel Derived Antireflective Coatings for Silicon, Solar Energy Materials, 52:159 (1981).

93. H. Vora and T.J. Moravec, Structural Investigation of Thin Films of Diamond-like Carbon, J. Appl. Phys. 52: 6151. (1981).

94. M.A. Lind ed. Proc. of the Second Solar reflective Materials Workshop Solar Energy Materials 3:1 (1980).

95. C.F. Jefferson, H. Myers and F. Russo, Solar Reflectors Made from Silver Metallo-Organic Resinates, Proc. Of SPIE 324:74 (1982).

96. R.B. Pettit and E.P. Roth in Solar Materials Science (L.Murr edt.) Academic Press, New York, 1980.

97. B.L. Butler and R.B. Pettit, Optical Evaluation Techniques for Reflecting Solar Concentrators, Proc. of SPIE 114:43 (1977).

98. A. Goetzberger and V. Wittwer, Fluorescent Planar Collector-Concentrators: A Review, Solar Cells 4:3 (1981).

99. A. Goetzberger and W. Greubel, Solar Energy Conversion with Fluorescent Concentrators, Appl. Phys. 14:123 (1977).

100. V. Wittwer, W. Stahl and A. Goetzberger, Fluorescent Planar Concentrator, Solar Energy Materials 11:187 (1984).

101. W. Stahl and V. Wittwer, Highly Selective Narrowband Absorbers in Combination with Fluorescent Concentrators, Proc. of SPIE 428:187 (1983).

102. J. A. Levitt and W.H. Weber, Materials for Luminescent Greenhouse Solar Collectors, Appl. Optics 16:2684 (1977).

103. A. Bennett and L. Olsen, Analysis of Multiple Cell Concentrator- Photovoltaic Systems,Proc. of IEEE Photov. Spec. Conf. 13:868 (1978).

104. L. De Sandre, D.Y. Sing, H.A. Macleod and M.R. Jacobson, Thin Film Multilayer Filter Designs for Hybrid Solar Energy Conversion Systems, Proc. of SPIE 562:155 (1985).

105. M.A.C. Chendo, D.E. Osborn and R. Swenson, Analysis of Spectrally Selective Liquid Absorption Filters for Hybrid Solar Energy Conversion, Proc. of SPIE 562:160 (1985).

106. J. Jannson, T. Jannson, K.H. Yu, Solar Control Tunable Lippman Holowindows, Proc of SPIE 562:75 (1985).

107. G. Kienel, and W. Dachselt, Cold Light Mirrors, Ind. Res. and Develop. 22:135 (1980).

108. R. Winegarner, Greenhouse Selective Baffle Collector, Proc. of ISES American Section, p.33 (1977).

109. M.R. Brambley and M. Godec, Effectiveness of Low Emissivity Films for Reducing Energy Consumption in Greenhouses, Proc. of ASES, p. 25 (1982).

110. M. Rubin and C. M. Lampert, Transparent Insulating Silica Aerogels, *Solar Energy Materials* 11:1 (1984).

111. J. Fricke Edit., *Aerogels*, Springer-Verlag, Berlin, FRG, 1986.

112. J.H. Mazur and C. M. Lampert, High Resolution Electron Microscopy Study of Silica Aerogel Transparent Insulation, *Proc.of SPIE* 502:123 (1984).

113. R. Exelby and R. Grintner, Phototropy and Photochromism, *Chem. Rev* 64:247 (1964).

114. G. Brown and W. Shaw, Phototropism, *Rev. Pure Appl. Chem.* 11:2 (1961).

115. G.H. Dorion and A.F. Wiebe, *Photochromism,* Focal Press, London 1970.

116. G.P. Smith, Photochromic Glass: Properties and Applications, *J. Mat Sci.* 2:139 (1967).

117. N.Y.C. Chu, Photochromic Performance of Spiroindolinonaphthoxazines in Plastics, *Proc of SPIE* 562:6 (1985).

118. J.H. Day, Chromogenic Materials, in *Ency. of Chemical Technology*, J. Wiley, New York, 1977.

119. G. V. Jorgenson and J.C. Lee, Thermochromic Materials Research for Optical Switching Films, *Proc. of SPIE* 562:2 (1985).

120. J.C. Lee, G.V. Jorgenson and R.J. Lin, Thermochromic Materials Research for Optical Switching, *Proc. of SPIE* 692:3 (1986).

121. C.M. Lampert, Electrochromic Materials and Devices for Energy Efficient Windows, *Solar Energy Materials* 11:1 (1984).

122. W.C. Dautremont-Smith, Transition Metal Oxide Electromic Materials and Displays: A Review, *Displays* 4:3,67 (1982).

123. B.W. Faughan and R.S. Crandall, Electrochromic Displays Based on WO_3., in *Display Devices* (J.I. Pankove Edit.), Springer-Verlag, Berlin FRG. 1980.

124. J.S.E.M. Svensson and C.G. Granqvist, Electrochromic Coatings for "Smart Windows", *Solar Energy Materials* 12:391 (1985).

125. J. Nagai, T. Kamimori and Mizuhashi, Transmissive Electrochromic Device, *Proc. of SPIE* 562:39 (1985).

126. D.K. Benson and C.E. Tracy, Amorphous Tungsten Oxide Electrochromic Coatings for Solar Windows, *Proc. of SPIE* 562:46 (1985).

127. R.B. Goldner et al. Recent Research Related to the Development of Electrochromic Windows, *Proc. of SPIE* 562:32 (1985).

128. S.F. Cogan, E.J. Anderson, T.D. Plante and R.D. Rauh, Materials and Devices in Electrochromic Window Development, *Proc. of SPIE* 562:23 (1985).

129. C.M. Lampert, T.R. Omstead, P.C. Yu, Chemical and Optical Properties of Electrochromic Nickel Oxide Films, <u>Solar Energy Materials</u> 14:14 (1986).

130. P.C. Yu, G. Nazri and C.M. Lampert, Spectroscopic and Electrochemical Studies of Electrochromic Hydrated Nickel Oxide Films, <u>Proc of SPIE</u> 653:16 (1986).

131. S. Chandrasekhar, <u>Liquid Crystals</u>, Cambridge University Press, Cambridge, England (1977).

132. S. Sherr, <u>Electronic Displays</u>, J. Wiley, New York (1970).

133. J.L. Fergason, Polymer Encapsulated Nematic Liquid Crystals for Display and Light Control Applications, SID Digest 85:68 (1985).

SPECTRALLY SELECTIVE COATINGS
FOR ENERGY EFFICIENT WINDOWS

C.G. Granqvist

Physics Department, Chalmers University of Technology,
S-412 96 Gothenburg, Sweden

ABSTRACT

This paper gives an overview over coatings for energy efficient windows. Thin film optics, model dielectric functions, and techniques to evaluate dielectric functions are covered briefly. After this preamble, we introduce spectral selectivity, which is the key concept for energy efficiency. We make a distinction between coatings with static properties (with a further subdivision into coatings for solar control and for low thermal emittance) and coatings with dynamic properties. In the former class, we discuss metal-based and doped-semiconductor-based coatings in some detail. In particular we dwell on ion-plated gold films and on e-beam evaporated tin-doped indium oxide films in two case studies. Among the coatings with dynamic properties, we discuss electrochromic and thermochromic films.

1.   INTRODUCTION

The purpose of this paper is to present an overview over
coatings for energy efficient windows. Usually the desired property
is obtained by spectral selectivity. This term is taken to imply that
the radiative properties (i.e., the absorptance A, emittance E, reflectance R, and transmittance T) are qualitatively different within
different parts of the electromagnetic spectrum. The optical properties are connected to one another at each wavelength λ by the general
relations

$$A(\lambda) + R(\lambda) + T(\lambda) = 1, \tag{1}$$

$$E(\lambda) = A(\lambda). \tag{2}$$

Both relations follow from energy conservation. Equation (2), known
as Kirchhoff's law, states that at a given wavelength the absorptance
is equal to the emittance for matter in thermodynamical equilibrium.

Section 2 contains some basic material which must be covered
before we can turn to the window coatings. It introduces thin film
optics, models for dielectric functions of uniform and non-uniform
materials, and techniques to evaluate dielectric functions from optical measurements. Section 3 treats spectral selectivity in general
by presenting the radiative properties pertinent to our ambience and
giving formulas for computing suitably integrated optical quant-
tities. Section 4 introduces window coatings of different kinds and
covers the types of spectral selectivity demanded for hot and cold
climates, as well as a review over coatings for "solar control" and
for obtaining low thermal emittance. Section 5 gives a case study
of the optical properties of ion-plated Au films. Another case study
of e-beam evaporated In_2O_3:Sn films is presented in Section 6. The
case studies are not as restrictive as one may first believe but
illustrate the salient features of coatings based on noble metal films
and on heavily doped oxide semiconductor films. Section 7 contains
a survey over optical switching coatings of electrochromic and thermo-

chromic type. The reference list at the end of this article contains books, review articles and - to a limited extent - original scientific papers of particular relevance. General reviews over selective surfaces for various heating and cooling application are given in Refs. 1-6.

2.   NOTES ON THE OPTICAL PROPERTIES OF MATERIALS

2.1   Basic Thin Film Optics

We consider light incident towards the boundary between two media denoted i and j. The angle to the surface notmal if θ_i, as indicated in Fig. 1. The media are characterized by their complex

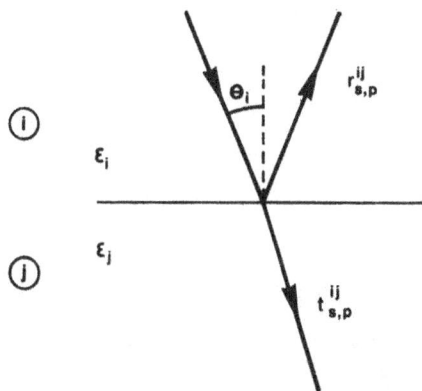

Fig. 1 Definition of symbols entering Fresnel's relations.

dielectric functions ε_i and ε_j. Part of the light is reflected at the boundary (r^{ij}) and part is transmitted (t^{ij}) through the boundary. We distinguish between light with s-polarisation (E vectors normal to the plane spanned by the incident, reflected and transmitted beams) and with p-polarisation (H vectors normal to the same plane). From Maxwell's equations, one can now obtain the well known Fresnel's relations for the reflected field amplitudes.

$$r_s^{ij} = \frac{(\varepsilon_i)^{1/2}\cos\theta_i - (\varepsilon_j - \varepsilon_i\sin^2\theta_i)^{1/2}}{(\varepsilon_i)^{1/2}\cos\theta_i + (\varepsilon_j - \varepsilon_i\sin^2\theta_i)^{1/2}}, \tag{3}$$

$$r_p^{ij} = \frac{(\varepsilon_i)^{1/2}(\varepsilon_j - \varepsilon_i\sin^2\theta_i)^{1/2} - \varepsilon_j\cos\theta_i}{(\varepsilon_i)^{1/2}(\varepsilon_j - \varepsilon_i\sin^2\theta_i)^{1/2} + \varepsilon_j\cos\theta_i}, \tag{4}$$

$$t_s^{ij} = \frac{2(\varepsilon_i)^{1/2}\cos\theta_i}{(\varepsilon_i)^{1/2}\cos\theta_i + (\varepsilon_j - \varepsilon_i\sin^2\theta_i)^{1/2}}, \tag{5}$$

$$t_p^{ij} = \frac{2(\varepsilon_i\varepsilon_j)^{1/2}\cos\theta_i}{(\varepsilon_i)^{1/2}(\varepsilon_j - \varepsilon_i\sin^2\theta_i)^{1/2} + \varepsilon_j\cos\theta_i}. \tag{6}$$

Fresnel's relations can be used to discuss the optical properties of a thin film on a substrate. We consider the geometry specified in Fig. 2 and let (2) denote the film (of thickness d) and (3)

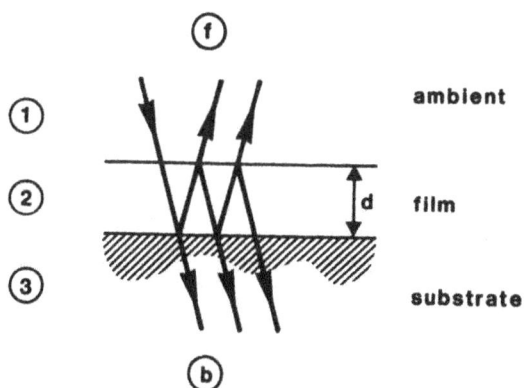

Fig. 2   Geometry for discussing the optics of a thin film on a
substrate.

the substrate. A medium (1) surrounds the coated substrate. Further
we let (f) signify light incident from the frontside and (b) signify
light incident from the backside. Equations (3) - (6) yield ex-
pressions for r^{12}, r^{23}, t^{12}, t^{21}, t^{23}, t^{32}, r^{31} and t^{31}. Neglecting
the effect of the backside of the substrate, we obtain the reflec-
tance and transmittance amplitudes for the film, r_2 and t_2, by

$$r_{2s}^{f} = \frac{r_s^{12} - r_s^{23} e^{2i\delta}}{1 + r_s^{12} r_s^{23} e^{2i\delta}} , \tag{7}$$

$$t_{2s}^{f} = \frac{t_s^{12} t_s^{23} e^{i\delta}}{1 + r_s^{12} r_s^{23} e^{2i\delta}} . \tag{8}$$

Relations of the same form are easily found for r_{2s}^b, t_{2s}^b, $r_{2p}^{f,b}$, $t_{2p}^{f,b}$ as well. In Eqs. (7) and (8), δ is the phase change of the light beam upon traversing the film. Specifically

$$\delta = \frac{2\pi}{\lambda}(\varepsilon - \sin^2\theta_1)^{1/2}d. \tag{9}$$

The measurable optical quantities are light intensities. These are the squares of the absolute values of r_{2s}^f etc., denoted by capital letters, i.e. for example R_{2s}^f.

If the substrate is metallic, then the transmittance is zero and only the reflectance has to be considered. Thus the reflectance for s-polarised light is obtained simply by squaring the absolute value of Eq. (7), and the reflectance for p-polarised light is obtained in the same manner from an expression analogous to Eq. (7). If the substrate is transparent, a more complicated situation exists, and multiple reflections in the substrate must be taken into account. They are incoherent for large substrate thicknesses and must be included through addition of the intensities of the multiply reflected beams. The final expressions for the reflectance and the transmittance are

$$R_s = R_{2s}^f + \frac{T_{2s}^f T_{2s}^b R_{3s}}{1 - R_{2s}^b R_{3s}}, \tag{10}$$

$$T_s = \frac{T_{2s}^f T_{3s}}{1 - R_{2s}^b R_{3s}}, \tag{11}$$

where $T_{3s,p} = \left| r_{s,p}^{31} \right|^2$ and $T_{3s,p} = \left| t_{s,p}^{31} \right|^2$. Analogous relations are obtained for the p-components.

Detailed descriptions of thin film optics may be found in the books by Heavens (Ref.7) and by Born and Wolf (Ref. 8). Among many other things, these books discuss how to treat multiple layer films through matrix techniques.

2.2   Dielectric Functions Of Uniform Model Materials.

The optical properties of a material are conveniently discussed in terms of its complex frequency-dependent dielectric function $\varepsilon(\omega) = \varepsilon_1(\omega) + i\varepsilon_2(\omega)$. Alternatively, one may use a complex refractive index $N \equiv n + ik = \varepsilon^{1/2}$. In the latter notation, n and k are known as the optical constants of the material.

In many cases one can regard ε as a sum of individual contributions originating from different elementary excitations (cf, for example, the book by Mahan (Ref.9)). In terms of susceptibilities $\chi^i \equiv \chi_1^i + i\chi_2^i$ (in SI units), one can write

$$\varepsilon = 1 + \chi^{VE} + \chi^{PH} + \chi^{FC}, \tag{12}$$

where VE = valence electrons, PH = phonons, and FC = free carriers (usually electrons). The various susceptibilities can easily be distinguished if their resonances fall in well separated wavelength regions. For a particular χ^i, the contribution far from its resonance is real and constant.

Figure 3 gives a schematic representation of the χ^is. The real and imaginary parts are consistent with the Kramers-Kronig relations.

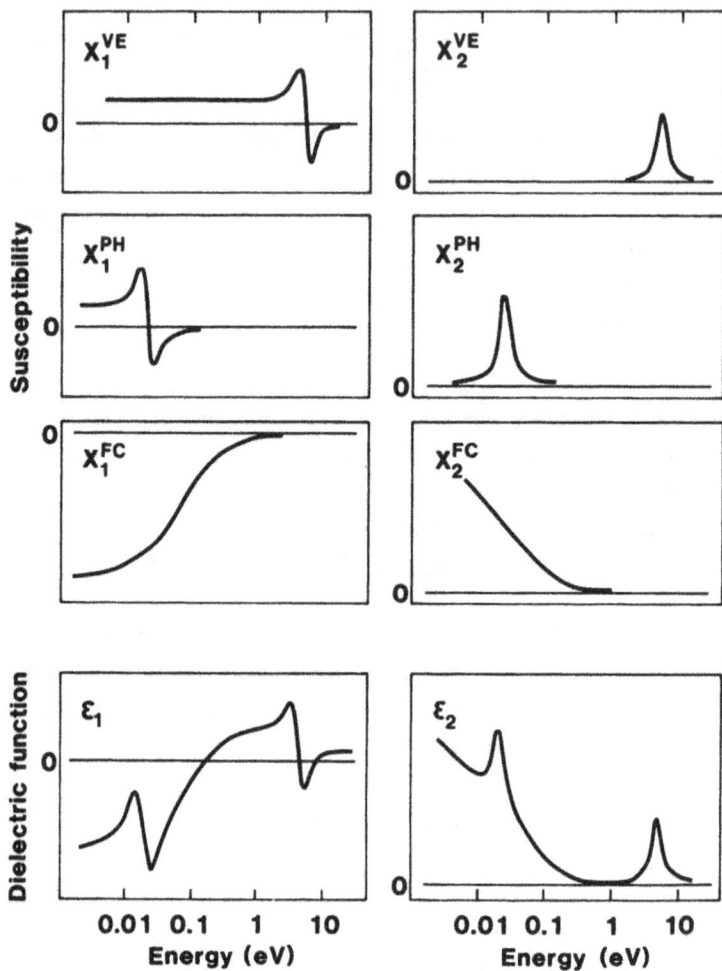

Fig. 3 Contributions to the dielectric function from susceptibilities
due to valence electrons (VE), phonons (PH), and free carriers
(FC).

For χ^{VE} and χ^{PH}, one can often represent the susceptibilities by a sum of damped Lorentz oscillators, i.e. by expressions of the kind

$$\chi^{Lorentz} = \frac{\Omega^2}{\omega_L^2 - \omega^2 - i\omega\Gamma}, \tag{13}$$

where Ω is the oscillator strength, ω_L is the resonance frequency and Γ represents the width of the resonance peak. For most good metals, χ^{VE} is resonant in the ultraviolet or the blue part of the visible spectrum, while χ^{PH} is resonant in the thermal infrared. For χ^{FC} one can make use of the Drude theory, at least for a first-order description. The susceptibility can be written

$$\chi^{Drude} = -\frac{\omega_p^2}{\omega^2 + i\omega\gamma} \tag{14}$$

where ω_p is the plasma frequency and γ represents the width of the resonance. More elaborate free-electron theories are required for quantitative assessments. It is seen that χ^{Drude} can be obtained by setting $\omega_L = 0$ in $\chi^{Lorentz}$. Below $\hbar\omega_p$, χ_1^{FC} becomes strongly negative while χ_2^{FC} becomes strongly positive, as apparent from Fig. 3. The location of $\hbar\omega_p$ depends on the free electron density. For a metal, $\hbar\omega_p$ is normally in the ultraviolet. For a doped semiconductor, $\hbar\omega_p$ can be in the infrared. In a metal, χ^{PH} is usually not apparent owing to the dominating influence of χ^{FC}. The lowest part of Fig. 3 illustrates the ensuing performance of ϵ for the case of a heavily doped semiconductor.

2.3   Dielectric Functions For Non-uniform Model Materials.

Surface coatings of interest for energy efficient windows, as
well as for many other applications, can be non-uniform on a scale
which is much smaller than the wavelengths of concern. Frequently,
the coatings comprise a mixture of metallic and insulating components.
Under these conditions, the optical performance of the coating is
governed by a spatial average of the properties of the two components.
The averaging, which is far from trivial for non-dilute composites, is
expressed in terms of an effective-medium theory (EMT), which yields an
effective dielectric function $\bar{\epsilon}$.

EMTs are of importance for basic scientific reasons and also
because they permit modeling and optimization of the optical performance of coatings by use of a computer. Many EMTs have been formulated
and investigated recently. Much of this work has been motivated by
its potential interest for efficient solar energy utilization. Reviews
are given in Refs. 10-14.

Figure 4 serves as a background for the formulation of different
EMTs. The averaging needed in the derivation of $\bar{\epsilon}$ depends crucially on
the specific microgeometry of the composite, and parts (a) and (b) show
two structures which are characteristic for many practically occurring
two-component composites. In order to formulate the EMTs, we want to
single out the essential feature of the structure; thus we need
simplifying assumptions which do not leave out the basic physics. This
is done by the introduction of random unit cells (RUCs), as shown in
Figs. 4 (c) and (d). The separed-grain structure is represented by a
sphere of "A" surrounded by a concentric shell of "B", and the aggregate structure is represented by a sphere having a probability f_A of
being "A" and $1-f_A$ of being "B".

A basic definition of an effective medium is that the RUC -
embedded in the effective medium - should not be detectable in an experiment using radiation confined to a specified wavelength range.
Thus the extinction of the RUC should be the same as if it were replaced

MICROSTRUCTURES

Separated-grain structure   Aggregate structure

(a)   (b)

Material "A"; filling factor f_A

Material "B"; filling factor $1-f_A$

RANDOM UNIT CELLS

Maxwell Garnett theory   Bruggeman theory

$\bar{\varepsilon}^{MG}$ $\bar{\varepsilon}^{Br}$

(c) $\varepsilon_B \varepsilon_A$ (d) $\varepsilon_A , \varepsilon_B$

Ratio of volumes determines f_A Probability f_A of being "A" Probability $1-f_A$ of being "B"

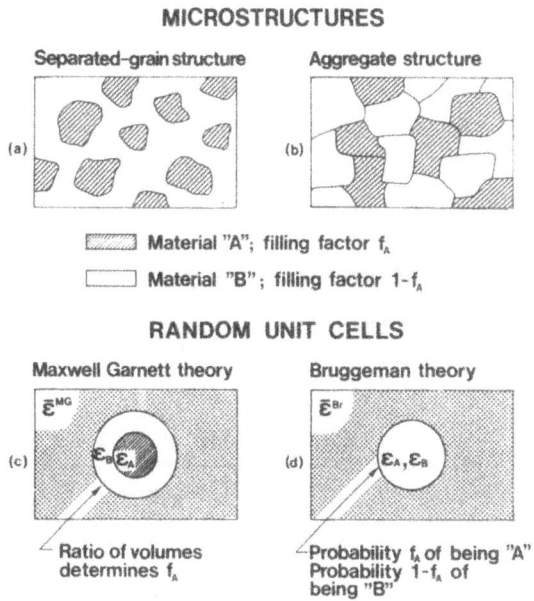

Fig. 4   Microstructures and random unit cells for two types of composite materials.

by a sphere whose dielectric function is $\bar{\epsilon}$. An optical theorem for absorbing media states that

$$C_{ext} = 4\pi \ \text{Re} \ \left[S(0)/k^2 \right], \tag{15}$$

$$k = 2\pi\bar{\epsilon}^{-1/2}/\lambda, \tag{16}$$

where C_{ext} is the extinction of the RUC compared with that of the surrounding material and $S(0)$ is the scattering amplitude in the forward direction. Requiring now that $C_{ext} = 0$ yields

$$S(0) = 0, \tag{17}$$

which states the fundamental property of an effective medium.

The RUC of the separated-grain structure is a coated sphere (cs) which yields[15-17]

$$S^{cs}(0) = i(kb)^3 \ \frac{(\epsilon_B - \bar{\epsilon})(\epsilon_A + 2\epsilon_B) - f_A(2\epsilon_B - \bar{\epsilon})(\epsilon_A - \epsilon_B)}{(\epsilon_B + 2\bar{\epsilon})(\epsilon_A + 2\epsilon_B) + f_A(2\epsilon_B - 2\bar{\epsilon})(\epsilon_A - \epsilon_B)} \tag{18}$$

$$+ \mathcal{O}\left[(kb)^5 \right].$$

The "filling factor" is given by $f_A = (a/b)^3$, where a(b) denotes the radius of the inner (outer) sphere. For small particles, we set the leading term in $S^{cs}(0)$ equal to zero and obtain (with $\epsilon \equiv \epsilon^{MG}$)

$$\frac{\bar{\epsilon}^{MG} - \epsilon_B}{\bar{\epsilon}^{MG} + 2\epsilon_B} = f_A \ \frac{\epsilon_A - \epsilon_B}{\epsilon_A + 2\epsilon_B} , \tag{19}$$

which is the well known expression for the effective dielectric function of the Maxwell Garnett (MG) theory.[18]

The RUC of the aggregate structure is a sphere (s) with[15-17]

$$S^s(0) = i(kb)^3 \frac{\varepsilon - \bar{\varepsilon}}{\varepsilon + 2\bar{\varepsilon}} + O\left[(kb)^5\right],\tag{20}$$

where ε is either ε_A or ε_B. In the small size limit we obtain (with $\bar{\varepsilon} \equiv \varepsilon^{-Br}$)

$$f_A \frac{\varepsilon_A - \varepsilon^{Br}}{\varepsilon_A + 2\varepsilon^{-Br}} + (1 - f_A) \frac{\varepsilon_B - \varepsilon^{-Br}}{\varepsilon_A + 2\varepsilon^{-Br}} = 0.\tag{21}$$

This equation was first given by Bruggeman (Br).[19]

It is possible to extend the EMTs to non-spherical particles and also to set limits on the permissible particle diameters. These and other aspects are treated in Ref. 20. Finally, we note that there have appeared certain general results (Bergman-Milton bounds) for $\bar{\varepsilon}$ which hold irrespective of the detailed microgeometry.[21,22]

2.4 Evaluation Of Dielectric Functions From Optical Measurements

The extraction of optical constants of thin films from various types of optical measurements is a field of widespread interest, and a large number of schemes have been devised in the literature. The current quest for coatings for energy efficiency has accentuated the demand for routine techniques applicable in the full visible and infrared wavelength regions. Preferrably, such techniques should be experimentally straightforward and compatible with conventional equipment.

The most commonly used technique to determine the optical constants n and k of materials involves measuring the normal incidence transmittance T and the near-normal incidence reflectance R of a thin

film on a transparent substrate. We refer to this as the (T,R) method.
It usually includes spectrophotometry. The popularity is derived
mainly from the experimental simplicity. No angular adjustments,
polarisers, or other special equipment are needed, unintentional
polarisation effects in the optical system do not enter, and the sample
preparation is straightforward. An alternative technique, known as the
(T,R_m) method, combines normal incidence transmittance with near-
normal reflectance R_m against a metallized part of the substrate.
The latter technique is particularly useful for absorbing materials.
One obtains n and k, or ϵ_1 and ϵ_2, by fitting computed optical data-
reached by applying the Fresnel relations to the pertinent sample configurations - to corresponding experimentally recorded quantities.

The (T,R)- and (T,R_m) methods have their specific pros and cons,
which may be conveniently illustrated by use of contour maps. These
show curves, computed from the Fresnel relations, for constant R,T and
R_m in the (n,k) plane. Figure 5 shows a contour map for T(n,k) and
R(n,k) at wavelength λ for a thin film of thickness d, so that d/λ =
0.1. The film is backed by a transparent substrate with refractive
index n_s = 1.5. One notes that the transmittance contours constitute a very regular grid in the (n,k) plane with the curves running
virtually parallel to each other over a large region. The transmittance value is evidently determined almost solely by the imaginary part
of the refractive index, whereas the dependence on the real part of the
refractive index is weak. In an accurate procedure for determining
n and k, the transmittance data should be combined with an experiment
yielding a regular grid of contours oriented approximately perpen-
dicularly to the transmittance grid. The accuracy of the ensuing optical constants is governed by two factors: the spacing of the grids and
their angle of intersection. Small spacing and perpendicular orientation give the highest accuracy. Clearly the (T,R) combination is far
from ideal. As shown in Fig. 5, the R contours make a suitably
oriented grid only in very limited regions of the (n,k) plane; specifically, only for large k and small n (k > 1, n < 1), and for very small
k values near the n axis, do the contours intersect under suitable
angles. For almost any point in the (n,k) plane at which n > k, the

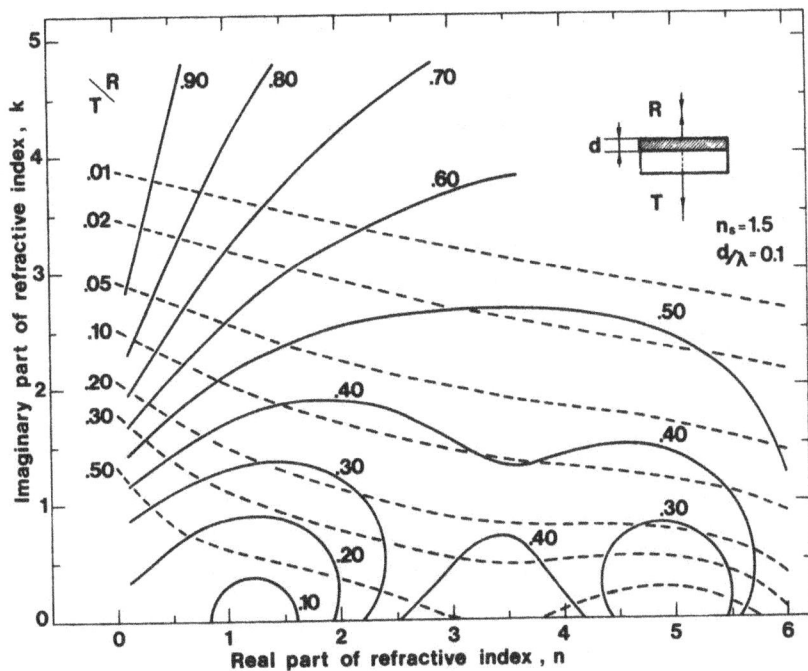

Fig. 5 Plot of the contours of constant normal transmittance T and
reflectance R of a thin film on a transparent substrate.
The shown magnitudes of d, λ and n_s were used.

angle of intersection is quite small thus leading to increased errors of the (T,R) method. Furthermore, in a region close to the line n = k + 1, the transmittance and reflectance contours run parallel to each other and hence even a minor experimental error in this region leads to gross errors in the extracted optical constants (particularly in the real part). A further problem of the (T,R) method is that several solutions (n,k) exist to each set (T,R). This multivaluedness, which is an inherent property of any technique using two intensity measurements, produces a problem in the (T,R) case in regions near the contour tangency points, which are branch points between different solutions. A minor experimental error in this region can cause a branching from one solution (n_1,k_1) to another solution (n_2,k_2) which is physically incorrect.

Figure 6 gives a contour plot for the (T,R_m) case. Contours of constant transmittance T of a thin film ($d/\lambda = 0.1$) on a transparent substrate ($n_s = 1.5$) are combined with contours of reflectance R_m for a thin film on a metallic substrate. The optical constants of the substrate metal, $n_m + ik_m$, were set to $n_m = 20$ and $k_m = 70$, which is typical of aluminium in the mid-infrared. The exact choice of optical constants for the metal plays a very subordinate role for this plot. The (T,R_m) contour map evidently differs from the (T,R) map in several important ways. Particularly noteworthy is the fact that the R_m contours form a regular grid oriented almost perpendicular to the transmittance contours in a large region of the (n,k) plane (for $n \leq 2.4$). This means that the (T,R_m) combination makes an excellent technique for the determination of thin film optical constants in this part of the (n,k) plane. It is interesting to note that this good region of the (T,R_m) technique includes the inherently troublesome region of the (T,R) technique. The pole structure apparent in Fig. 6 can be moved by having a different thickness of the film. This feature extends the applicability of the (T,R_m) technique. A more detailed discussion of the (T,R) and (T,R_m) methods, including error analyses, can be found in Ref. 23.

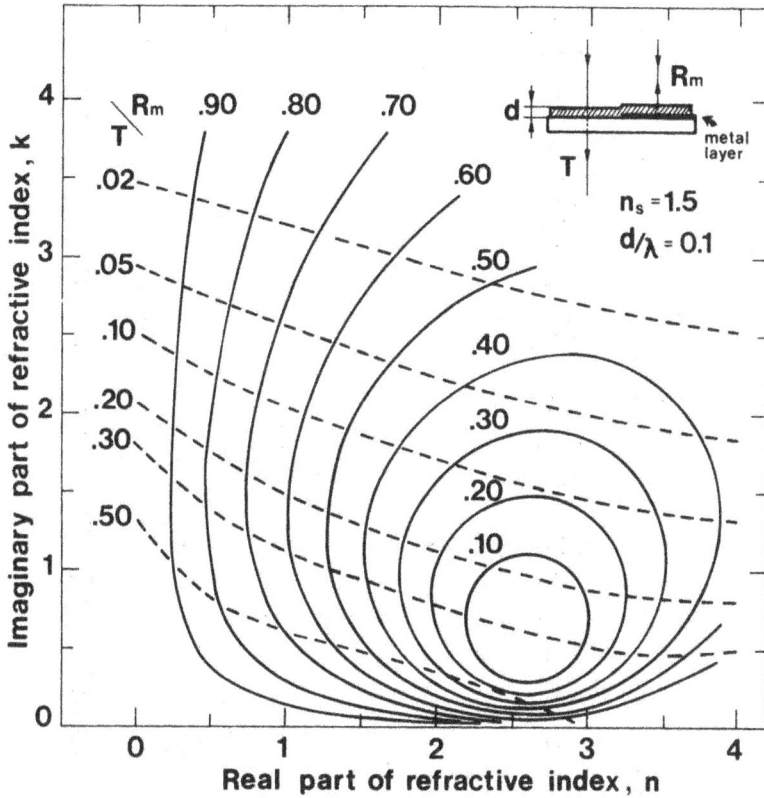

Fig. 6 Plot of the contours of constant normal transmittance T and reflectance R_m of a thin film on a partly metallized substrate. The shown magnitudes of d, λ and n_s were used. The metal substrate had a refractive index of 20 + i70.

The above mentioned optical techniques for evaluating ε_1 and ε_2 or n and k are by no means the only ones. A class of techniques is based on measurements of polarimetric ratios. It can yield very high precision, mainly in the visible wavelength range. It relies for its accuracy on excellent wide angle polarisers and a very well collimated beam. These are demands that are not easily met, particularly in the infrared. Specialized equipment (ellipsometers) is also needed to take full advantage of the technique. Optical constants can also be extracted from a single optical measurement (for example of trans- mission or reflection) and subsequent use of the casuality relations (Kramers-Kronig analysis). This approach has a drawback in that ex- perimental data are needed over an extremely wide wavelength range; in principle, all the way from zero to infinite wavelength. A very direct technique utilizes Fourier transform interferometers, which measure amplitue as well as phase in a transmission or reflection ex- periment. This method has so far been restricted mainly to the (far) infrared region and also demands sophisticated and specialized equip- ment. A recent exhaustive survey over techniques to determine ε_1 and ε_2 or n and k is given in Ref. 24.

3.   SPECTRAL SELECTIVITY: A GENERAL INTRODUCTION

3.1   Ambient Radiative Properties

Spectral selectivity, as a means to achieve energy efficiency, rests on the radiative properties of our ambience. These are illus- trated in Fig. 7 with a common logarithmic wavelength scale on the abscissa. Part (a) depicts blackbody spectra for four temperatures of practical interest. The vertical scale gives power per unit area and wavelength increment. The curves appear bell-shaped. Practically no energy is emitted for wavelength shorter than 2 µm. Blackbody spectra pertaining to room temperature are peaked at ~ 10 µm. The thermal radiation from a real body is obtained by multiplying the blackbody radiation by a numerical factor - the emittance - which is less than unity. In general, the emittance is wavelength dependent.

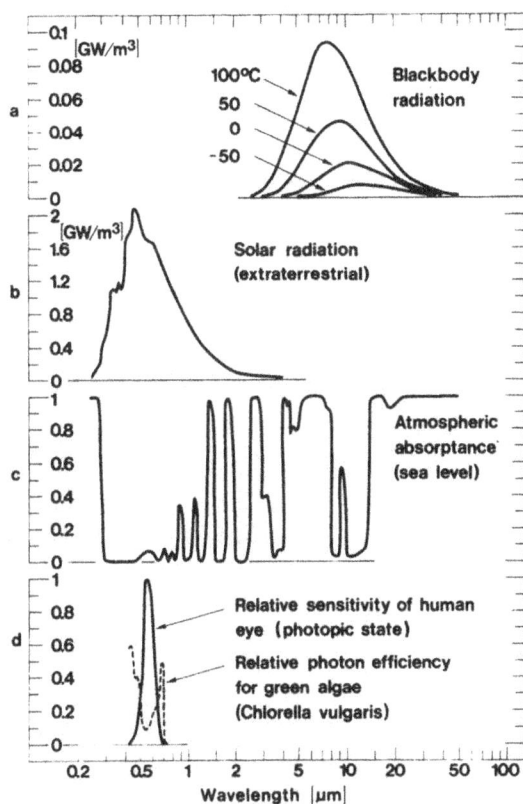

Fig. 7 Spectra for (a) blackbody radiation pertaining to four temperatures, (b) solar irradiation outside the earth's atmosphere, (c) typical absorptance across the full atmospheric envelope, (d) relative sensitivity of the human eye and an example of the relative photon efficiency for photosynthesis.

Figure 7 (b) shows the spectrum for solar radiation incident on the atmospheric envelope. The curve agrees rather well with a black-body spectrum defined by the sun's surface temperature (~ 6000°C). We observe that the solar spectrum is confined to the 0.25 - 3-μm interval, so that there is almost no overlap with the thermal spectra in part (a). Hence one can have surfaces whose properties are entirely different with regard to thermal and solar radiation, as we discuss at length for window coatings in subsequent sections of this article.

The windows of interest are located near ground level, and hence it is important to consider the role of the atmosphere. Figure 7(c) shows typical absorptance vertically across the atmospheric envelope under clear weather conditions. The pronounced peaks are due to molecular absorption in water vapour, carbon dioxide, and ozone. It is seen that most of the solar radiation can be transmitted through the air. In contrast, the thermal ratiation from an upwards-facing surface is absorbed to a large extent at 8 to 13 μm where the absorptance is weak (provided that the atmospheric humidity is not too large). The thermal radiation can be large in the 8-13-μm interval, and hence we conclude that a non-negligible part of the emitted energy can go straight through the atmosphere. This phenomenon constitutes the basis for radiative cooling, which is sometimes of importance for upwards-facing windows.

Figure 7 (d) presents two biological constraints which are relevent to window applications. The solid curve shows that the human eye is sensitive only in the 0.4-0.7-μm range. Hence a large part of the solar energy comes as infrared radiation, which is an important fact for window coatings. The dashed curve indicates that photosynthesis in green plants uses radiation with wavelengths shorter than 0.7 μm, which is of obvious interest for greenhouses.

3.2   Integrated Optical Properties

For assessing the performance of surfaces with regard to energy efficiency, we must know one or several suitably integrated optical properties. The integration could be over the eye's sensitivity curve to get the luminous performance, over a typical solar spectrum, over a blackbody spectrum to get the thermal performance, or over the atmospheric irradiance spectrum to find the radiative cooling performance. Luminous, solar and thermal quantities are discussed here. Spectral selectivity implies that not all of these are equal. Colour specifications also use integrated optical quantities.

Integrated luminous (lum), solar (sol), and thermal (therm) quantities are conveniently obtained from the relation

$$X_\gamma = \int d\lambda \emptyset_\gamma(\lambda) X(\lambda) / \int d\lambda \emptyset_\gamma(\lambda). \tag{22}$$

Depending on which quantity we want to compute, we set $\emptyset_\gamma$ equal to $\emptyset_{lum}$, $\emptyset_{sol}$ or $\emptyset_{therm}$. For the function $\emptyset_{lum}$ it is proper to use the standard luminous efficiency function for photopic vision as specified vy CIE (Ref. 25). For $\emptyset_{sol}$ one could use tabulated spectra appropriate to a certain air mass (AM) and turbidity (Ref. 26). For example one could use AM0 (extra-terrestial conditions), AM1 (sun at zenith), AM2 (sun at 30° above the horizon) etc. For thermal radiation, the weighting function can be written

$$\emptyset_{therm} = c_1 \lambda^{-5} \left[\exp(c_2/\lambda\tau) - 1 \right]^{-1} \tag{23}$$

with $c_1 = 3.7418 \times 10^{-16} \left[Wm^2 \right]$, $c_2 = 1.4388 \times 10^{-2} \left[mK \right]$, and τ signifying temperature. X in Eq. (22) denotes the spectral property; for example normal transmittance, normal reflectance, or hemispherical emittance E. The emittance can be obtained from

$$E(\lambda) = 1 - T_h(\lambda) - R_h(\lambda), \tag{24}$$

where T_h is the hemispherical transmittance given by

$$T_h(\lambda) = \tfrac{1}{2} \int_o^{\pi/2} d(\sin^2\theta)\left[T_s(\theta,\lambda) + T_p(\theta,\lambda)\right], \qquad (25)$$

and an analogous relation holds for the hemispherical reflectance R_h. The subscripts s and p in Eq. (25) refer to different states of polarisation.

4. A SURVEY OF WINDOW COATINGS

4.1 The Different Types Of Spectral Selectivity

There is a large number of window coatings and in order to have an organized presentation we choose first to consider those with static properties, i.e., coatings whose optical performance remains the same irrespective of external conditions. This group can be conveniently subdivided into coatings for "solar control" and for providing low thermal emittance. The second group of coatings, whose study is presently emerging as a self-contained research topic, have dynamic properties and encompasses optical switching coatings of different types.

With the purpose of introducing the "solar control" coatings, we note that in a hot climate it is frequently the case that the solar energy which enters through a window and is absorbed in the room causes overheating. Hence, there is a need for cooling to obtain a comfortable indoor temperature. Conventional air conditioning uses a lot of electrical energy, part of which can be saved by the use of proper window coatings. From the upper part of Fig. 8, which contains basically the same information as in Fig. 7, it is seen that only some of the solar radiation comes as visible light. It is clearly effective to have a coating which is transparent only for $0.4 < \lambda < 0.7$ μm while it is reflecting for $0.7 < \lambda < 3$ μm so that the infrared part of

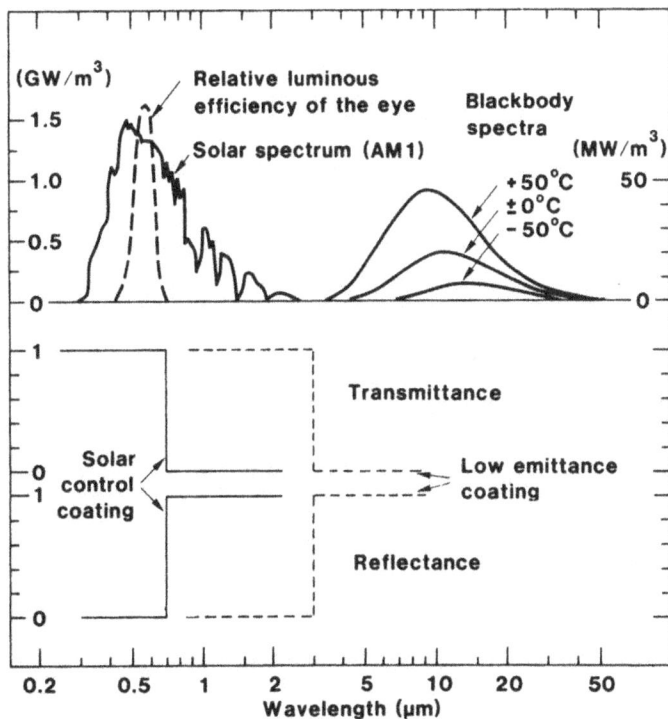

Fig. 8 Upper part shows the luminous efficiency of the high-adapted eye, a typical solar spectrum for one air mass, and blackbody spectra corresponding to three temperatures. Lower part shows transmittance and reflectance for idealized window coatings designed for "solar control" (solid lines) and for low thermal emittance (dashed lines).

the solar radiation is excluded. This spectral profile is shown by
the solid lines in the lower part of Fig. 8. The improvement caused
by such a coating depends to some extent on the weather, but as a
rule of thumb it is conceivable to decrease the solar heating by
almost 50 % at negligible loss of visible transmittance.

We now turn to the coatings with low thermal emittance and note
that in a cold climate a window is usually causing an undesired loss
of energy. A large part of this loss is by thermal radiation to the
ambience. It is clear from the upper part of Fig. 8 that energy
efficiency can be gained by coatings which are transparent for
$0.3 < \lambda < 3$ µm, so that maximum use is made of the solar energy, while
the emittance of thermal radiation is minimized. The required optical
properties are hence high transmittance for $0.3 < \lambda < 3$ µm and high
reflectance for $3 < \lambda < 100$ µm. This performance is indicated by the
dashed lines in the lower part of Fig. 8. The advantage of low-emittance
window coatings can be expressed in various ways: one is to state that
a similar thermal insulation can be obtained in a window with double
glazing and coating as in a window with triple glazing and no coating.
The lower weight of the former alternative is an important benefit.

The previous two types of coating are of great value for energy
efficient windows. However, they have a fundamental limitation in
their properties being static, which make them incapable of adjusting
according to the variable demands on heating and lighting during the
day or season. For example, it may be that a coating which has low
thermal emittance, and is adequate during the winter in a midlatitude
region, causes overheating during the summer. It is obvious that a
"smart window" should exhibit dynamic radiative properties, which can
be automatically tuned in such a way that a suitable amount of solar
energy or visible light is introduced. The pertinent coatings can be
of different kinds: photochromic (i.e., change their properties in
response to the irradiation); thermochromic (i.e., change their properties in response to the temperature); and electrochromic (i.e.,
change their properties in response to the strength and direction

of an applied electric field. Here we focus mainly on the electrochromic ones but we also touch upon thermochromism. Research on optical switching coatings for energy efficient windows is presently
being started in different parts of the world.

Reviews of window coatings with static properties are found in
Refs. 27-43. Some work on electrochromic coatings is reviewed in
Ref. 44. Useful information of window coatings is found also in
Refs. 1-6.

4.2 Metal Based Coatings

Very thin metal films can be used to provide spectrally selective transmittance. Best results are obtained for the free electron
metals Cu, Ag and Au. Most other metals exhibit a pronounced and
normally undesirable interband absorption within the solar range.
The thickness of the coatings has to be above a certain threshold
value, since otherwise they tend to be broken up in a two-dimensional
array of discrete islands which do not exhibit metal-like properties.
Figure 9, taken from Ref. 45, illustrates the salient properties of a
growing metal film. Going from left to right, the figure presents
growth stages, film structures, thickness scales and theoretical
models for the optical properties. The growth stages are well
known:[46,47] when metal vapour condenses onto a surface nuclei will be
formed at certain sites; continued deposition makes the nuclei grow
via surface diffusion and direct impingement; the metal islands thus
formed undergo coalescence growth so that larger and more irregular
islands appear; the growing film then goes through large-scale coalescence so that a metallic network extends through the sample; subsequently the voids get smaller and more regular; and finally a uniform metal layer may be formed. The corresponding structures (with
metal shown as black) are sketched. The thickness scales pertain to
the deposition of gold onto glass by use of conventional evaporation
as well as by ion-plating. One should note the compressed thickness
scale valid for ion-plating. A similar compression - albeit perhaps

not equally strong - is expected for sputtered films, since this deposition technique involves bombardment with energetic ions. This observation is important because magnetron sputtering is widely used today for the production of window coatings. Generally, the thickness scales are dependent on many parameters such as the deposited species, substrate material, substrate temperature, vacuum conditions, the presence of electric fields, etc. However, even if the thickness scales may display a varying degree of compression and expansion, it is expected that the growth stages are traversed in a certain order and hence that the optical performance varies with thickness in a definite way. The corresponding theoretical models for the optical properties are indicated in the right-hand part of Figure 9. The discontinuous metal film can be understood from the Maxwell Garnett theory (cf. Sec. 2.3), provided that particle shape and interaction - both of which become increasingly complicated with increasing thickness - are accounted for properly. Around large-scale coalescence, the film is characterized by an interpenetrating maze structure and we conjecture that effects of continuum percolation may be observable at optical frequencies. The non-uniform film can be understood to some degree from the Bruggeman theory (cf. Sec. 2.3) with due regard to void shapes. Finally, the continuous film can be understood to a considerable degree from the Drude theory (cf. Sec. 2.2). For a window coating to be practially useful, it should have a thickness well above the one required for large scale coalescence.

The optical properties of growing films, which were treated in principle in Fig. 9, are substantiated in Fig. 10. This figure is reproduced from a recent study[42] of selective transmission in conventionally evaporated films of Cu, Ag, Au, Cr, Fe, Co, Ni and Al. The shown data pertain to Ag coatings of different thicknesses onto glass. The thickest film (thickness 36 nm; curves denoted a) display an essentially bulk-like behaviour with high reflectance - and thus low transmittance - across the main part of the shown spectral range. For gradually thinner films, there is an increase in short-wavelength

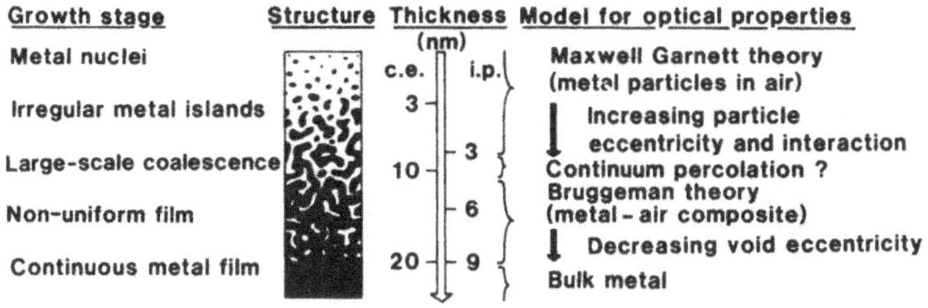

Growth stage	Structure	Thickness (nm)	Model for optical properties

Fig. 9   Survey on gold films coated onto glass by use of conventional evaporation (c.E.) and ion-plating (i.p.).

transmittance, while the reflectance for thermal radiation is not significantly affected. Thus there appears a gradual onset of selective transmission. For the thinnest film (thickness 6 nm; curves denoted e), a qualitatively different behaviour is encountered which is indicative of a discrete nature of the film. We return to the optical properties of metal films in case study one reported on in Sec. 5.

The limited transmission in continuous noble-metal films is governed largely by reflection at the film boundaries rather than intrinsic absorption. Therefore it is possible to improve the transmittance by additonal coating(s) which act so as to antireflect the metal. We are then led to consider dielectric/metal and dielectric/metal/dielectric multilayers. Dielectrics with high refractive indices (such as Bi_2O_3, In_2O_3, SnO_2, TiO_2, ZnO and ZnS) are useful for creating the desired induced transmission. By selecting proper thick-

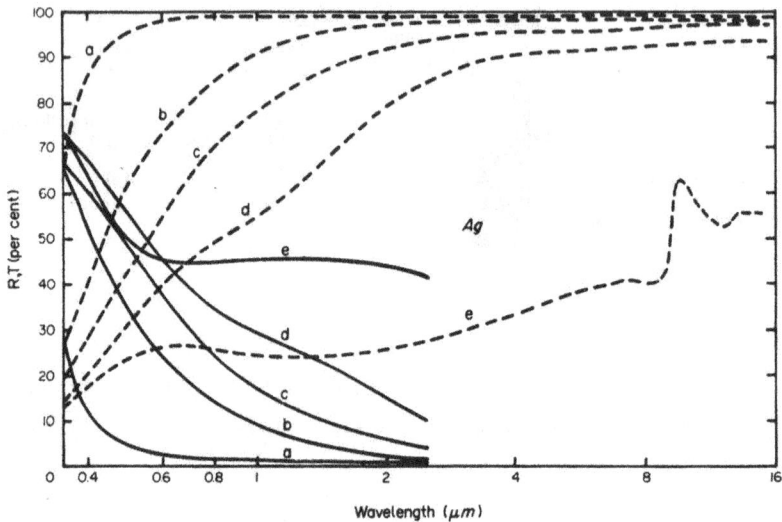

Fig.10 Spectral transmittance (solid curves) and reflectance (dashed curves) of selectively transmitting silverfilms on glass substrates. The films have the following thickness: (a) 36 nm, (b) 16 nm, (c) 12 nm, (d) 9 nm, (e) 6 nm. This figure is reproduced from Ref. 42.

nesses, one can optimize the coating either for solar control or for low thermal emittance. The optimization is conveniently done on a computer. Figure 11 reports on a high-performance selectively trans-mitting coating[29] suitable for solar control (according to the ideal property indicated by the dotted lines). The three-layer coating con-sists of titanium dioxide/silver/titanium dioxide, where each layer is 18 nm thick. The spectral transmittance and reflectance, shown by the solid curves, approach the ideal behaviour and, in particular, the visible transmittance is as large as about 80 %. The experimental data are in fair agreement with a computation (dashed curves) based on the known optical properties of the constituent materials. The ensuing differences are probably caused by a slightly non-homogeneous character of the silver layer.

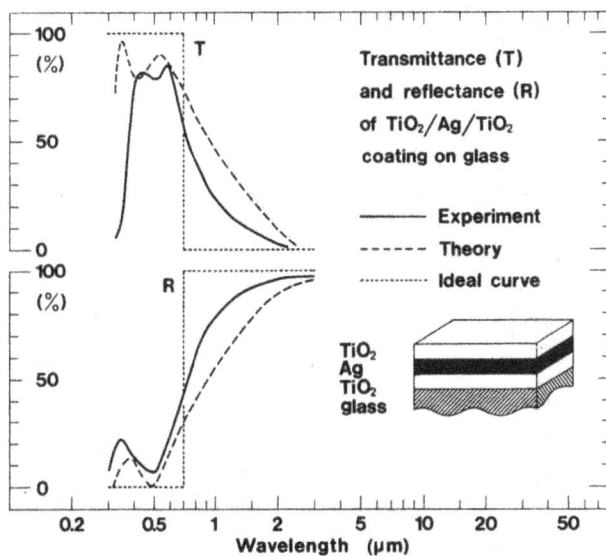

Fig.11   Spectral transmittance and reflectance of a $TiO_2/Ag/TiO_2$
coating on glass. Experimental and theoretical data are
reported and compared with an ideal spectral profile. Inset
depicts the coating structure.

4.3   Doped-Semiconductor-Based Coatings

Doped-semiconductor coatings offer a principally superior alternative to the metal-based coatings. The energy gap must be sufficient to allow good transmission down to 0.3 μm and the structure
must permit doping to a level so that the coatings exhibit free-
electron-like behaviour beyond a plasma wavelength of the order of a
few μm. It is important that the doping does not introduce substantial absorption at wavelengths shorter than the plasma wavelength, and
that the energy gap does not shift towards the visible range as a consequence of the doping. The useful materials are oxide semi-conduc-
tors based on Zn, Cd, In, Sn and combinations of these. The required
carrier densities can be achieved by non-stoichiometry (intrinsic
doping) as well as by inclusion of foreign elements (extrinsic doping).
Among the most widely studied materials we note SnO_2 doped by F or Sb,
In_2O_3 doped via non-stoichiometry or by Sn, ZnO doped by Al, and
Cd_2SnO_4.

The interesting oxide semiconductors have refractive indices of
the order of 2, and hence the short wavelength transmittance is
limited by reflection losses to an undesirable extent. One is then
led to regard dielectric/doped-semiconductor coatings, for which the
dielectric with low refractive index acts as an antireflecting layer.
Figure 12 shows a specific case having particularly good performance:
an Sn-doped In_2O_3 layer antireflected by MgF_2. It is found that the
transmittance is high for short wavelengths (particularly in the
visible) and that the reflectance is high for long wavelengths. These
properties are rather good for a low-emittance window coating; they
would have been even better if the onset of high reflectance were
moved towards the infrared, as can easily be done by decreasing the
Sn doping. We return to In_2O_3:Sn coatings in case study two in Sec.6
below. We note that most low-refractive index coatings cannot be
applied by high-rate sputtering, which is a drawback for practical
production. An exception to this rule seems to be offered by certain
novel metal-oxyfluoride coatings (Refs. 48, 49).

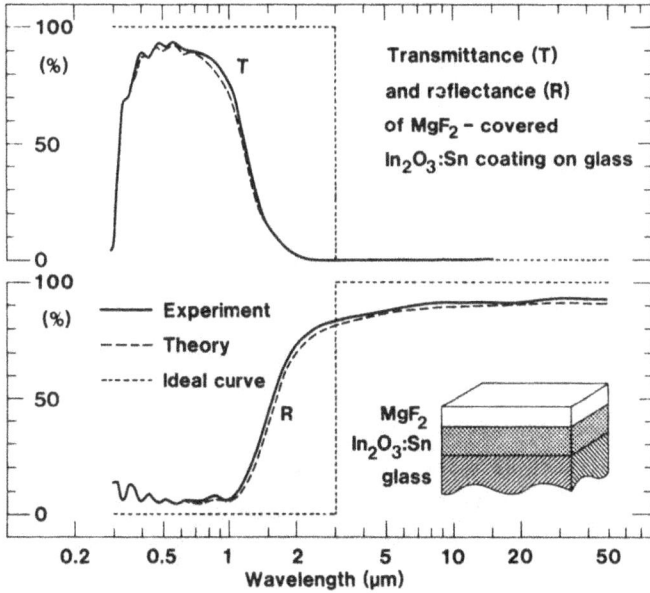

Fig.12 Spectral transmittance and reflectance of an MgF$_2$-coated
        In$_2$O$_3$:film on glass. Experimental and theoretical data are
        reported and compared with an ideal spectral profile. Inset
        depitcs the coating structure.

The final approach to selectively transmitting window coatings, which we mention just in passing, is to exploit some of the new materials, which are appearing today. Interesting results have been found for TiN, LaB_6 and $AuAl_2$.

5.   CASE STUDY ONE: SELECTIVE OPTICAL PROPERTIES OF ION-PLATED Au FILMS (REF.45).

This first case study reports on the optical properties of thin gold films made by ion-plating and compares them with results obtained by conventional evaporation. Film thicknesses on either side of the magnitude required for microstructural uniformity are of interest. Several novel effects of potential importance for improved window coatings are identified and discussed in terms of theoretical models. There can be little doubt that these findings would apply equally well to films of Cu and Ag.

5.1   Film Production And Characterization

Figure 13 is a sketch of the film deposition system. Gold was evaporated onto unheated glass substrates and simultaneously onto grids for electron microscopy. The evaporation rate was maintained at a certain value r by use of a vibrating quartz microbalance connected to a microprocessor-based deposition controller which regulated the power delivered from the evaporation power supply. Most of the films were produced by ion-plating, i.e., by continuous bombardment of the growing film with energetic ions. The ion current, I, was recorded via a small probe in the ion beam.

Ion-plating was performed as follows. After evacuation of the vacuum chamber an Ar flow was introduced so that the pressure became 10^{-4} Torr. The ion source was set to produce 500 eV Ar^+ ions with a certain flux $s(Ar^+)$, which was determined by measuring I. The gold flux $s(Au)$ was inferred from r. Typical experimental parameters were

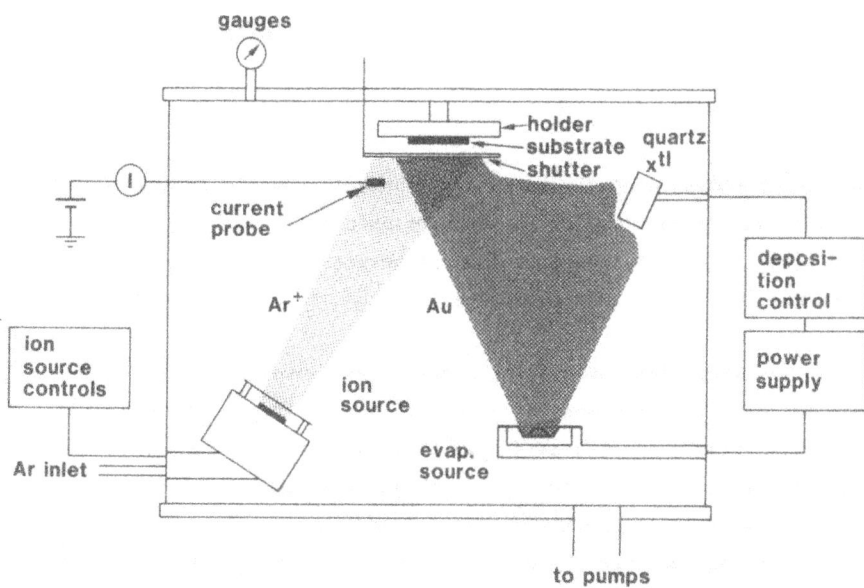

Fig.13   Schematic view of the ion-plating system and its process
controls.

$8 < I < 33 \text{ A/cm}^2$ and $0.1 < r < 0.2 \text{ nm/s}$. The flux ratio, defined by

$$g \equiv s(Ar^+)/s(Au),\qquad\qquad\qquad (26)$$

should be chosen properly in order to benefit from ion-beam-assisted film growth without having too much resputtering of deposited material. Empirically, we obtained good film properties with $g \approx 0.1$. Films with mass thicknesses in the range $4.5 < t < 18 \text{ nm}$ were produced.

The microstructure of the coatings was investigated by transmission electron microscopy. Figure 14 shows a typical bright-field micrograph for an ion-plated (g=0.1) film with t = 5.5 nm. A characteristic network structure with an areal coverage > 75 % is observed. There is a striking similarity to the sketch in Fig. 9. Ion-plated films, made under favourable conditions, had network configuration at $4 \lesssim t \lesssim 9 \text{ nm}$ and appeared uniform at $t \gtrsim 9 \text{ nm}$. Without ion impingement (g = 0), we obtained network films at mass thicknesses as large as ~ 15 nm.

5.2   Optical Data

Figure 15 shows spectral transmittance T and reflectance R in the visible and near-infrared. The upper part refers to ion-plated films obtained with g = 0.1. The thickest film (t ≈ 9 nm) displays an increase in R and a concomitant decrease in T for wavelengths $\lambda > 0.5 \text{ }\mu\text{m}$. This is characteristic for bulk-like Au, and we infer that the film is approaching a uniform structure. The lowering of T for $\lambda < 0.5 \text{ }\mu\text{m}$ is caused by interband absorption. Thinner films are qualitatively different: for $6 < t < 9 \text{ nm}$ there are plateaus of T and R in the near-infrared, which are followed by a weak decrease in T and increase in R at the longest wavelengths. It will be found below that the specific optical properties of films with t < 9 nm can be understood from considerations of non-uniform film structures. The lower part of Fig. 15 pertains to thicker films produced by conventional

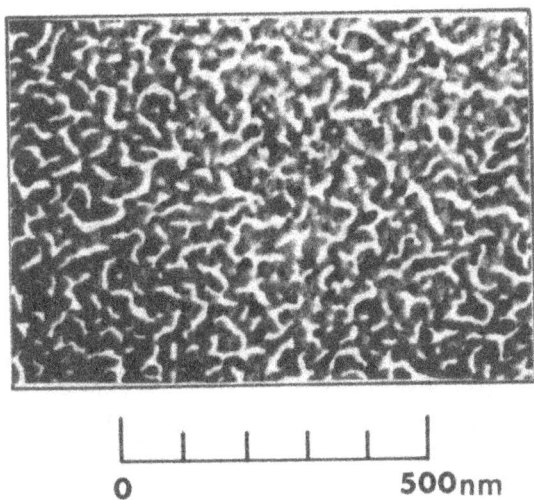

Fig.14   Transmission electron micrograph of a non-uniform ion-plated
Au-film.

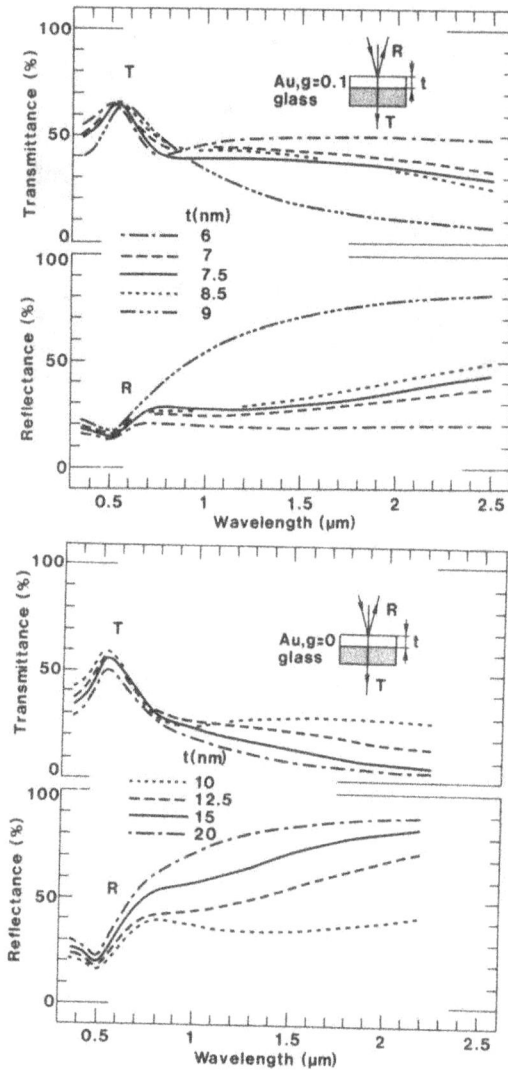

Fig.15 Spectral normal transmittance and near-normal reflectance for thin Au films produced by ion-plating (upper part) and conventional evaporation (lower part). Data are shown for films with different mass thicknesses.

evaporation (g=0). We note that near-infrared plateaus appear for
t ≲ 15 nm, which is well beyond the thickness where the ion-plated
films seem to be entirely bulk-like. Curves resembling those in
Fig.15 have been reported also by Valkonen et al.[42] in the paper from
which Fig. 10 was reproduced.

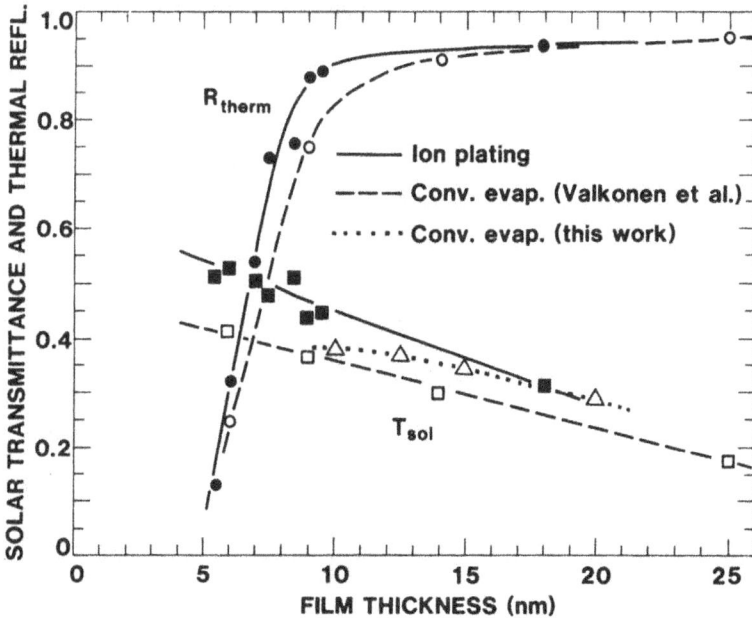

Fig.16 Normal solar transmittance and near-normal thermal reflectance
versus mass thickness for thin Au films produced by ion-
-plating (filled symbols) and by conventional evaporation
(open symbols). Results from measurements by Valkonen et al.
(Ref. 42) are included. Solid, dashed and dotted curves were
drawn solely to guide the eye.

Figure 16 reports T_{sol} and R_{therm} versus t for Au films produced
by ion-plating (g = 0.1) and by conventional evaporation (g = 0).
These quantities were computed by use of the formulas given in Sec.3.2
and τ = 300 K. The most important result concerns T_{sol}, which can be
significantly enhanced as a result of the ion impingement. For example,
one can combine R_{therm} = 76 % with T_{sol} = 51% in an ion-plated film
but only with T_{sol} = 37% using conventional evaporation. Hence ion-
plating can give significant gain in transmitted solar energy. Ion-
plating also leads to some improvement in the thermal properties.
All of the ion-plated films with t < 9 nm had 56 < T_{lum} < 58%. Large
enhancements of T_{sol} and/or T_{lum} can be achieved by embedding the gold
films between dielectric layers with high refractive indices, as we
have already commented on.

5.2   Analysis Of Uniform Films

We first analyze the thicker of the Au films - those with
t ≳ 9 nm in the case of ion-plating. These films showed no evidence
for non-uniformity, and it is meaningful to apply the Drude theory
(Sec. 2.2) at λ > 0.5 μm. It was verified that, in fact, the Drude
theory gives a very good representation of the extracted dielectric
function for uniform Au films. The only feature worth further comment
concerns the spectral dependence of the relaxation frequency γ
(cf. Eq. 14) which is intriguingly strong. It was found that the experimental data was consistent with a relation

$$\gamma = \gamma^{o} + \beta\omega^{2},$$
(27)

where γ^{o} and β are constants. An ω^{2}-dependence of the relaxation frequency has been reported before for the metal films, but the present magnitude of β is considerably larger than the one found in earlier work on thicker films. Several mechanisms could yield a relation such as the one indicated in Eq. 27; we can mention electron-electron interaction, electron-phonon interaction, a two-carrier mechanism accounting for crystalline grains and disordered inter-granular material, and surface electrodynamics. Recent work[50] showed convincingly that surface electrodynamics plays an important role. Our observation of enhanced βs in very thin ion-plated Au films is in accord with this view. The analysis also yielded a magnitude of γ^o which is consistent with the film thickness and crystallite size.

5.4   Analysis Of Non-Uniform Films

We now turn to the analysis of the non-uniform films consisting of a mixture of gold and air-filled voids. Since the basic structural units cannot be taken as round, the Maxwell Garnett and Bruggeman EMTs must be generalized so as to account for appropriate values of the depolarisation factors. Specifically, we took the two-dimensional images of the films to comprise a metallic phase represented by circular units (with depolarisation factor $L_m = 1/3$, pertaining to spheres) and a dielectric phase represented by elliptical units (with depolarisation factors L_1 and L_2, so that $L_1 + L_2 = 1$). The generalization of Eq. (19) now reads

$$\frac{\bar{\varepsilon}^{MG} - \varepsilon_m}{\bar{\varepsilon}^{MG} + 2\varepsilon_m} = \frac{f}{6} \sum_{i=1}^{2} \frac{1 - \varepsilon_m}{\varepsilon_m + L_i(1-\varepsilon_m)} , \tag{28}$$

and the generalization of Eq. (21) reads

$$\frac{f}{2} \sum_{i=1}^{2} \frac{1 + \bar{\varepsilon}^{Br}}{\bar{\varepsilon}^{Br} + L_i(1-\bar{\varepsilon}^{Br})} + (1-f) \frac{\varepsilon_m - \bar{\varepsilon}^{Br}}{\bar{\varepsilon}^{Br} + (\varepsilon_m - \bar{\varepsilon}^{Br})/3} = 0 , \tag{29}$$

where f is void fraction and ε_m refers to the dielectric function of
gold modified by standard techniques[51] so as to encompass a short mean
free path for the free electrons (set to 6.5 nm). It is then
straight-forward to compute $\bar{\varepsilon}^{Br}$ and $\bar{\varepsilon}^{MG}$ versus λ, t, f, L_1 and L_m.
Finally, we obtained T and R from Fresnel's equations (Sec. 2.1) and
with the underlying substrate represented by the dielectric function
for amorphous SiO_2 (Ref. 52).

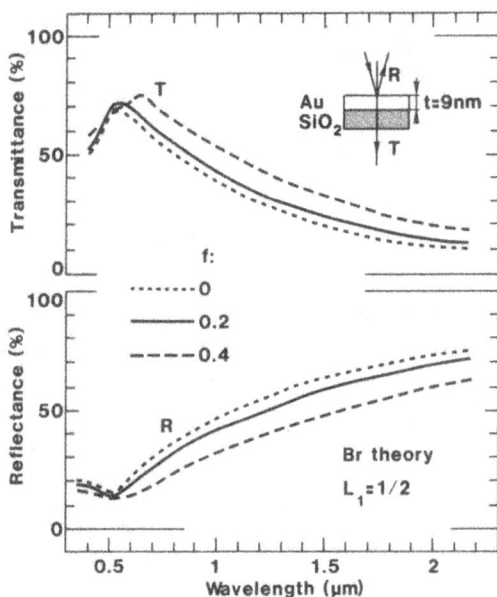

Fig. 17   Spectral normal transmittance and reflectance computed for a
non-uniform Au film on SiO_2. The Bruggeman effective medium
theory, appropriate to circular structural units, was used
together with the shown values of mass thickness and void
fraction.

Figure 17 shows spectral T and R for gold films characterized
by different void fractions. We used t = 9 nm and the Bruggeman
theory for circular structural units ($L_1 = L_2 = 1/2$). The dotted
curves pertain to f = 0, i.e., to a uniform film. The calculated
data is in good agreement with the measured results for the thickest
films reported on Fig. 15. An increase of the void fraction is seen
to have little effect on the luminous performance but to yield a considerable increase of T and decrease of R in the infrared. Figure 18
illustrates several important results. It considers a gold film
with t = 7.2 nm and f = 0.2. The solid curve refers to the Bruggeman

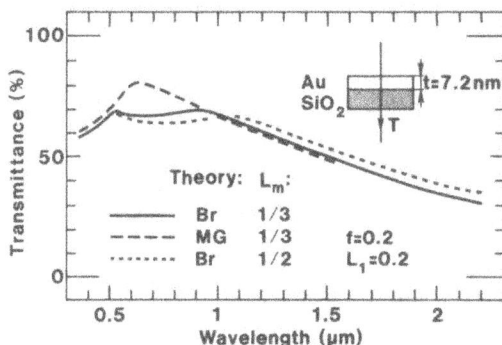

Fig.18 Spectral normal transmittance computed for a non-uniform Au
film on SiO_2. The Bruggeman and Maxwell Garnett effective
medium theories were used together with the shown values of
mass thickness, void fraction, and depolarisation factors for
voids and metallic units.

theory for non-spherical voids ($L_1 = 0.2$). A comparison with the upper solid curve in Fig. 17 demonstrates that the departure from circular void shapes leads to a characteristic plateau across the 0.5 - 1 - μm range. Corresponding results based on the Maxwell Garnett theory do not produce a plateau but are otherwise in good agreement with the Bruggeman data, as seen from the dashed curve in Fig 18. Setting $L_m = 1/2$, as in the computation giving the dotted curve, does not lead to significant departures from the curve obtained with $L_m = 1/3$. The plateaus in the infrared transmittance are important since they affect the solar transmittance. The plateaus are further investigated in Fig. 19, where we vary L_1 from 0.5

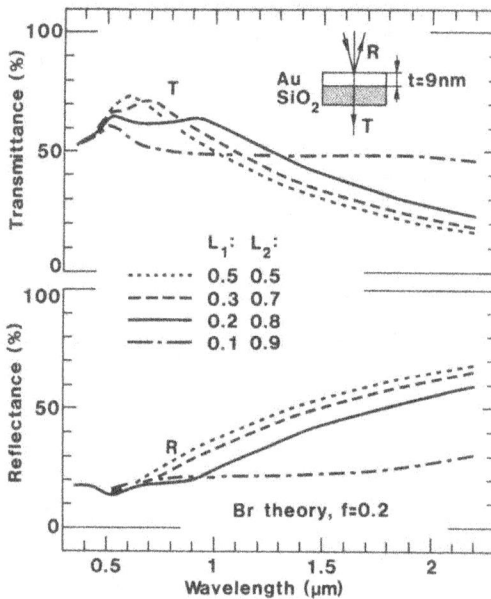

Fig.19 Spectral normal transmittance and reflectance computed for a non-uniform Au film on SiO_2. The Bruggeman effective medium theory was used together with the shown values of mass thickness, void fraction, and depolarisation factors for the voids.

(circle) to 0.1 for films with t = 9 nm and f = 0.2. The Bruggeman
formalism is used. It is found that the plateaus become increasingly
wide when L_1 is decreased, i.e., when the eccentricity of the elliptical voids is increased.

We now consider the computed spectra in some detail and discuss
to what extent we can use them for understanding the spectrophotometric data. We focus on the conspicuous plateaus in the infrared
spectra. They were clearly seen in the measured transmittance and
were apparent also in computations based on the Bruggeman effective
medium theory applied to non-spherical voids. Three specific points
demand attention. First, we observed that the measured plateaus were
widest in the thinnest film. Theoretically, this is to be expected
since the most elongated voids occur closest to large scale coalescence (or the percolation threshold). Second, the measured plateaus
were less distinct than in the computations. This is also very
reasonable when one remembers that a broad range of depolarisation
factors is needed to characterize a real non-uniform film. Third, we
observed that the transmittance maxima on the short-wavelength side
of the plateaus were more prominent than expected from the Bruggeman
theory. This is by no means unexpected, though, since the complex
structure of the experimental samples makes one believe that features
of both the Bruggeman and Maxwell Garnett theories are present.
Indeed, combining the solid and dashed curves in Fig. 18 makes the
short-wavelength peak dominant. In conclusion, we find that most
features of the transmittance plateaus can be reconciled with an
effective medium treatment based on the pertinent microstructure.
This theory also explains the increased solar transmittance of properly produced non-uniform metal films.

6.   CASE STUDY TWO: SELECTIVE OPTICAL PROPERTIES OF E-BEAM EVAP-
      ORATED In_2O_3:Sn FILMS (REFS.53-59)

Coatings of In_2O_3:Sn (also known as indium-tin-oxide or ITO) have
been investigated in depth during the last years. The theoretical
understanding seems to be more advanced for this oxide semiconductor
than for the other ones mentioned in Sec. 4.3. Hence we think it is
worthwhile to present a detailed analysis of this particular material
here.

6.1   Film Preparation And Optical Properties

In_2O_3:Sn coatings with selective optical properties can be produced by a variety of techniques including evaporation (resistive,
e-beam; non-reactive, reactive) sputtering (diode, magnetron, ion-
beam; planar, cylindrical; non-reactive, reactive), ion plating,
chemical vapour deposition, spray pyrolysis, dipping etc. In the
work to be discussed below we used reactive e-beam deposition of
In_2O_3 + SnO_2 onto heated substrates. This is a convenient technique
for laboratory-type investigations, and it is also used for industrial
production of In_2O_3:Sn coatings with high performance. The inset of
Fig. 20 shows a sketch of the film production unit comprising a
vacuum chamber with an e-beam source (whose input power is feed-back
controlled from a vibrating quartz microbalance), a heated substrate
holder, and facilities for controlled inlet of gases. In principle,
this is the same deposition unit as the one shown in Fig. 13. We
found that the best optical performance was obtained with substrate
temperature ~ 300° C, constant evaporation rate in the 0.2-0.3 nm/s
interval, and constant oxygen pressure in the $5-8 \times 10^{-4}$ Torr interval.

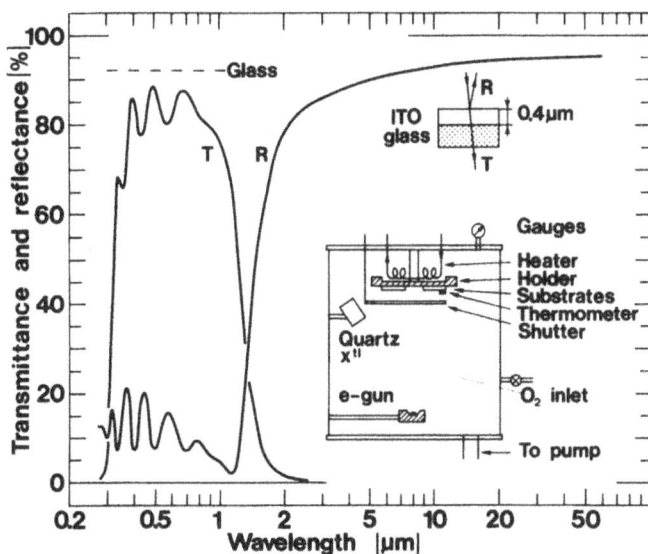

Fig.20   Spectral transmittance and reflectance of a 0.4 μm thick
In$_2$O$_3$:Sn film on glass. Inset shows a sketch of the film
deposition unit.

Main part of Fig. 20 shows spectrophotometric measurements of
transmittance and reflectance for a 0.4-μm-thick coating on glass.
The spectral selectivity is similar to the one shown in Fig. 12,
which is the expected result. The deposition parameters are rather
crucial, which is true not only for evaporation but also for other
techniques such as sputtering.[60,61] This is illustrated in Fig. 21,
which shows the spectral absorptance across the visible range for
0.3-μm-thick In$_2$O$_3$:Sn films deposited at different temperatures.
Only the film made at the highest temperature has very good performance. In our best films we have found an absorptance of only ~ 0.2 %
at λ = 0.5 μm.

In Fig.20 it is seen that the visible transmittance exhibits
pronounced peaks, which are indicative of optical interference. This

will lead to a slight colouring of the coatings. Further, if small thickness variations exist across the coated surface – which may be hard to avoid in a practical window coating – it will show iridescence, which is definitely an unwanted effect for most applications. An improvement in the colour performance is achieved by antireflection coats. Comparing Figs. 12 and 20 it is evident that the MgF_2 layer significantly cuts down the iridescence. The actual MgF_2 thickness is 90 nm in this example. The luminous transmittance can be ~ 95 % (including the substrate) while the thermal emittance is $\leq$ 20 % for 0.3-μm-thick In_2O_3:Sn films antireflection coated with 90 nm of MgF_2.

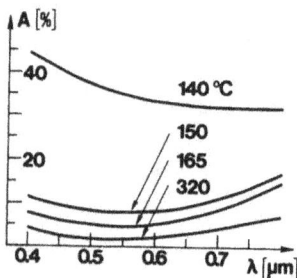

Fig.21   Spectral absorptance (A) for In_2O_3:Sn films (on glass) deposited at four different substrate temperatures.

Colour properties have not often received proper attention in scientific works on selectively transmitting window coatings. This is somewhat surprising, since colorimetric analysis is a well known subject treated in several books.[25,62-64] As an example of the quantitative results one can reach by colorimetric analysis, we present in Fig. 22 the x and y coordinates referred to the CIE 1931 standard colorimetric system and a daylight-type illuminant for normal transmission of light through glass coated with In_2O_3:Sn (thickness range 250 to 400 nm) and MgF_2 (thickness, t, being 0, 90, 95, 100, 105 nm).

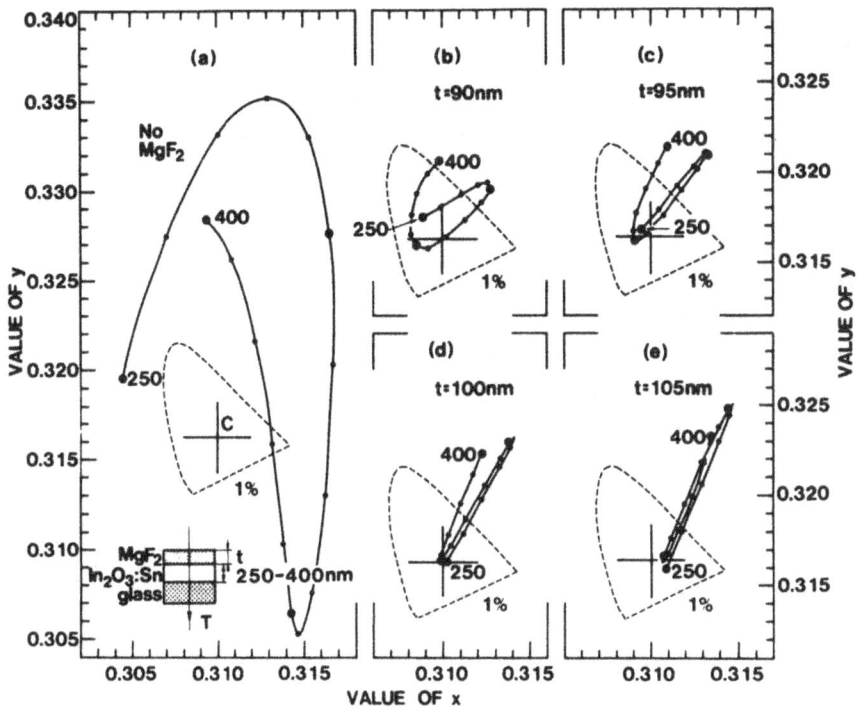

Fig.22   Solid curves show the CIE 1931 (x,y) chromaticity
         coordinates for normal transmission of daylight (standard
         illuminant C) through an In_2O_3:Sn film antireflected by MgF_2
         on glass.   Chromaticities corresponding to 10-nm thickness
         increments of In_2O_3:Sn are marked on the curves.

The data were computed by use of the dielectric functions of In_2O_3:Sn, MgF_2 and glass. The distance between a certain point on the curve and the point C describes the excitation purity of the colour (the graph corresponding to 1 % excitation purity is indicated). A full interpretation of the data in Fig. 22 is beyond the scope of this article; it is sufficient to observe that the excitation purity (or perceived colour) is significantly decreased when an MgF_2 layer is applied. Specifically, it was found that an excitation purity of < 1 % in normal transmission and < 10 % in normal reflection could be achieved with In_2O_3:Sn thicknesses in the 220-260- or 335-365 nm ranges and MgF_2 thicknesses in the 90-105-nm range.

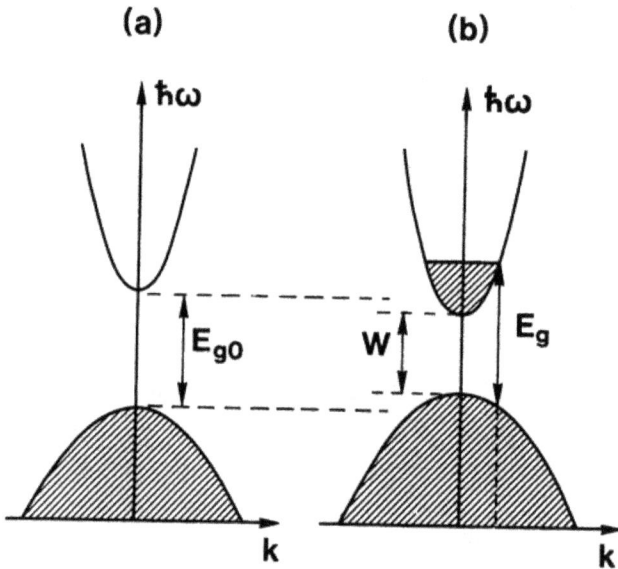

(a)   **(b)**

Fig.23 Part (a) shows the assumed bandstructure of undoped In_2O_3 in the vicinity of the top of the valence band and the bottom of the conduction band. k is wavevector amplitude and $\hbar\omega$ is photon energy. Part (b) describes the effect of Sn doping. A shift of the bands is indicated. Shaded areas denote occupied states.

6.2   Theoretical Analysis

We now turn to a theoretical analysis of the optical properties of In_2O_3:Sn. To this end we regard In_2O_3 as a host lattice to which the effect of the Sn dopant is added. The bandstructure of In_2O_3 - which forms the natural basis for a theory of the optical properties - is not known in any detail, but we believe that Fig. 23 (a) gives a good working model. It shows parabolic and isotropic bands separated by a direct semiconductor bandgap E_{g0}, which is 3.75 eV. It has been proposed that the conduction band is mainly from indium 5s electrons and the valence band is from oxygen 2p electrons. The Fermi energy E_F lies halfway between the energy bands. We next introduce a low density of donor atoms. Under these condictions, donor states are formed just below the conduction band, and E_F lies between the donor level and the conduction band minimum. For increased donor density, the donor states merge with the conduction band at a certain "critical" density n_c, whose magnitude can be estimated using Mott's criterion[65]

$$n_c^{1/3} a_o^* \approx 0.25. \tag{30}$$

The effective Bohr radius a_o^* is ~ 1.3 nm for In_2O_3. Above n_c, the Sn atoms are singly ionized. E_F is determined by the highest occupied states in the conduction band. Figure 23 (b) depicts the bandstructure of heavily doped In_2O_3. A partial filling of the conduction band as well as shifts in energy of the bands, relative to their location in In_2O_3, are indicated.

As earlier discussed in Sec.2.2, it is often possible to consider the complex dielectric function as a sum of contributions due to valence electrons, free electrons, and phonons (cf. Eq.12). Below we regard the susceptibilities for these contributions for the case of In_2O_3:Sn. We begin our discussion of χ^{VE} by noting that the fundamental bandgap of a heavily doped oxide semiconductor becomes wider as the electron density n_e is increased. The shift occurs as the net result of two competing mechanisms: a bandgap widening due to a blocking of the lowest states in the conduction band (so that absorption occurs for photon energies exceeding E_g rather than W in Fig. 23), and a bandgap narrowing due to electron-electron and electron-ion scattering (so that $W < E_{g0}$ in Fig. 23). Hence bandgap widening occurs as a consequence of the Burstein-Moss effect,[66] which is pronounced because of the small effective mass for the conduction electrons in In_2O_3:Sn. The empirical data follow approximately the relation $E_g - E_{g0} \propto n_e^{2/3}$. There is also a smearing of the bandgap which increases in proportion to n_e and which may also be dominated by effects of electron scattering (rather than of nonzero temperature).

If we confine our attention to the solar range, it is sufficient to regard the low-energy tail of the bandgap. Figure 24 shows absorption coefficient, evaluated from spectrophotometric measurements, versus photon energy for films of In_2O_3 and In_2O_3:Sn. The samples denoted A, B, C, D have the thicknesses 0.075, 0.115, 0.11, and 0.41 μm, and the electron densities 0.4, 1.7, 6.2 and 8.0 x 10^{20} cm^{-3}, respectively. The n_es were obtained from optically observed screened plasma energies. Below 4 eV, the data in Fig. 24 is seen to follow approximately the relation

$$\alpha = \alpha'' \exp \left[a \ (\hbar\omega - b) \right], \qquad (31)$$

where α'', a and b are parameters. Such logarithmic bandedges are often referred to as Urbach tails.[67] We may rewrite Eq. (31) in terms of a valence electron susceptibility, applicable in the bandedge (be) range, as

$$\chi_{be}^{VE} = \varepsilon_\infty - 1 + i \ (\varepsilon_\infty^{1/2}/2\pi)\lambda\alpha'' \exp \left[a(\hbar - b) \right], \qquad (32)$$

where ε_∞ is the extrapolated high-frequency dielectric constant.

The bandgap shift can be represented by

$$b = cn_e^{2/3} + d, \qquad\qquad (33)$$

with c and d being parameters. The empirical results in Fig. 24 can be reproduced by setting $\alpha'' = 1 \text{ cm}^{-1}$, $a = 3.99 \text{ (eV)}^{-1}$, $c = 0.376 \times 10^{-14} \text{ cm}^2 \text{ eV}$, and $d = 0.979 \text{ eV}$. Urbach tails may be connected with electric microfields due to the ubiquitous ionized defects,[68] thermal fluctuations of the bandgap,[69] etc.

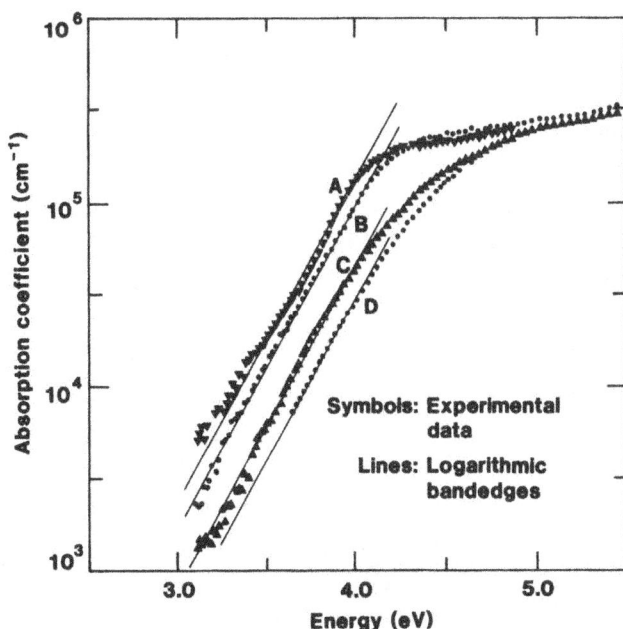

Fig.24 Absorption coefficient versus photon energy for films of In₂O₃ and In₂O₃:Sn. Symbols refer to measurements at room temperature; the scatter illustrates the difficulty of using thin films to determine small absorption coefficients. The parallel lines indicate logarithmic band-edges.

We now turn to the free-electron properties and note that the spectral selectivity required for a window coating rests to a large extent on these. The analysis of free-electron properties is conveniently performed in terms of a complex dynamic resistivity $\rho \equiv \rho_1 + i\rho_2$, which is connected with the susceptibility via the general relation

$$\chi = i/(\varepsilon_o \omega \rho), \tag{34}$$

where ε_o is the permittivity of free space. This formalism is useful[70] since ρ_1 is directly related to the type of scattering which prevails among the free electrons. The dots and solid curves in Fig. 25 show empirical results for three heavily doped In_2O_3:Sn films with somewhat different electron densities. The data is based on spectrophotometric measurements of transmittance and reflectance in the solar and thermal ranges. We find that ρ_1 is rather constant below the plasma energy $\hbar\omega_p$ and joins smoothly with the measured dc resistivity ρ_c. Above the plasma energy, ρ_1 falls off according to a power law. With due consideration of experimental uncertainties, it is found that a relation

$$\rho_1 \propto \omega^s, \text{ with } -1.3 \geq s \geq -2, \tag{35}$$

is obeyed. At the highest energies shown in Fig. 25, ρ_1 levels off and starts to increase at the approach of the semiconductor bandgap. $-\rho_2$ also follows a power law with an exponent between 0.85 and 1.0. Additional data for $\hbar\omega < 0.1$ eV were consistent with the results in Fig. 25.

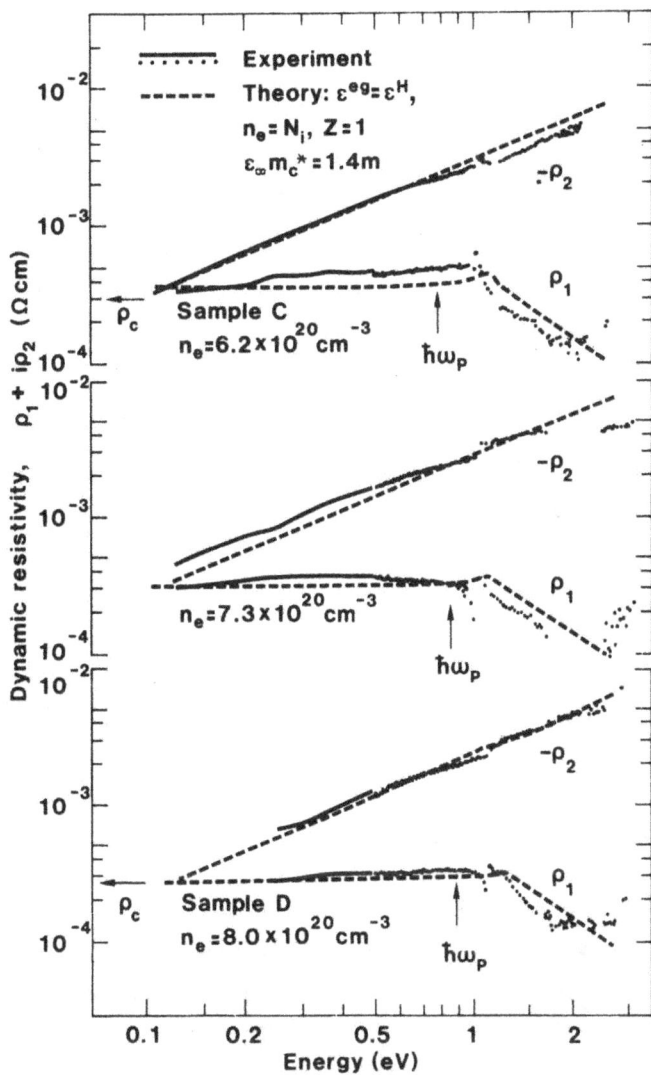

Figure 25.

Fig.25 Real and imaginary parts of the dynamic resistivity for
In$_2$O$_3$:Sn films. Dots and solid curves were evaluated for
three films. The actual spread in the data is apparent from
the dots; the lines are shown for the energies where the
spread is insignificant. Dashed curves were computed from
the theory of ionized impurity scattering with the shown
parameters. The detailed plasmon features around the "knee"
in the ρ_1 data are not resolved. The pertinent values of
plasma energy $\hbar\omega_p$ and dc resistivity ρ_c are indicated by
arrows.

We now wish to present a quantitative theoretical model for ρ.
To this end we consider ionized impurity scattering, which is unavoidable for $n_e \gg n_c$. Using the Gerlach-Grosse theory[70] appropriate to
a Coulomb potential we can write

$$\rho(\omega) = i\ \frac{Z^2 N_i}{6\pi^2 \epsilon_o n_e^2 \omega} \int_0^\infty k^2 dk \left[\frac{1}{\epsilon^{eg}(\vec{k},\omega)} - \frac{1}{\epsilon^{eg}(\vec{k},0)}\right] - i\ \frac{\omega}{\epsilon_o \epsilon_\infty \omega_p^2},$$

(36)

where Z is the charge of the impurities, N_i is their density, and
ϵ^{eg} is the dielectric function of the free-electron gas. For heavily
doped In$_2$O$_3$:Sn, we may set $Z = 1$ and $N_i = n_e$, thus assuming that the
free electrons stem from singly ionized Sn atoms. The remaining part
of the theory regards ϵ^{eg}. We first consider the Random Phase
Approximation[9,72] (RPA). For electron collision processes it is the
longitudinal part of the dielectric function that matters, and in the
limit of small damping we have for the degenerate electron gas
(with $\epsilon^{eg} \equiv \epsilon^{RPA}$)

$$\varepsilon^{RPA}(k,\omega) \equiv \varepsilon_\infty + vP(k,\omega) = \varepsilon_\infty + \frac{3\omega_p^2}{k^2 v_F^2} \, (f_1 + if_2), \qquad (37)$$

where v_F if Fermi velocity. Explicit expressions for f_1 and f_2 may be found in Ref. 56. RPA is known[9] to give a good description of the free-electron gas particularly when the parameter

$$r_s^* = a_o^* \, (4\pi n_e/3)^{1/3} \qquad (38)$$

is small. For heavily doped In_2O_3:Sn we can have $r_s^* \sim 1$. This is in fact smaller than for metals, which attests to the applicability of the RPA. Nevertheless, the RPA is not entirely satisfactory since it leaves out the effects of exchange and correlation in the electron gas, and it is of some interest to invoke more elaborate theories. One such extension of RPA, which includes exchange but not correlation, was introduced by Hubbard.[73] Within this theory, screening in the electron gas is represented (with $\varepsilon^{eg} \equiv \varepsilon^H$) by

$$\varepsilon^H(k,\omega) = \varepsilon_\infty + \left[1 - G^H(k)\right] vP(k,\omega), \qquad (39)$$

$$G^H(k) = k^2/2(k^2 + k_F^2). \qquad (40)$$

Here $k_F = (3\pi^2 n_e)^{1/3}$ denotes the Fermi wavevector.

The dashed curves in Fig. 25 were computed from the theory outlined above with no free parameter. We used empirical results for n_e, the Hubbard correction to the RPA, and

$$\varepsilon_\infty m_c^* = 1.4 \, m, \qquad (41)$$

where m_c^* is the effective conduction band mass and m is the free-electron mass. The relation in Eq. (41) was taken from the detailed work by Köstlin et al.[74] The correspondance between theory and ex-

periment is striking; not only the slopes of the curves but also their
magnitude in the whole spectral region – from the thermal infrared
and well into the visible – are in agreement. Specifically, the parameter s (cf. Eq. 35) obtained from the computation was equal to – 1.5,
which is as expected for ionized impurity scattering. We thus conclude that the Gerlach-Grosse theory of ionized impurity scattering,
with screening in the electron gas represented by the RPA or an extension thereof, provides a quantitative model for the free-electron
properties of high-quality In_2O_3:Sn films produced by reactive e-beam
deposition. The crystallite size of the analyzed coatings is ~ 50 nm,
which is an order of magnitude larger than the mean free path due to
ionized impurity scattering below the plasma energy. Above $\hbar\omega_p$, the
ions become less efficient scatterers, and in the visible spectral
range it may be that other types of scattering cannot be neglected.
To account for this possibility one may use the provisional formula

$$\chi^{FC} = \frac{i}{\varepsilon_0 \omega \, \rho(\omega)} + i\eta, \tag{42}$$

with ρ taken to represent ionized impurity scattering only. For the
coatings considered here, it appears that $\eta \approx 0.01$. In general, η
can be regarded as a quality factor whose magnitude depends on the
fabrication technique.

As a final ingredient in the theoretical model, we consider
phonon absorption and present in Fig. 26 spectrophotometric reflectance
data for an In_2O_3 film. The spectrum, given on a wavenumber axis,
displays three main peaks at 330, 365 and 412 cm^{-1}. Shoulders on these
peaks, as well as smaller features at higher wavenumbers, are apparent.
All of these structures are believed to correspond to damped transverse
phonon-polaritons, whose effect on the optical data is a reflectance
maximum. The longitudinal polariton coincides with the plasmon resonance, which is a consequence of the free electrons in In_2O_3. In

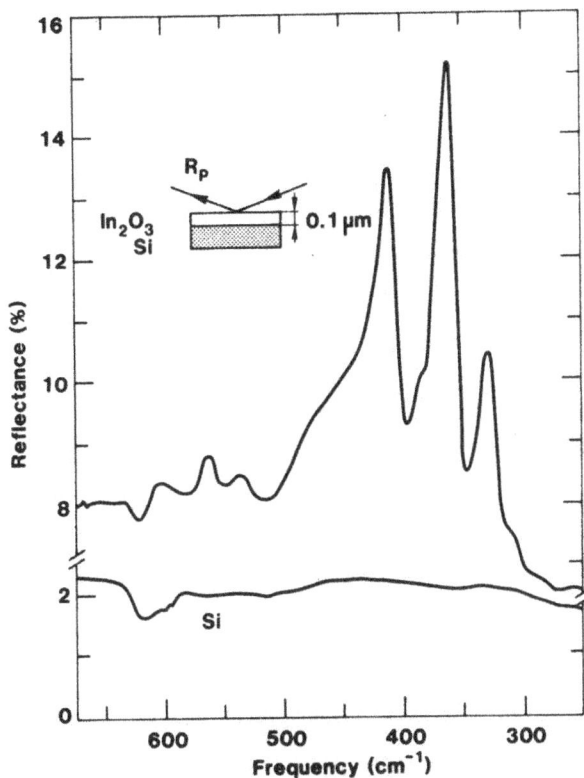

Fig.26 Spectral reflectance of a bare Si substrate (lower curve) and
after coating it with a 0.1 μm thick film of In_2O_3 (upper
curve) whose electron density was ~ 0.5 x 10^{20} cm^{-3}. The
optical measurement used 70° angle of incidence and p-
-polarised light. The experimental configuration is shown
in the inset.

In_2O_3:Sn, the added free electrons effectively screen out the transverse polaritons, so that hardly any observable structure remains in the spectra. According to White and Keramidas,[75] a factor group analysis predicts 16 infrared active modes in In_2O_3. It appears that 9 of these can be identified in the spectra. We may represent χ^{PH} by a sum of three damped Lorentz oscillators (cf. Eq. 13) according to

$$\chi^{PH} = \sum_{j=1}^{3} \frac{\Omega_j^{PH}}{(\omega_j^{PH})^2 - \omega^2 - i\omega\Gamma_j^{PH}} . \tag{43}$$

For $j = 1,2,3$, ω_j^{PH} takes the values 412, 365, and 330 cm^{-1}, Ω_j^{PH} takes the values 330, 600, and 450 cm^{-1}, and Γ_j^{PH} takes the values 5,12, and 16 cm^{-1}.

6.3   Computed Optical Properties For Idealized Window Coatings.

Equations (12), (32), (42) and (43) specify the dielectric function of In_2O_3:Sn, from which one can compute transmittance, reflectance, emittance, etc. In the first calculations reported below we took the In_2O_3:Sn films to be backed by a medium characterized by a constant refractive index equal to 1.5. This simulates a substrate of glass with low Fe_2O_3 content in the visible and near-infrared, or a foil of a polymer such as polyester, polyethylene, etc. In subsequent calculations we represented the substrate by the dielectric function of amorphous SiO_2.[52] Figure 27 shows spectral normal transmittance and reflectance for a $0.2 - \mu m -$ thick film with $10^{20} \leq n_e \leq 3 \times 10^{21}$ cm^{-3}. The bandgap widening, the onset of strong reflectance at a wavelength which scales with n_e, and a gradual disappearance of the phonon-induced structure are all clearly seen. At $n_e = 6 \times 10^{20}$ cm^{-3}, we find high solar transmittance combined with high thermal-infrared reflectance, and we conclude that such coatings are of interest as low emittance window coatings. Setting $n_e = 3 \times 10^{21}$ cm^{-3} we have high luminous transmittance and high near-infrared reflectance, which is the desired property for a solar-control coating. However, coatings with $n_e \geq 10^{21}$ cm^{-3} cannot be produced. Excessive doping yields unwanted absorption due to various Sn-based complexes,[71] so that the theoretical model breaks down for the highest electron densities considered in Fig. 27.

We have now a sufficient background to assess the In$_2$O$_3$:Sn films for use on energy-efficient windows. To this end we need integrated luminous, solar and thermal quantities, which were derived from the spectral data by use of the formulas given in Sec. 3.2. Figure 28 shows T_{lum} and R_{lum} for In$_2$O$_3$:Sn coatings with four thicknesses and $10^{20} \leq n_e \leq 3 \times 10^{21}$ cm^{-3}. At $n_e < 10^{21}$ cm^{-3} we find $75 \leq T_{lum} \leq 85\%$ and $10 \leq R_{lum} \leq 20$ %; the actual magnitudes of these quantities are strongly influenced by optical interference. As the electron density goes up, there is a tendency for R_{lum} to decrease while T_{lum} varies in a more erratic manner. Figure 29 is the counterpart for T_{sol} and R_{sol}. It is seen that T_{sol} decreases and R_{sol} increases when n_e goes up. This can be interpreted as a result of the plasma wavelength which gradually moves into the solar spectrum from the infrared side. At $n_e \leq 5 \times 10^{20}$ cm^{-3} we have $R_{sol} \approx 13$ % irrespective of film thickness. Figure 30 reports on hemispherical thermal emittance. The solid curves were computed for In$_2$O$_3$:Sn films backed by a non-emissive material having a refractive index of 1.5. In order to make valid assessments for a glass window, we also studied a configuration with an intermediate SiO$_2$-layer which was sufficiently thick so as to simulate bulk-like properties. The dotted curves in Fig.30 pertain to this latter configuration. Generally, there is a decrease in the emittance as the electron density and the film thickness increase, which is a direct consequence of the increased number of free electrons. We conclude from Fig.30 that substrate emission is important for $n_e \leq 5 \times 10^{20}$ cm^{-3}, but negligible above this electron density provided that the In$_2$O$_3$:Sn films are thicker than ~ 0.2 μm. These latter films have $E_{therm} \leq 15$ % for $n_e \approx 10^{21}$ cm^{-3}.

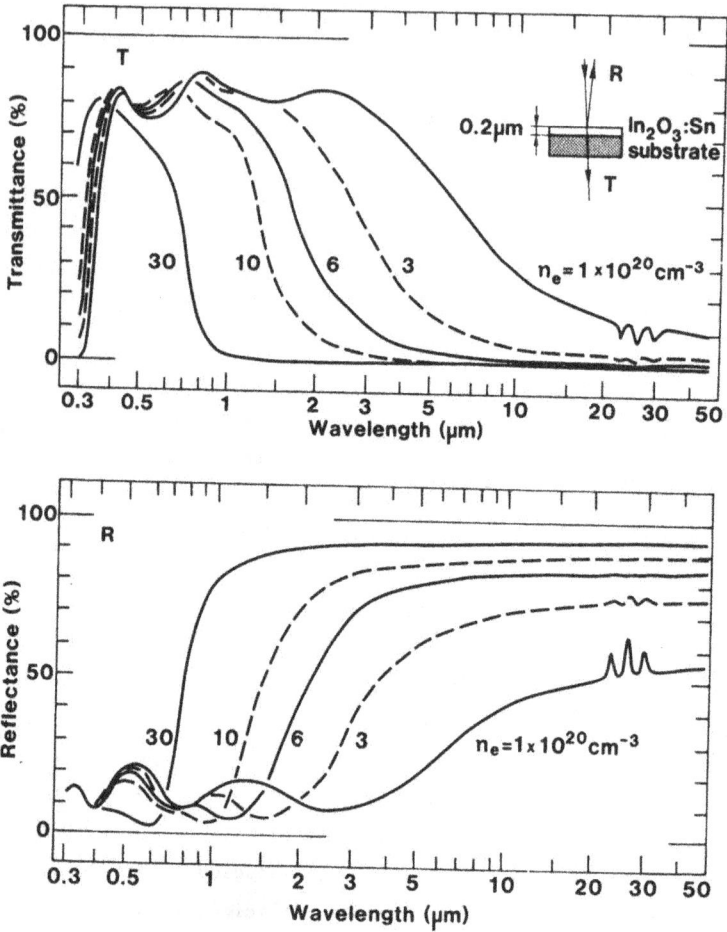

Fig.27 Spectral normal transmittance (upper part) and reflectance
(lower part) as computed from the model for the optical properties of In_2O_3:Sn. The shown values of electron density and
film thickness were used.

Figures 28-30 show that In_2O_3:Sn films can combine high transmittance of luminous and solar radiation with low thermal emittance,
and consequently they are useful for energy efficient windows. A
fully quantitative optimization is outside the scope of this article,
but it is nevertheless possible to draw some general conclusions. To
this end we first fix E_{therm} at a low value. It is not meaningful
to diminish it to the extreme, since conductive and convective heat
transfer cannot be eliminated in an ordinary gas-filled unit.
Specifically, we took E_{therm} to be 15, 20 and 25 %. The dotted curves
in Fig.30 yield corresponding electron densities for the investigated film thicknesses, and the pertinent values of T_{sol} can be
read from Fig. 29. Figure 31 displays solar transmittance as a function of film thickness with E_{therm} as parameter. Requiring E_{therm} =
15% one cannot exceed $T_{sol} \approx 70$ %, which is undesirably low for many
applications. On the other hand, limiting the requirement to E_{therm} =
20 %, one can have $T_{sol} \sim 78$ %. At still larger values of E_{therm} it
is possible to reach a marginally improved solar transmittance.
Optimum performance can be obtained with 0.2 - μm - thick In_2O_3:Sn
films having $4 \lesssim n_e \lesssim 6 \times 10^{20}$ cm^{-3}. Spectral data for such a film
was shown in Fig.27. The optimization of the luminous transmittance
and the colour properties can be regarded as a separate problem,
whose solution rests on antireflection treatments. These aspects
were touched upon earlier.

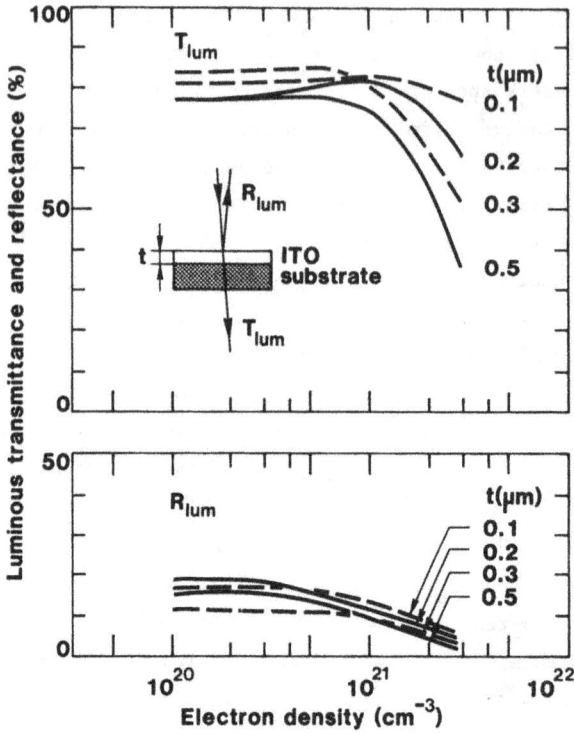

Fig.28 Luminous normal transmittance (upper part) and reflectance
(lower part) versus electron density as computed from the
model for the optical properties of In_2O_3:Sn. Results are
shown for four film thicknesses.

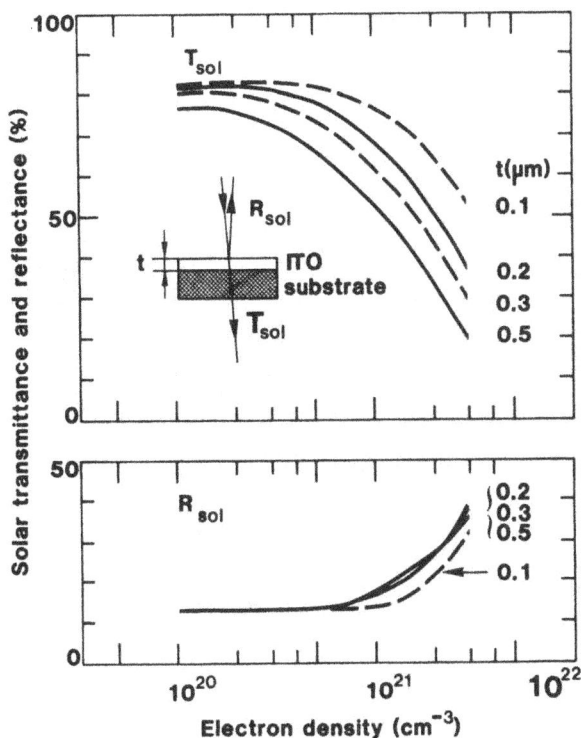

Fig.29 Solar normal transmittance (upper part) and reflectance (lower part) versus electron density as computed from the model for the optical properties of In_2O_3:Sn. Results are shown for four film thicknesses. The curves for R_{sol} corresponding to 0.2, 0.3 and 0.5 µm thick films are almost overlapping.

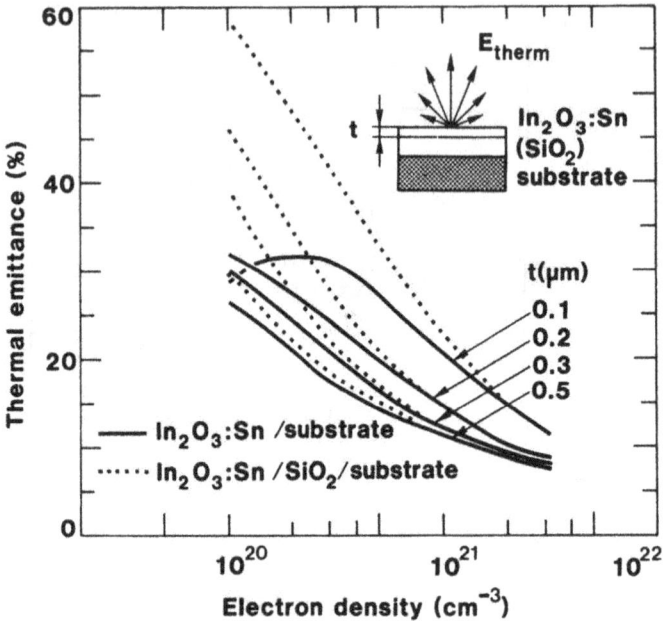

Fig.30 Thermal hemispherical emittance versus electron density as
computed from the model for the optical properties of
In_2O_3:Sn. The substrate is taken to be non-emitting with a
refractive index equal to 1.5 (solid curves) or having properties given by the dielectric function of amorphous SiO_2
(dotted curves). The shown film thicknesses were used.

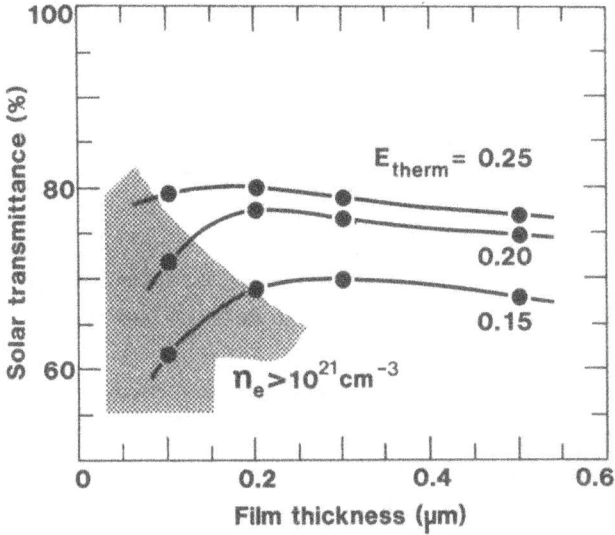

Fig.31 Normal solar transmittance versus film thickness for three
values of hemispherical thermal emittance, as computed from
the model for the optical properties of In_2O_3:Sn and with
the substrate represented by the dielectric function for
amorphous SiO_2. Curves are drawn only to guide the eye.
The shaded area refers to films with $n_e > 10^{21} cm^{-3}$, which
cannot be produced experimentally.

7.   OPTICAL SWITCHING COATINGS FOR SMART WINDOWS

7.1   General Aspects On Electrochromic Coatings

Electrochromism, i.e. the persistent change of the optical
properties under the action of an electric field, is wellknown in
oxides based on W, Mo, Ir, Ni and others, as well as in numerous
organic materials. Earlier work, mainly oriented towards non-emissive
information displays, is surveyed in Refs. 44 and 76-79.

An electrochromic coating for use on a "smart window" (i.e., a
window whose optical performance can be altered in accordance with
dynamic needs for heating and cooling) must encompass several materials. Figure 32 sketches a basic configuration which is convenient
for discussing the operating principles. It comprises two transparent
conducting layers, required for applying the electric field, and
intervening materials serving as active electrochromic layer, ion
conductor, and ion storage. The optical properties of the electrochromic material change under insertion or withdrawal of ions. Either
of the transparent electrodes can be a heavily doped oxide semiconductor or, possibly, a semitransparent metal film. These latter
materials were discussed at length above. The ions needed in the
electrochromic reaction are provided by the ion storage and are injected into, or withdrawn from, the electrochromic layer via the
ion conductor. The detailed design can be of different types: for
devices using liquid electrolytes the electrolyte can serve both as
storage and conductor for the ions, while in all-solid-state devices
the ion conductor can be an appropriate dielectric and the ion
storage can be another electrochromic layer (preferably anodic if the
base electrochromic layer is cathodic, or vice versa). One may also
combine the conductor and storage media into one layer. Further it is
possible to exclude the ion storage medium and instead rely on a re-
plenishment of H^+ ions, originating from the dissociation of water

molecules diffusing in from an ambience with controlled humidity; obviously, this requires a substantial atomic permeability of the outer
electrode.

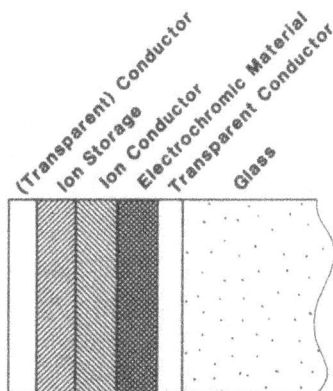

Fig.32 Basic design of an electrochromic coating for "smart windows".

The radiative properties of a "smart window" can be varied
gradually and reversibly between different states. The nature of this
switching can be different depending on the basic requirement of the
window, i.e., whether the main goal is to achieve control of the energy
flowing through the window aperture, or if daylighting and glare
control are of prime importance. Under certain conditions it may also
be possible to have variable thermal insulation. In Fig. 33 we
illustrate idealized cases for energy control. For this application
it is important to remember that the solar spectrum extends over the
0.3-3-μm range, whereas the eye is sensitive only in the 0.4-0.7-μm
interval. The pertinent spectra are shown by the shaded areas in
Fig.33; they were illustrated also in Figs.7 and 8. Almost 50 % of
the total solar energy comes as infrared radiation, and hence it is
possible - at least in principle - to change the energy throughput

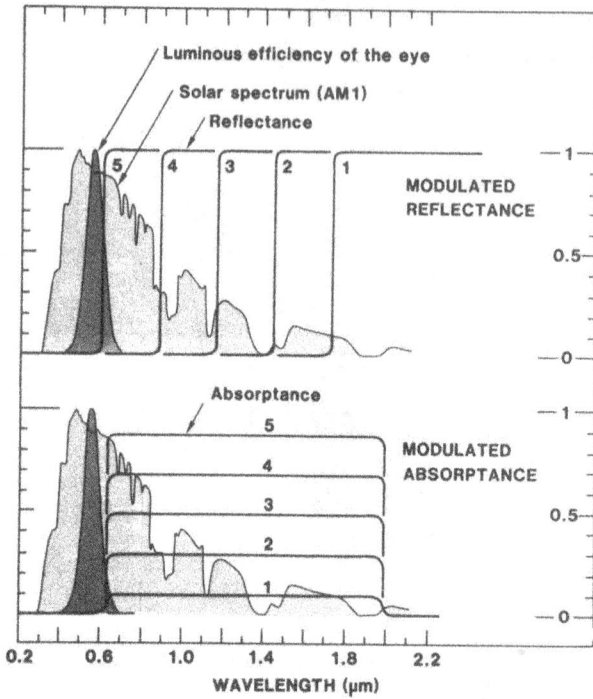

Fig.33 Optical properties of idealized switching coatings for controlled energy throughput in "smart windows". The performance with modulated reflectance and modulated absorptance as working principles are shown (the spectral position of the absorption band is somewhat arbitrary). Shaded areas denote the luminous efficiency of the eye and a typical solar irradiance spectrum.

within rather wide limits without affecting the luminous transparency. The changed infrared transmittance can be achieved in two ways: by modulating the infrared reflectance as indicated in the upper part of Fig.33, or by modulating the infrared absorptance as indicated in the lower part. A gradual decrease in the transmitted energy is obtained as the reflectance edge moves towards shorter wavelength, or as the absorptance goes up. Both types of modulation can occur in electrochromic WO_3 as we will see shortly: the properties of crystalline WO_3 films are conveniently treated in terms of modulated reflectance, whereas amorphous WO_3 displays modulated absorptance. The thermal properties of the "smart windows" are governed by their reflectance in the 3-100-μm range. The all-solid-state concept, which is preferable in a window, includes an extended outer transparent conductor which provides low thermal emittance. In many cases this is the desired property. If instead a high thermal emittance is required, this can be obtained by applying a suitable external coating to the electrochromic multilayer structure.

7.2   Theoretical And Experimental Results For Electrochromic
       Coatings

We first consider crystalline electrochromic WO_3. The reversible colouring and bleaching is generally believed[76] to be the result of a double injection of positive ions (M^+) and electrons (e^-) according to the overall reaction

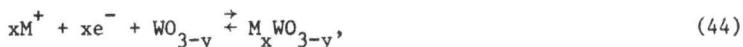

$$xM^+ + xe^- + WO_{3-y} \overset{\rightarrow}{\leftarrow} M_x WO_{3-y},$$
(44)

with $0 < x \leq 0.5$ and $y \leq 0.03$. The optical properties can be understood at least qualitatively from the formation of an electron gas. Since each injected ion is associated with one electron, it is possible to compute the limiting performance of an electrochromic window coating including crystalline WO_3 by considering only the inevitable ionized impurity scattering. We further disregard the role of all layers except the electrochromic one. With this premise one can apply precisely the same theory as the one used to treat the free-electron properties of In_2O_3:Sn in Sec. 6.2. We set $\varepsilon_\infty = 4.8$, $Z = 1$ and $m_c^* = 0.5$ m. Figure 34, reproduced from Ref. 80, shows some typical data for computed spectral reflectance. The results pertain to a 0.2 μm thick electrochromic WO_3 film with $10^{21} \leq n_e \leq 10^{22}$ cm^{-3}, backed by a glass substrate whose refractive index was put to 1.53. With increasing n_e, the onset of strong reflectance is shifted towards

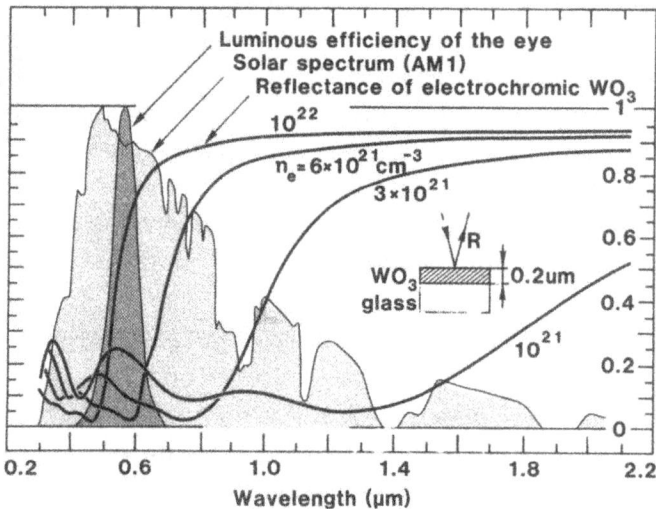

Fig.34 Computed spectral reflectance for crystalline electrochromic WO_3 films with different electron densities. Solar and luminous spectra are shown.

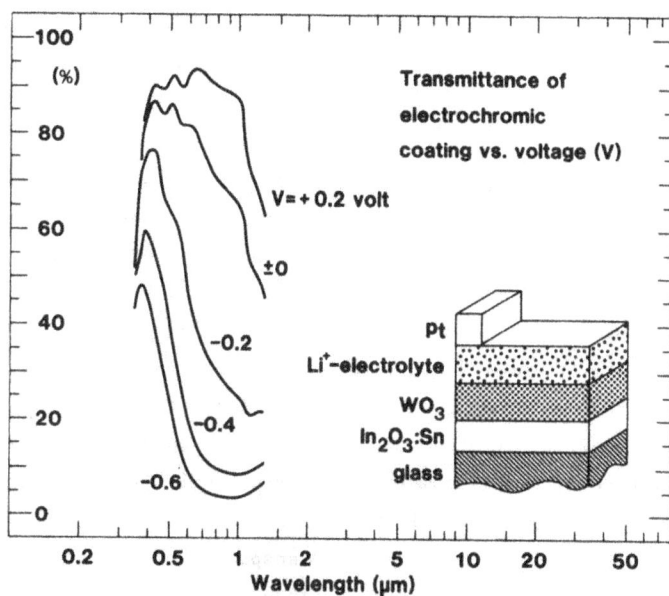

Fig.35 Normal spectral transmittance as a function of voltage for an electrochromic configuration with a glass substrate, coated with In_2O_3:Sn and WO_3, immersed in a liquid Li^+-electrolyte.

shorter wavelength and the transition becomes progressively sharper. This follows from an altered plasma wavelength. The results may be appreciated by comparing them with characteristic spectra for sunlight and for the luminous efficiency of the eye (shaded areas). The potential for varying the reflection and transmission of infrared sunlight is apparent; quantitative data are given in Ref. 80. Experimental data on crystalline WO_3 have been published in Refs. 81 and 82.

We now turn to amorphous WO_3 films and discuss some recent experiments.[83,84] Also in this case, the colouring can be understood from the double injection model in Eq. (44). The first design comprises a glass substrate, made electrically conducting by an In_2O_3:Sn film, coated with a 0.3-µm-thick electrochromic layer prepared by reactive e-beam deposition of WO_3. This sample was mounted so that it formed the working electrode in a transparent electrochemical cell containing a small Pt counter-electrode and a liquid electrolyte of 1 M $LiClO_4$ in $C_4H_6O_3$. The sample was connected in a standard three-electrode potentiostatic configuration with a saturated calomel reference electrode. The whole cell was put in the sample compartment of a double-beam spectrophotometer. An analogous cell, containing electrolyte but no glass sample, was put in the reference beam. Transmittance was recorded in the 0.35-1.3-µm-range. In the combined electrical and optical measurements we first applied a voltage between the working electrode and reference electrode until maximum transmittance was observed. The WO_3 layers were then coloured, by gradually forming Li_xWO_3, through a reversal of the voltage. The colouring was halted at intermediate levels in the colour-bleach cycle and corresponding transmittance spectra and voltages (V; measured between the working electrode and the reference electrode) were recorded. Figure 35 shows a typical set of curves. Large changes of the optical properties are seen to occur as a result of a low voltage. The devices exhibit open-circuit memory. The integrated solar transmittance obtained from Fig.35 lies between 86 and 12 %.

All-solid-state devices are preferrable for practical purposes. Here we describe some work of ours on

this subject.[84] The specimen consists of - in order -
In_2O_3:Sn, 0.15 μm of WO_3 produced by evaporation in the presence of O_2,
0.05 to 0.1 μm of MgF_2 produced by evaporation in the presence of
water vapour, and Au. The top electrode is ~ 15 nm thick and hence
semitransparent at short wavelengths. As a voltage (V) is applied
between the In_2O_3:Sn and Au films, the transmittance is changed. This
may be caused by dissociation of H_2O molecules and subsequent formation
of H_xWO_3. Figure 36 shows some initial data on variable transmittance
and reflectance in the 0.35-2.2-μm interval. The corresponding integrated solar transmittance varies between 25 and 3 % when the voltage
is altered. The durability of these coatings is not yet satisfactory
for practical use, and storage in dry air tends to deteriorate the
electrochromic property.

Generally speaking, durability may be a problem for WO_3-based
devices, and there is a need to have alternative electrochromic
materials. One such alternative is hydrated nickel oxide.[85,86] The
inset of Fig. 37 shows a sample configuration recently investigated by
us.[86] A glass plate with an In_2O_3:Sn film was coated with NiO_x by
reactive rf-magnetron sputtering. The samples were bleached and
coloured during immersion in 1M KOH. It appeared that the electrochromic reaction[87] could be written, schematically, as

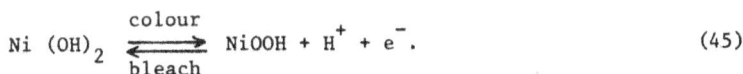

$$Ni\,(OH)_2 \underset{\text{bleach}}{\overset{\text{colour}}{\rightleftharpoons}} NiOOH + H^+ + e^-. \tag{45}$$

Thus colouring is associated with hydrogen extraction (rather than
hydrogen insertion as for WO_3). Figure 37 shows normal transmittance T
and near-normal reflectance R in the 0.35 - 2.5-μm interval. The
initial bleached film had T ~ 75 % and R ~ 15 % in the mid-luminous
range (λ ~0.55 μm), as indicated by the dotted curves. The performance
remained almost unchanged after ~ 10^4 colour-bleach cycles, as apparent
from the solid curves. Typical properties of a heavily coloured sample
are given by the dashed curves, which were obtained for a sample from
which 150 mC had been extracted. Now T ~ 12 % and R ~ 6 % at

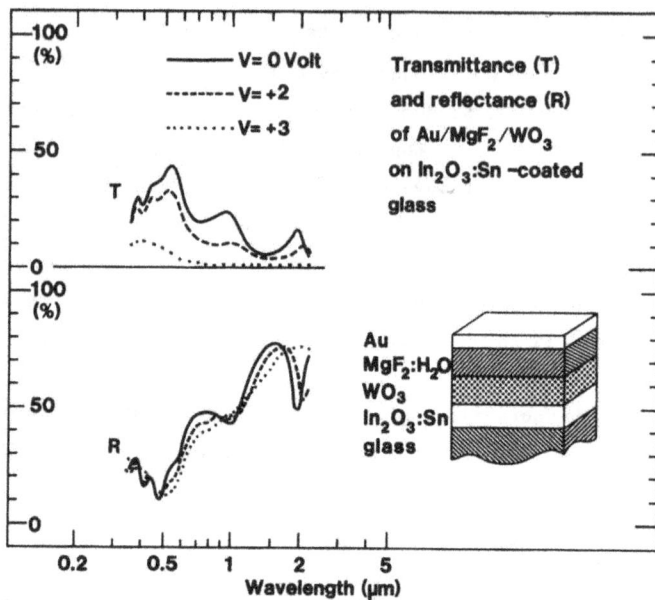

Fig.36   Normal spectral transmittance and reflectance as a function of
voltage for a glass substrate covered with a multilayer stack
of In_2O_3:Sn, WO_3, MgF_2 and semitransparent Au.

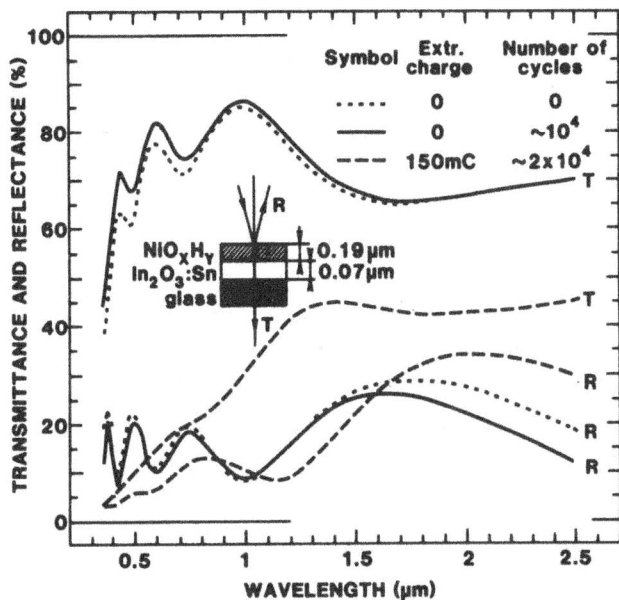

Fig.37 Spectral transmittance at normal incidence and reflectance at
        10° angle of incidence for an electrochromic NiO_x-based layer
        on In_2O_3:Sn-coated glass. The curves refer to different
        magnitudes of extracted charge and number of color-bleach
        cycles. The sample configuration of sketched in the inset.

$\lambda \sim 0.55$ µm. The large optical modulation as well as the extreme durability of the coatings are noteworthy. The window performance of the coatings was assessed by evaluating the integrated solar and luminous transmittance by the formulas given in Sec.3.2. Figure 38 reports T_{sol} and T_{lum} versus extracted charge as obtained from the spectral normal transmittance data. A monotonic decrease of the integrated transmittance is found with T_{sol} going from $\sim$ 75 % to $\sim$ 20 % and T_{lum} going from $\sim$ 75 % to $\leq$ 10 % when the charge extraction is increased to 200 mC. The rate of change for T_{sol} and T_{lum} is largest for small amounts of extracted charge.

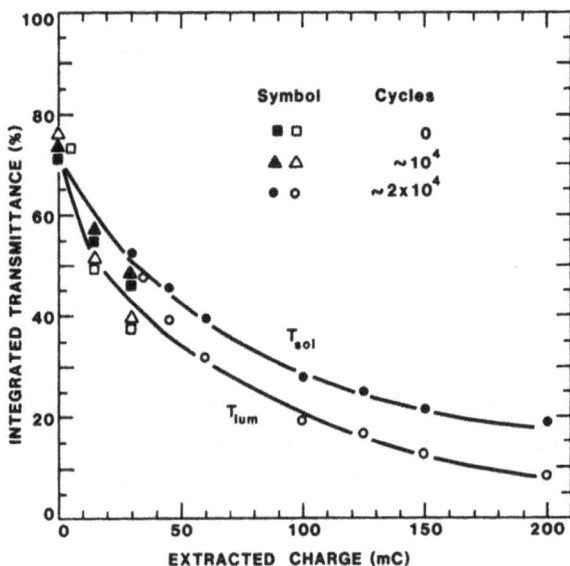

Fig.38 Integrated solar (filled symbols) and luminous (open symbols) transmittance for electrochromic NiO_x-based coatings according to the inset of Fig.37. Squares, triangles and circles refer to 0, $\sim 10^4$, and $\sim 2 \times 10^4$ color-bleach cycles, respectively. The curves are drawn only as a guide to the eye. A further decrease of the transmittance was observed for 500 mC extracted charge.

7.3   Thermochromic Coatings

We end this article with a few initial results from our laboratory on thermochromic VO_2 coatings.   Some earlier work on thermochromic coatings is surveyed in Ref. 88.   It is wellknown that VO_2
undergoes a reversible metal-to-semiconductor transition when 68°C is
passed.   The transition temperature can be depressed by alloying as
well as by several other means.   Figure 39 shows spectral transmittance in the 0.35-2.5-µm range for a 0.2 µm thick VO_2 film above
and below the transition, as taken from Ref. 89.   The sample was
prepared by e-beam deposition of V metal at a substrate temperature
of 310°C and at a rate of 0.2 nms^{-1}, followed by an anneal in 0.2 Torr
of air at 400°C for 30h.   It is seen that the solar transmittance is
considerably higher at a low temperature than at a high temperature,
which is the desired property for temperature control.   Thus it
appears to be feasible to produce interesting window coatings based
on VO_2.

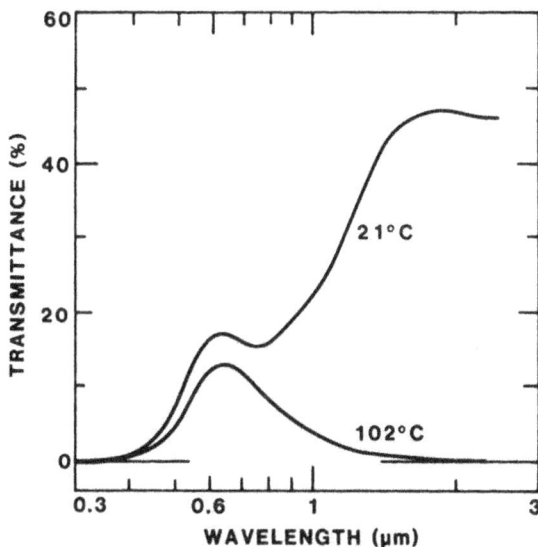

Fig.39   Spectral transmittance of a thermochromic VO_2 coating at the
         indicated temperatures.

ACKNOWLEDGEMENT

This paper is based, to a large extent, on work carried out by the author together with numerous colleagues and students. In particular, I wish to acknowledge the important contributions by T.S. Eriksson, I. Hamberg, A. Hjortsberg, G.A. Niklasson, G.B. Smith, and J.S.E.M. Svensson. Our work was supported by grants from the Swedish Natural Science Research Council, the National Swedish Board for Technical Developments and other funding agencies.

REFERENCES

1. Granqvist, C.G., Appl. Opt. <u>20</u>, 2606 (1981).

2. Lampert, C.M., Proc. SPIE <u>324</u>, 1 (1982).

3. Granqvist, C.G., Proc. SPIE <u>401</u>, 330 (1983).

4. Lampert, C.M., Opt. Engr. <u>23</u>, 92 (1984).

5. Granqvist, C.G., The Physics Teacher <u>22</u>, 372 (1984).

6. Granqvist, C.G., Physica Scripta <u>32</u>, 401 (1985).

7. Heavens, O.S., "Optical Properties of Thin Solid Films" (Butterworth, London, 1955).

8. Born, M., and Wolf, E., "Principles of Optics", 6th edition (Pergamon Oxford, 1980).

9. Mahan, G.D., "Many Particle Physics" (Plenum, New York, 1981).

10. Landauer, R., AIP Conf. Proc. <u>40</u>, 2 (1978).

11. Sievers, A.J., in "Solar Energy Conversion: Solid State Physics Aspects," Vol. 31 of "Topics in Applied Physics", edited by Seraphin, B.O., p. 57 (Springer, Berlin, 1979).

12. Granqvist, C.G., J. Phys. (Paris) <u>42</u>, C1-247 (1981).

13. Aspnes, D.E., Am. J. Phys. <u>50</u>, 704 (1982). Thin Solid Films <u>89</u>, 249 (1982).

14. Niklasson, G.A. and Granqvist, C.G., in "Contributions of
     Clusters Physics to Materials Science and Technology",
     edited by Davenas, J. and Rabette P., (Martinus Nijhoff,
     Dordrecht, 1986).

15. Kerker, M., "The Scattering of Light and Other Electromag-
     netic Radiation" (Academic, New York, 1969).

16. van de Hulst, H.C., "Light Scattering by Small Particles"
     (Dover, New York, 1981).

17. Bohren, C.F., and Huffman, D.R., "Absorption and Scattering
     by Small Particles"(Wiley, New York, 1983).

18. Garnett, J.C.M., Philos. Trans. Roy. Soc. London 203, 385
     (1904); 205, 237 (1906).

19. Bruggeman, D.A.G., Ann.Phys. (Leipzig) 24, 636 (1935).

20. Niklasson, G.A., and Granqvist, C.G., J. Appl. Phys.55, 3382
     (1984).

21. Bergman, D.J., in "Macroscopic Properties of Disordered Media",
     Vol. 154, in "Lecture Notes in Physics", edited by
     Burridge, R., Childress C., and Papanicolaou G., (Springer,
     Berlin, 1982); p.10.

22. Bergman D.J., Ann.Phys. (NY) 138, 78 (1982).

23. Hjortsberg A., Appl. Opt. 20, 1254 (1981).

24. Palik, E.D., (editor), "Handbook of Optical Constants"
     (Academic, New York, 1985).

25. Wyszecki, G., and Stiles, W.S., in "Color Science", 2nd edition (Wiley, New York, 1982) p.256.

26. Thekaekara, M.P., in "Solar Energy Engineering", edited by Sayigh, A.A.M., (Academic, New York, 1977), p. 37.

27. Ritter, E., in "Progr. Electro-Optics", edited by Camatini, E., (Plenum, New York, 1975); p.181.

28. Berman, S.M., and Silverstein, S.D., editors, AIP Conf. Proc. 25, 286 (1975).

29. Fan, J.C.C., and Bachner, F.J., Appl. Opt. 15, 1012 (1976).

30. Vossen, J.L., Phys. Thin Films 9, 1 (1977).

31. Haacke, G., Ann. Rev. Mater. Sci. 7, 73 (1977).

32. Lampert, C.M., Solar Energy Mater. 6, 1 (1981).

33. Lampert, C.M., Ind. Chem. Prod.Res.Rev. 21, 612 (1982).

34. Köstlin, H., Festkörperprobleme 22, 229 (1982).

35. Haacke, G., Proc. SPIE 324, 10 (1982).

36. Manifacier, J.C., Thin Solid Films 90, 297 (1982).

37. Jarzebski, Z.M., Phys. Stat. Sol. A 71, 13 (1982).

38. Frank, G., Kauer, E., Köstlin, H., and Schmitte, F.J., Proc. SPIE 324, 58 (1982).

39. Lampert, C.M., Energy Res. 7, 359 (1983).

40.   Berning, P.H., Appl. Opt. $\underline{22}$, 4127 (1983).

41.   Chopra, K.L., Major, S., and Pandya, D.K., Thin Solid Films
      $\underline{102}$,1 (1983).

42.   Valkonen, E., Karlsson, B., and Ribbing, C.-G., Solar Energy
      $\underline{32}$, 211 (1984).

43.   Granqvist, C.G., Hamberg, I., and Svensson, J.S.E.M., Ind.
      Chem. Prod. Res. Dev. $\underline{24}$, 93 (1985).

44.   Lampert, C.M., Solar Energy Mater. $\underline{11}$, 1 (1984).

45.   Smith, G.B., Niklasson, G.A., Svensson, J.S.E.M., and
      Granqvist, C.G., J. Appl. Phys. $\underline{59}$, 571 (1986).

46.   Chopra, K.L., "Thin Film Phenomena" (McGraw-Hill,
      New York, 1969).

47.   Neugebauer, C.A., in "Handbook of Thin Film Technology",
      edited by Maissel, L.I., and Glang, R., (McGraw-Hill,
      New York, 1970) Ch.8.

48.   Harding, G.L., Solar Energy Mater. $\underline{12}$, 169 (1985).

49.   Harding, G.L., Hamberg., I., and Granqvist, C.G., Solar Energy
      Mater. $\underline{12}$, 187 (1985).

50.   Gugger, H., Jurich. M., Swalen, J.D., and Sievers, A.J.,
      Phys.Rev. B$\underline{30}$, 4189 (1984).

51.   Norrman, S., Andersson, T., Granqvist, C.G., and Hunderi O.,
      Phys. Rev. B$\underline{18}$, 674 (1978).

52.   Eriksson, T.S., Jiang, S.-J., and Granqvist, C.G., Appl. Opt.
       24, 745 (1985).

53.   Hamberg, I., Hjortsberg, A., and Granqvist, C.G., Appl. Phys.
       Lett. 40, 362 (1982); Proc. SPIE 324, 31 (1982).

54.   Hamberg, I., and Granqvist, C.G., Appl. Opt. 22, 609 (1983).

55.   Hamberg, I., and Granqvist, C.G., Thin Solid Films 105, L 83
       (1983).

56.   Hamberg, I., and Granqvist, C.G., Proc. SPIE 428, 2 (1983);
       Appl. Phys. Lett. 44, 721 (1984).

57.   Hamberg, I., Granqvist, C.G., Berggren, K.-F., Sernelius, B.E.,
       and Engström, L., Phys. Rev. B 30, 3240 (1984); Proc. SPIE
       502, 2 (1984).

58.   Hamberg, I., and Granqvist, C.G., Appl. Opt. 24, 1815 (1985);
       Proc. SPIE 562, 137 (1985).

59.   Hamberg, I., and Granqvist, C.G., J. Appl. Phys., to be published.

60.   Howson, R.P., Ridge, M.I., and Suzuki, K., Proc. SPIE 428, 14
       (1983), and references therein.

61.   Jiang, S.-J., and Granqvist. C.G., Proc. SPIE 562, 129 (1985).

62.   Hardy, A.C., "Handbook of Colorimetry" (Massachusetts
       Institute of Technology, Cambridge, 1963).

63.   Grum, F., and Bartleson, C.J., (editors), "Optical Radiation
       Measurements", Vol. 2, Color Measurement (Academic, New York,
       1980).

300

64.   MacAdam, D.L., "Color Measurement: Theme and Variations
        (Springer, Berlin, 1981).

65.   Mott, N.F., "Metal-Insulator Transitions" (Taylor and Francis,
        London, 1974).

66.   Burstein, E., Phys. Rev. 93, 632 (1954); Moss, T.S., Proc.
        Phys. Soc. London B 67, 775 (1954).

67.   Kurik, M.V., Phys. Stat.Sol. A8, 9 (1971).

68.   Redfied, D., Phys. Rev. 130, 914, 916 (1963).

69.   Skettrup, T., Phys. Rev. B18, 2622 (1978).

70.   Gerlach, E., and Grosse, P., Festkörperprobleme 17, 157 (1977).

71.   Frank, G., and Köstlin, H., Appl. Phys. A 27, 197 (1982).

72.   Lindhard, J., Kgl. Danske Videnskab. Selskab, Mat.-Fys. Medd.
        28. No. 8 (1954).

73.   Hubbard, J., Proc. Roy. Soc. London, Ser. A. 243, 336 (1957).

74.   Köstlin, H., Jost, R., and Lems, W., Phys. Stat. Sol. A 29, 87
        (1975).

75.   White, W.B., and Keramidas, V.G., Spectrochim. Acta 28A, 501
        (1972).

76.   Faughnan, B.W., and Crandall, R.S., in "Display Devices",
        edited by Pankove, J.I., "Topics in Applied Physics", Vol. 40
        (Springer-Verlag, Berlin, Heidelberg, 1980), p.181.

77. Beni, G., and Shay, J.L., Adv. Image Pickup and Display $\underline{5}$, 83 (1982).

78. Dautremont-Smith, W.C., Displays, January 1982, p.3; April 1982, p.72.

79. Svensson, J.S.E.M., and Granqvist, C.G., Proc. SPIE $\underline{502}$, 30 (1984); Solar Energy Mater. $\underline{12}$, 391 (1985).

80. Svensson, J.S.E.M., and Granqvist, C.G., Appl. Phys. Lett. $\underline{45}$, 828 (1984).

81. Goldner, R.B., Medelsohn, D.H., Alexander, J., Hendersson,N.R., Fitzpatrick, D., Haas, T.E., Sample, H.H., Rauh, R.D., Parker, M.A., and Rose, T.L., Appl. Phys. Lett. $\underline{43}$, 1039 (1983), Goldner, R.B., and Rauh, R.D., Proc. SPIE $\underline{428}$, 38 (1983); Solar Energy Mater. $\underline{11}$, 177 (1984).

82. Goldner, R.B., Norton, P., Wong, K., Foley, G., Goldner, E.L., Seward, G., and Chapman, R., Appl. Phys. Lett. $\underline{47}$, 536 (1985).

83. Svensson, J.S.E.M., and Granqvist, C.G., Solar Energy Mater. $\underline{11}$, 29 (1984).

84. Svensson, J.S.E.M., and Granqvist, C.G., Thin Solid Films $\underline{126}$, 31 (1985).

85. Lampert, C.M., Omstead, T.R., and Yu, P.C., Solar Energy Mater. $\underline{14}$, 161 (1986).

86. Svensson, J.S.E.M., and Granqvist, C.G., Appl. Phys. Lett., to be published.

87.   Oliva, P., Leonardi, J., Laurent, J.F., Delmas, C.,
        Braconnier, J.J., Figlarz, M., Fievet, F., and de Guibert, A.,
        J. Power Sources $\underline{8}$, 229 (1982).

88.   Jorgenson, G., Lawrence Berkeley Laboratory Report LBL –
        18299 (August 1984), unpublished.

89.   Babulanam, S.M., Eriksson, T.S., Niklasson, G.A., and Granqvist,
        C.G., Proc. SPIE $\underline{692}$, to be published.

Solar Cell Modelling And Diffusion Length Determination

Klaus G.Bücher

Institut f. Werkstoffe der Elektrotechnik
Technische Universität Berlin
Jebensstr. 1, D-1000 Berlin 12, W-Germany

1. Summary

A model that describes the contribution of the emitter region, space charge
region and base region to the current density collected in a solar cell of
finite thickness (not semiinfinite) is presented. Analytical solutions are
given, so the different terms are open to physical interpretation. As the performance of a solar cell has to be optimized, we varied the parameters descri-
bing the solar cell. The cell depth, base doping rate and the diffusion length
in the base strongly influence spectral response and white light performance.
A new method to determine the diffusion length from spectral response measurements is given. We use logarithmic derivatives of $S(\lambda)$, therefore we need only
relative measurements of spectral response, reflectivity versus λ. The evaluation is less sensitive to measurements errors and unknown cell parameters
(emitter thickness) than previous schemes where very large uncertainties
could arise.

2. Solar Cell Model

A cross-section of solar cells under consideration is given in fig. 1. Photons
pass the antireflective coating from the left but for the reflected fraction
$R(\lambda)$. They are absorbed in the semiconductor either in the emitter, the space
charge region (scr) or the base region. Assuming a finite thickness of the
cell, a certain amount of photons reaches the back contact and is reflected
back into the base (metallic reflectivity r). The carriers which are not
collected either recombine in the volume or at the surface of the cell (surface recombination velocity s). Further cell parameters are the dopant density
N_D, N_A and the diffusion length L_p, L_n in the emitter and the base region
respectively.

This model of a solar cell has been analytically solved /1/, including the
effects of the emitter and space charge region and the effects of a finite
thickness of the cell.

The transport equation

$$(1) \qquad \frac{d^2N}{dx^2} - \frac{N}{L^2} = - \frac{\eta\alpha\lambda}{Dhc} \cdot E(x,\lambda)$$

describes the minority carrier excess concentration N (Δp or Δn) in the
emitter resp. base. L is the diffusion length, D the diffusion coefficient,
η the quantum efficiency of electron-hole pair generation (assumed to be equal
unity here). Including internal reflection, the power density penetrating into

the material is

$$E(x;\lambda) = E(x = -d;\lambda).$$

(2)
$$\cdot \ [\exp(-\alpha(d+x)) + r \cdot \exp(-\alpha(2D+d-x))\ ]$$

To solve the differential equation, one has to subject it to the following boundary conditions:

   i) All carriers diffusing to the front surface recombine there.

(3a)
$$x = -d : \ qs \cdot \Delta p\Big|_{x=-d} = -q \cdot D_p \cdot \frac{d\Delta p}{dx}\Big|_{x=-d}$$

   ii) No excess carrier concentration is present at the back contact (ohmic contact).

(3b)
$$x = +D : \ \Delta n\Big|_{x=+D} = 0$$

   iii) At the margins of the space charge region, the carrier concentrations are given by the Boltzmann factor $\Delta N = N_o \ (\exp \ (V/V_T \ ) - 1 \ ).$

(3c)
$$x = -w_n : \ \Delta p\Big|_{x=-w_n} = p_o [\exp(\frac{V}{V_T})-1] \cdot \frac{A}{C} \cdot (\frac{n_{i,eff}}{n_i})^2$$

(3d)
$$x = +w_p : \ \Delta n\Big|_{x=+w_p} = n_o [\exp(\frac{V}{V_T})-1] \cdot \frac{B}{C} \ ,$$

Equations (3c,3d) take account of high majority carrier concentrations using the generalized Boltzmann equilibrium between the two margins of the space charge region with

(4)
$$A = 1 + (\frac{n_{i,eff}}{p_{po}})^2 \cdot \exp(\frac{V}{V_T}) \ ,$$

(5)
$$B = 1 + (\frac{n_i}{n_{no}})^2 \cdot \exp(\frac{V}{V_T}) \ \ \text{and}$$

(6)
$$C = 1 - \exp(- \ 2(V_D - V)/V_T) \ .$$

High doping effects in the emitter, leading to bandgap narrowing and a higher intrinsic carrier concentration n_{ieff} have been allowed for the emitter region in equation (3c,4),

(7)
$$n_{i,eff} = n_i \cdot \exp(\frac{\Delta W}{2kT}) \cdot \sqrt{F_{1/2}(n_{fn})/\exp(n_{fn})}$$

where $F_{1/2}$ equals $(2/\sqrt{\pi})$ times the Fermi-Dirac integral with the normalized energy difference n_{fn} between conduction band and Fermi level as argument.

(8)
$$n_{fn} = \frac{W_C - W_F}{kT}$$

The solution of eq.(2) yields the concentration of excess minority carriers in the base and emitter. Its slopes at the boundary of the scr give the diffusion currents of minority carriers. They are both summed up to form a dark current and a photogenerated current. This photocurrent is obviously still a spectral one since eq.(2) has been formulated for monochromatic irradiance.

The dark current originating from the space charge region is calculated using the model of Sah, Noyce, Shockley and Hall/2,3/. The excess recombination rate R_{scr} is

(9)
$$R_{scr} = \frac{n \ p - n_i^2}{\tau_{p_e}(n+n_i) + \tau_{n_e}(p+n_i)}$$

We have assumed an exponential relation between the carrier concentrations n,p

and the potential which has been chosen for an abrupt junction. The photocurrent inside the scr is calculated from the optical generation rate $g(x,\lambda)$
and then integrated over the entire wavelength range $\Delta\lambda$.

These results yield the total current density of a solar cell:

$$j((V) = (j_{no} + j_{po}) \; [\exp \; (\frac{V}{V_T}) - 1] \quad - \int_{\Delta\lambda} [j_{nph} \; (\lambda) + j_{pph}(\lambda)] \; d\lambda$$
$$(10) \qquad\qquad\qquad + q \int_w \; [R_{SCR}(x) - \int_{\Delta\lambda} g(x,\lambda)d\lambda]dx \quad .$$

3. Influence of Cell Parameters on Cell Performance

Using the model presented in section 2, the influence of solar cell properties
on spectral response and white ligth I-V characteristics has been determined.
Unless otherwise mentioned, the following standard values of crystalline cells
have been chosen:

$D= 400$ μm, $N_A = 1.5 \; 10^{16} cm^{-3}$, $L_n = 100$ μm, $r=0.6$, $D_n = 20 cm^2 s^{-1}$

$d = 0.3 \mu m$, $N_D = 10^{19} cm^{-3}$, $L_p = 5 \mu m$, $s = 10^4$ cm/s , $D_p = 3 cm^2 s^{-1}$.

The base thickness D, diffusion length L of electrons in the base and the dopant density of the base strongly influence the response, while other parameters are of minor importance only.

Fig. 2 shows the contributions of emitter, scr and base to the total spectral
response $S = j/E$ of a solar cell. In the visible part of the spectrum significant contributions are generated in the emitter and scr (even at wavelengths as
high as 800 nm. As the absorption coefficient is large in this part of the
spectrum, more photons are absorbed in the emitter and scr leading to a higher
current originating there. Nevertheless only little changes in spectral
response are found on the red shoulder of the curve so this data can be used
to determine the diffusion length in the base.

The effect of this important parameter, L_n , is shown in fig. 3. As L_n increases
more carriers can diffuse to the junction and be collected there, so the
spectral response increases and is shifted to longer wavelength.

If the base thickness is reduced it contributes less carriers as only a part
of the photons is absorbed. The maximum of the spectral response is lowered
and shifted, approaching the stand alone contributions of emitter and scr (fig.4).

After integration of the current densities with respect to the spectral
distribution of AM1 irradiation, fig. 5 shows the influence of the parameters
on white ligth I-V characteristics. The current density collected, i.e. fill
factor and efficiency, strongly depends on the base diffusion length as
important characterising parameter of the cell.

4. Diffusion Length Measurement

As can be seen from fig.5 it is important to know and improve the diffusion
length. In a sligthly simplified description as in the previous sections i.e.
neglection of the scr and provision of a semiinfinite base we find for j:

$$j = \left[\frac{qD_n}{L_n} n_o + \frac{qD_p}{L_p} p_o \frac{\frac{D_p}{L_p} \sinh\frac{d}{L_p} + s \cosh\frac{d}{L_p}}{\frac{D_p}{L_p} \cosh\frac{d}{L_p} + s \sinh\frac{d}{L_p}} \right] \left[e^{\frac{V}{U_T}} - 1 \right] + \frac{qL_n I_o e^{-\alpha d}}{1 + \alpha L_n} + \frac{qL_p I_o e^{-\alpha d}}{1 + \alpha^2 L_p^2} \left[\frac{-(\alpha D_p + s) e^{\alpha d} + \frac{D_p}{L_p} \sinh\frac{d}{L_p} + s \cosh\frac{d}{L_p}}{\frac{D_p}{L_p} \cosh\frac{d}{L_p} + s \sinh\frac{d}{L_p}} + \alpha L_p \right]$$

Considering actual dimensions of a cell: $d/L_p \to 0$, $\alpha D \gg 1$, $D/L_n \gg 1$, we find

(12)
$$S_{base}(\lambda) = \frac{q\lambda}{hc} \cdot \alpha(\lambda) \cdot e^{-\alpha(\lambda)d} \cdot \frac{1}{L_n^{-1} + \alpha(\lambda)}$$

which directly shows the influence of diffusion length on carrier collection. Reflection, emitter/scr contribution, contact grid shading are still neglected.

There are several methods to extract L_n from eq. (12) which shall be discussed first. The last part of this sections reviews the approximations leading to eq.(12).

If the absolute spectral response has been measured for one or several values of λ at the red shoulder, eq. (12) can be solved for L

(13)
$$L_n = \frac{1}{\alpha(\lambda) \; [S(\lambda)^{-1} \frac{q\lambda}{hc} \exp(-\alpha(\lambda)d) \; -1]}$$

If several data points have been determined, approximation routines, e.g.least squares fits or a graphic method /4/ can be used. Nevertheless there are some major drawbacks using this method for silicon devices:
  i) One needs a measurement of the absolute value of $S(\lambda)$.
 ii) The emitter thickness d is generally not known exactly.
iii) One subtracts two quantities of approximately the same size (~ 1) in
      the denominator, leading to large errors after taking the inverse of
      it. For this reason even reasonable measurement errors can lead to
      wrong or negative values for L_n.
Therefore we suggest the following method of obtaining the diffusion length: We calculate the logarithmic derivative of $S(\lambda)$,

(14)
$$\frac{d \ln S}{d\lambda} = \frac{1}{S_{abs}} \frac{dS}{d\lambda} abs \equiv \frac{1}{S_{rel}} \frac{dS}{d\lambda} rel$$

and solve for L_n

(15)
$$L_n = - \frac{1}{\alpha} \{ \frac{\frac{1}{\alpha} \frac{d\alpha}{d\lambda}}{\frac{1}{\lambda} - \frac{1}{S_{rel}} \frac{dS}{d\lambda} rel - d\frac{d\alpha}{d\lambda}} + 1 \}$$

We now need only measurements of the relative change of the spectral response with λ , which are easier and more accurate to prepare. As the derivatives are negative on the red shoulder of $S(\lambda)$, all values in the denominator are now added (cf. iii) above). $d\frac{d\alpha}{d\lambda}$ is small and can often be neglected as compared to the other summands. This is not so in other methods which are sensitive to the exponential relation on d ($\exp(-\alpha d)$), e.g. the well known graphic method proposed by Stokes and Chu /5/. There the neglection of para- meters as emitter thickness, reflectivity, grid shading can lead to errors as high as 40% so that results should be interpreted as 'effective' diffusion length in comparison with other cells only. In addition that method implies that L_n is a constant which need not be true for different light bias conditions /6/.

If necessary, a lower limit for d can be estimated by the assumption, that at low wavelength all carriers in the base will be collected, as now the diffu- sion length is much larger than the absorption depth of photons. In this simple model the solar cell consists of two layers: The top layer doesn't contribute any carriers to the current (e.g. because of surface recombination) while all

carriers in the bottom layer are collected /7/. Now $S(\lambda)$ is simply

(16)
$$S(\lambda) = e^{-\alpha(\lambda)d} \; \frac{q\lambda}{hc}$$
and

(17)
$$\ln S(\lambda) \; h \; c/q\lambda = -\alpha(\lambda) \; d$$

under the assumption $\alpha^{-1} \ll L_n$ (low wavelength, 400 ... 650 nm). So a logarithmic diagram of spectral response $S(\lambda)$ versus $\alpha (\lambda)$ renders the thickness of the inactive layer as the slope of the line. Again relative mea- surements of $S(\lambda)$ are sufficient and this lower limit for d can be used in the evaluation of L_n.

Errors in the determination of L_n as large as 40% in the standard methods are introduced for a number of reasons:
1) We have neglected the part of the current originating in the emitter and scr. Depending on λ the base part $P(\lambda)$ of the total current is only 95 ..100% of $j(\lambda)$ on the red shoulder.
2) The exatvalue of the absorption coefficient depends on temperature and the history of the material (e.g. stress relieved/stressed /8/).
3) The spectral reflectance $R(\lambda)$ off the surface of the cell is 10 ... 30% of the total incident light.
4) The effective area of the cell is reduced by the shading of the contact grid, $a_{eff} \sim 80 ...95$ % .

$R(\lambda)$ and a_{eff} are given by measurements, $P(\lambda)$ predicted by the model of section 2. In the standard methods these corrections can change L_n by a large amount, e.g. from $L = 128 \mu m$ to the much smaller value of $L \doteq 28 \mu m$ (see the sensitivity of the evaluation to i) - iii) above). Eq.(12) is now written as

$$(18) \qquad L_n = \frac{1}{\alpha(\lambda)\{S(\lambda)^{-1} P(\lambda)^{-1} a_{eff} \exp{(-\alpha(\lambda)d)} (1- R(\lambda)) \frac{q\lambda}{hc} - 1 \}}$$

Using the correction factors in our method of logarithmic derivatives,

$$(19) \qquad L_n = - \frac{1}{\alpha(\lambda)} \{ \frac{\frac{1}{\alpha(\lambda)} \frac{d\alpha(\lambda)}{d\lambda}}{\frac{1}{\lambda} + \frac{1}{1-R(\lambda)} \frac{d(1-R(\lambda))}{d\lambda} - \frac{1}{P(\lambda)} \frac{dP(\lambda)}{d\lambda} - d \frac{d\alpha(\lambda)}{d\lambda} - \frac{1}{S(\lambda)} \frac{dS(\lambda)}{d\lambda}} +1 \}$$

We need only the relative change of all correction factors with λ. If they change only slowly with λ as compared to $\frac{1}{S} \frac{dS}{d\lambda}$,they can be neglected at all. Therefore we need not know the effective area of the cell. Using relative spectral response data, L_n can be extracted from single data points (e.g.for quick reference) as well as from sets of data points (as in graphic methods).

5. Conclusions

We have presented the analytic solution of a model of crystalline solar cells. The influence of all cell parameters on solar cell performance has been simulated. The diffusion length in the base, the base thickness and doping concentration turned out to be of major importance.
In addition a new method to extract the base diffusion length from relative spectral response data has been proposed. It greatly reduces uncertainties in the determination of this important parameter.

6. References

/1/ H.C.Scheer, H.G.Wagemann, Archiv f. Elektrotechnik 66(1983), 327
/2/ C.T.Sah, R.N.Noyce, W.Shockley, Proc. IRE 45(1957), 1228
/3/ R.N.Hall, Proc. IRE Suppl. B17(1959), 923
/4/ N.D.Arora, S.G.Chamberlain, D.J.Roulston, Appl. Phys. Lett. 37(1980), 325
/5/ E.D.Stokes, T.L.Chu, Appl. Phys. Lett. 30(1977), 425
/6/ J.Metzdorf, H.Kaase, Proc. 6th EC Photov. Solar Energy Conf. (1985), 160
/7/ B.Reinicke, K.Zander, Hahn-Meitner-Institut Berlin, Bericht HMI-B 60 (1967)
/8/ W.E.Phillips, Sol. State Electronics 15(1972), 1097

Fig. 2

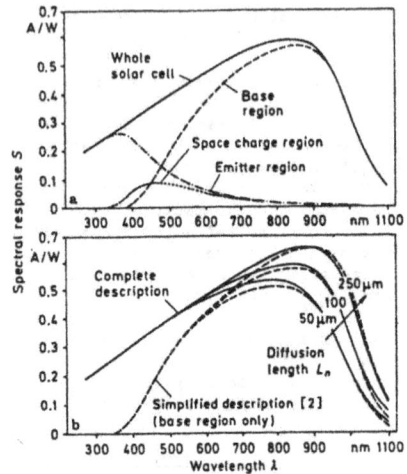

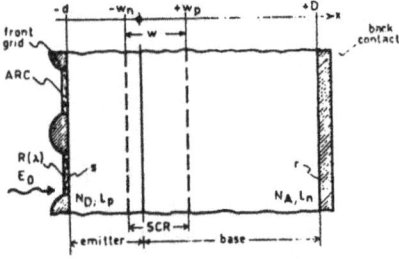

Fig. 1

Fig. 3

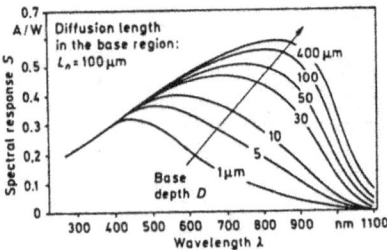

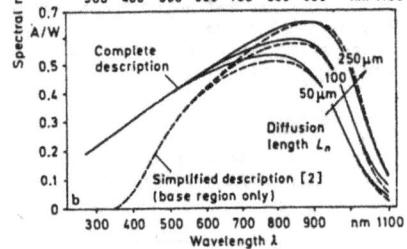

Fig. 4

Fig. 5
a~c

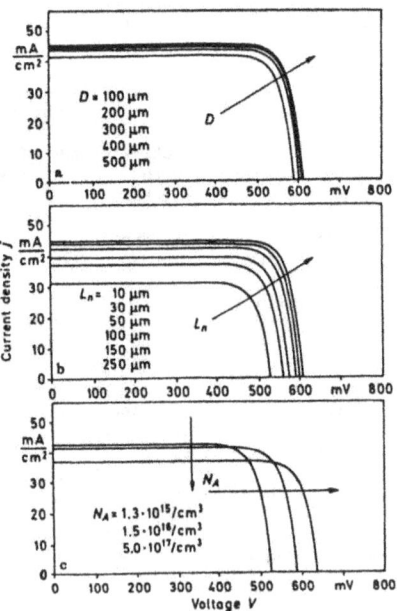

7. Figure Captions

Fig. 1 Cross section of a solar cell
Fig. 2 Contribution of emitter, scr and base to spectral resonse
Fig. 3 Variation of spectral response with diffusion length L_n
Fig. 4 Variation of spectral response with base depth
Fig. 5 White light characteristics of solar cells

SOLAR-ENERGY RELATED TOPICS

MATERIALS FOR PHOTOELECTROCHEMICAL CELLS

Bruno Scrosati

Dipartimento di Chimica, Universita' of Rome, Italy.

Pinciples of operation of photoelectrochemical cells.

A photoelectrochemical cell (PEC) is based on the junction between a semiconductor (either or n or p type) and an electrolyte, generally liquid, containing a suitable redox couple. The energy schemes for n-type and p-type semiconductors in junction with a redox couple in solutions are illustrated in Figure 1. At the equilibrium, the Fermi level of the semiconductor (which is the electrochemical potential of electrons) equalizes the Fermi level of the redox couple (which is the Nernst potential). This induces band bending with a formation for a depletion layer at the interface. (1-5)

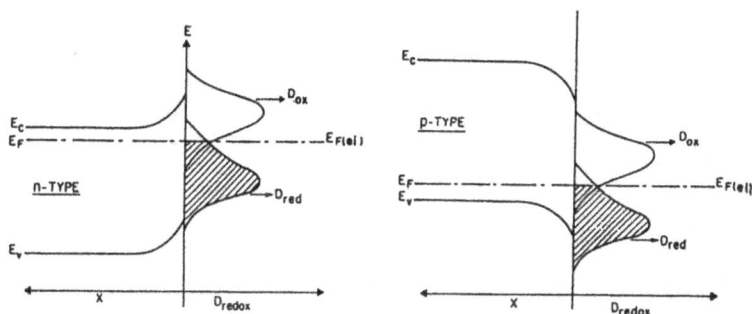

FIGURE 1- The energy scheme of a n-type and a p-type semiconductor in junction with a redox couples.

E_c= conduction band; Ev= valence band;

E_F= Fermi level of the semiconductor; $E_{F(el)}$ = Fermi level of the redox couple (Nernst potential).

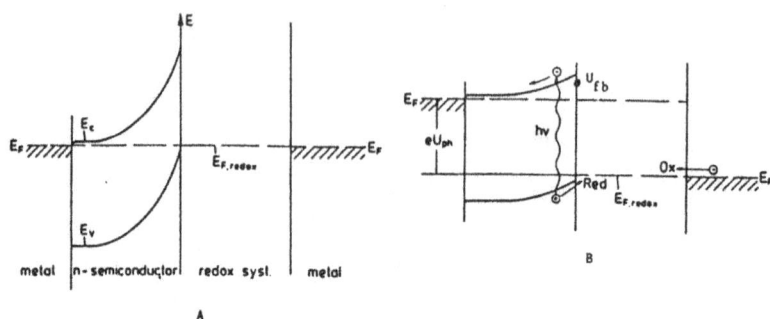

FIGURE 2- Energy scheme of a regenerative photoelectrochemical cell (PEC) using a n-type semiconductor electrode at the equilibrium (A) and under illumination (B).

Let us now consider a sequence formed by a semiconductor (supposely n-type), immersed in an electrolyte containing a redox couple (schematized as Ox/Red e.g. Fe^{3+}/Fe^{2+}) and a metal (e.g. platinum or carbon). At equilibrium in the dark the Fermi levels of the three components equilize, as shown in Figure 2A. When the semiconductor is illuminated with an energy $h\nu$ equal or greater than the bandgap, Ebg, electron-hole couples are generated with a decrease in band-bending and a rise in the Femi level (Figure 2B). This situation gives rise to photopotential, $eUph$, equal to the difference between the Fermi level of the illuminated semiconductor and that of the redox couple in solution. Under these circumstances, the maximum photovoltage corresponds to the condition where the bands are totally unbended. Therefore, the corresponding value of potential, called flat-band potential, Ufb, is of obvious relevance in PEC characterization. Furthermore, the photogenerated couples are readily separated by the field in the depletion layer. The minority carriers are injected into the solution where they drive the following oxidation process:

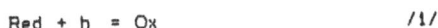

$$Red + h = Ox \qquad\qquad /1/$$

e.g.:

$$Fe^{2+} + h = Fe^{3+}$$

The majority carriers move towards the bulk of the semiconductor and, via an external load, can travel to the counter electrode, from which they are injected into the solution to drive the following reduction process:

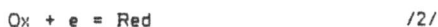

$$Ox + e = Red \qquad\qquad /2/$$

e.g.:

$$Fe^{3+} + e = Fe^{2+}$$

which is just the opposite of process /1/.

In this fashion, it is possible to obtain a photovoltage and a photocurrent, and thus electric power, through the system, without changes in the electrolyte composition.

The one described is the operation of the so called 'regenerative' PEC, where, as again schematized in Figure 3, the light (solar) energy can be directly transformed into electric energy.

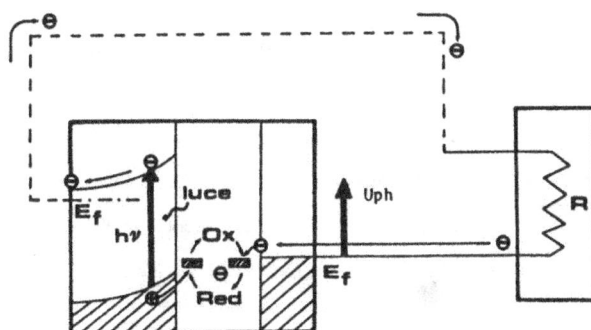

FIGURE 3- Operational scheme of a PEC.

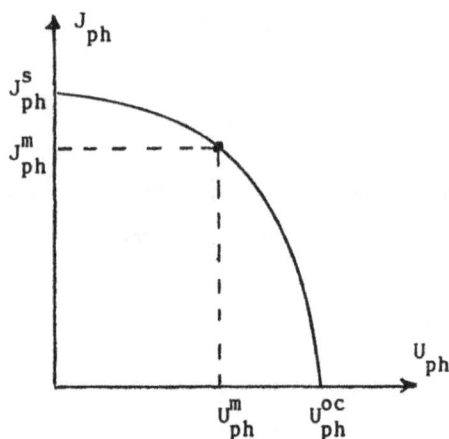

FIGURE 4- Output characteristic of a PEC.

In fact, under load, a photovoltage, Uph, establishes between the semiconductor and the counterelectrode and a corresponding photocurrent density, Jph, flows. The properties of a PEC are characterized by a current-voltage plot, as typically shown in figure 4.

The Jph x ph product is the power output at any point and its highest value Pm represents the maximum output power available. The ratio between Pm and the sunlight power incident on the electrode area, Po, is the maximum solar energy efficiency, ηm, of the PEC. The limiting power of a PEC is obviously the product between the open circuit voltage, $\overset{oc}{U}$ph, and the short circuit current, Jph. The ratio Pm/($\overset{\bullet\bullet}{U}$ph/Jph) is defined as the fill factor, FF, of the PEC.

The energy scales

Energy schemes in semiconductor studies are generally related to a "physical
scale", referred to the electrons at rest in vacuo (0 eV), while electrochemical
studies are based on the scale referred to the standard potential of the hydrogen
electrode (0 V).   The   interpretation   of   the   energy   schemes   of
semiconductor/electrolyte junctions requires a relationship between the two

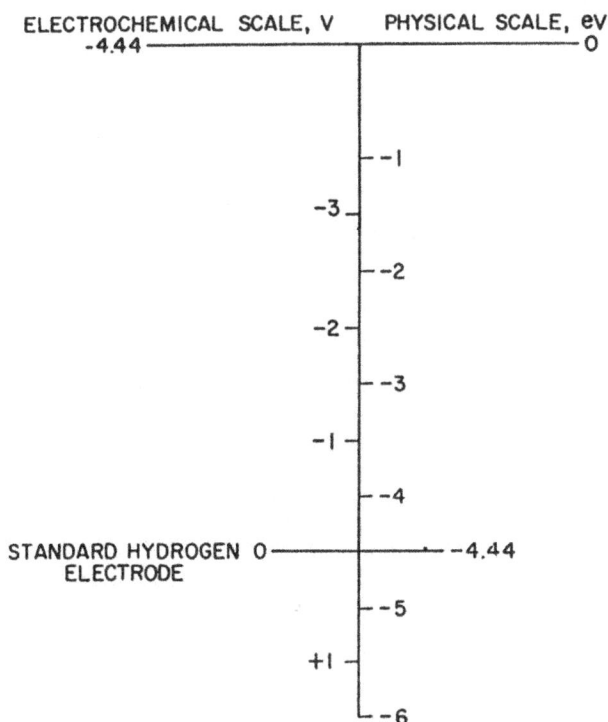

ELECTROCHEMICAL SCALE, V   PHYSICAL SCALE, eV

FIGURE 5- Physical and electrochemical scales.

scales. Figure 5 shows how the two scales can be related: the reference point
remains the standard potential of the hydrogen electrode, to which is attributed
0 V and -4.44 e V.

<u>The problem of photocorrosion.</u>

The concept of operation of PEC is technologically appealing since the solid/liquid junction is easily obtained (just by immersion of the semiconductor electrode into the liquid electrolyte) and by the selection of semiconductors having suitable bandgap values, reasonable high efficiencies may be on principle foreseen.

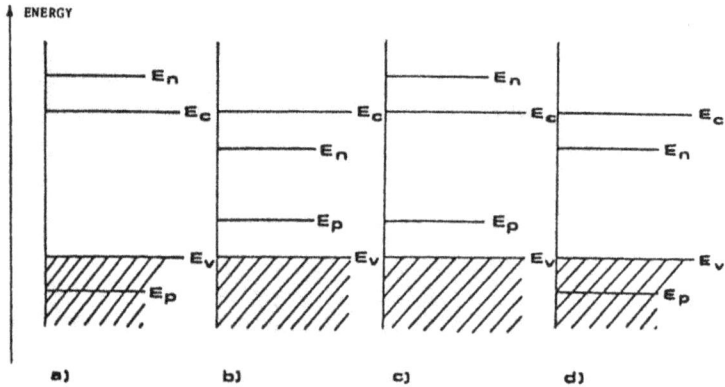

FIGURE 6- Correlations between band edges and the potential levels of the anodic (E_p) and cathodic (E_c) decomposition processes of the semiconductor electrode (6).

a) thermodynamic stability; b) both anodic and cathodic decompposions are possible; c) anodic decomposizion possibe; d) cathodic decomposition possible.

However, the photogenerated minority carriers, rather than drive the disired electrochemical reaction in solution, may oxidize (n-type) or reduce (p-type) the semiconductor itself, this leading to photocorrosion processes which rapidly degrade the life of the PEC. Figure 6 schematizes the various cases which may verify at the semiconductor interface, by illustrating the correlation between valence band, E_v, and conduction band, E_c, edges and the potential levels of the oxidation process (E_p=anodic decomposition) and of the reduction process (E_n=cathodic decomposition) of the semicoductor electrode (6).

In case (a), En lies above Ec and Ep lies below Ev and thus neither the reduction nor the oxidation process may take place; therefore, case (a) represents the thermodynamic stability of the semiconductor but, unfortuantely, not very likely to verify in practise. Case (b) is just the opposite: the semiconductor electrode is thermodynamically unstable both towards anodic and cathodic decomposition; also this case is practically improbable.

The cases most commonly met are (c) and (d), where the oxidation and the reduction, respectively, of the semiconductor are thermodynamically possible. Under these circumstances, two opposite situations may be envised at a semiconductor/electrolyte junction.

These are illustrated in Figure 7 in the case of a n-type semiconductor.

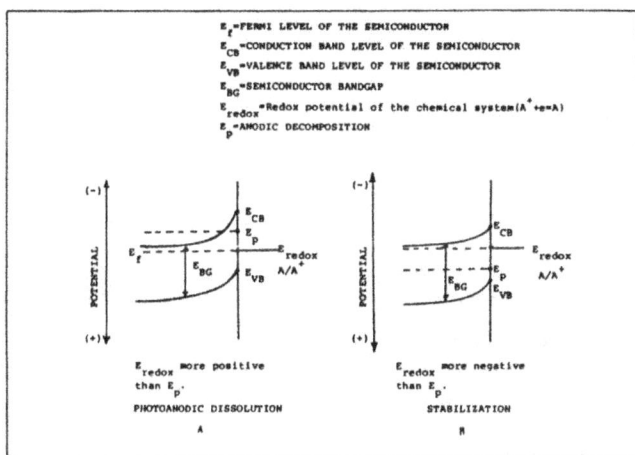

FIGURE 7- The two cases of photoanodic dissolution and of stabilization in the case of n-type semiconductor.

Figure 7A shows that the anodic decomposition potential lies above the energy level of the desired redox process: the decomposition is thus thermodynamically favoured and the semiconductor is readily photo-oxidized. Figure 7B shows the opposite situation, where the electrochemical process is thermodynamic favoured and thus stabilization of the semiconductor is achieved.

These two critical situations are further illustrated in Figure 8 and 9 with a practical example related to the junction between CdS and two different redox couples in aqueous solutions (7).

In the first case (Figure 8), where the ferro/ferricyanide redox couple is used, the semiconductor decomposition reaction, Ep:

$$CdS + 2h = Cd^{2+} + S \qquad\qquad /3/$$

is thermodynamically favoured over the electrochemical reaction, Eredox:

$$/Fe(CN)_6/^{4-} + h = /Fe(CN)_6/^{3-} \qquad /4/$$

Therefore, upon illumination, the semiconductor readily decomposes with the dissolution of Cd^{2+} ions in solution and the precipitation of S on its surface.

In the second case (Figure 9), where the sulfur/polysulfide couple is used, the electrochemical reaction, Eredox:

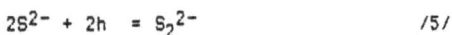

$$2S^{2-} + 2h = S_2^{2-} \qquad\qquad /5/$$

is favoured over the decomposition process:

$$Cd + 2h = Cd^{2+} + S \qquad\qquad /3/$$

Therefore, upon illumination, the semiconductor remains stable and the electrochemical reaction proceeds.

The Kinetic approach.

Figure 8 and 9 clearly shows that the Eredox and the Ep levels are very close each other. Therefore, shifts in level positions may easily verify (e.g. by changes in redox couple concentrations) to the point of reversing the situation of a stability condition into that of a photocorrosion.

To prevent this, two main routes can be followed, one based on a kinetic approach and the other based on a material selection approach.

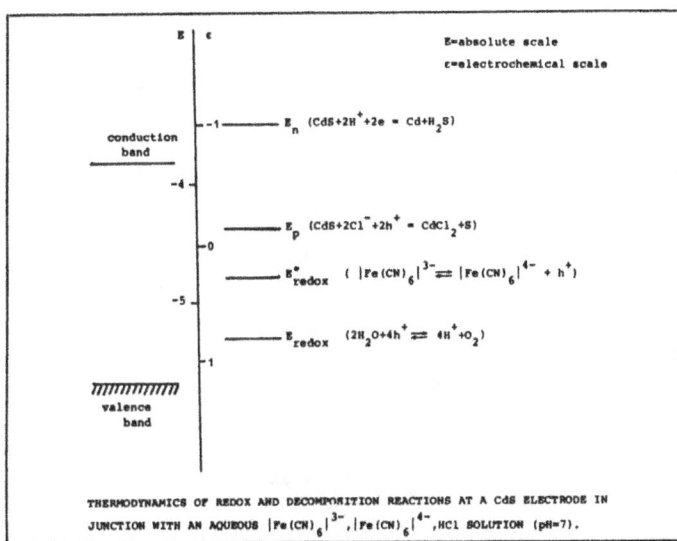

FIGURE 8- Thermodynamic condition of non-stability of CdS photoanode in junction with the ferro/ferricyanide redox couple (7).

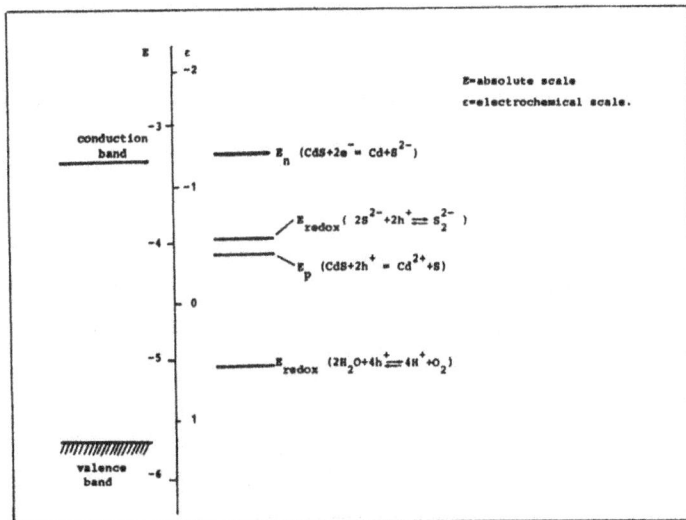

FIGURE 9- Thermodynamic condition of stability of a CdS photoanode in junction with the sulfur/polysulphide couple (7).

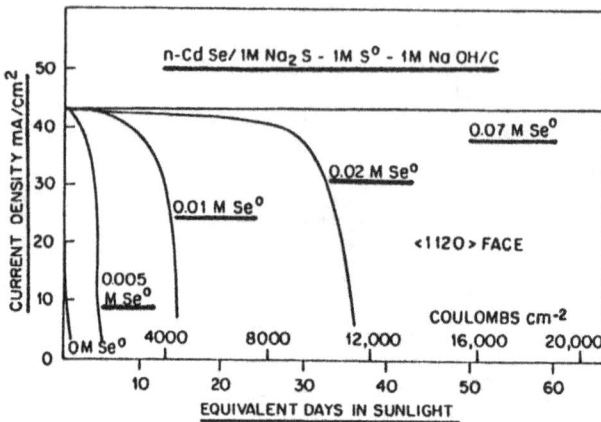

FIGURE 10- Stabilization of a CdSe PEC by additions of selenium to the S^{2-}/S electrolyte (8).

The first route is pursued with the choice of suitable electrolyte compositions in order to make the rate of the transport of the photogenerated minority carriers to the redox couple much faster than that of phtocorrosion. In such a way, even if the photocorrosion process is thermodynamically favoured, its rate is made negligibly small so that effectively the semiconductor electrode becomes kinetically stable.

This kinetic approach has been followed in the case of PEC based on 'classical' semiconductors, such as CdS, CdSe and GaAs. For example, Figure 10 illustrates the effect of electrolyte composition in the stabilization of a CdSe semiconductor in junction with a S^{2-}/S redox couple (8). In normal conditions (i.e. with no modification of the electrolyte), the photocurrent decays rapidly to zero, this indicating very fast and destructing photocorrosion phenomena). However, small additions of selenium Se, to the electrolyte, improves dramatically the stability of the semiconductor, to the point that at concentrations the order of 0.07M, the photocurrent output becomes stable and the photocorrosion is fully overcome. With this type of modified electrolyte, stable, PEC have been realized using single-crystal CdS electrodes.

Similar stabilization effects have been obtained in the case of GaAs in junction with the Se^{2-}/Se_2 couple. Addition of selenium improves consistently the semiconductor stability. Consequently, stable and highly-efficient (9%) PEC have been obtained with single-crystal GaAs photoanodes (9). However, the practical utilization of these PECs is somewhat restrained by the unsafely of the selenium-based electrolytes.

The material approach.

This approach is directed to the selection of semiconductor materials whose optical transitions do not involve bonding orbitals, this on principle giving promises of good stabiltiy against photocorrosion.

This condition appears to be fulfilled by layered semiconductors (e.g. molybdenum and tungsten chalcongenides) and by ternary semiconductors (e.g. copper indium selenide).

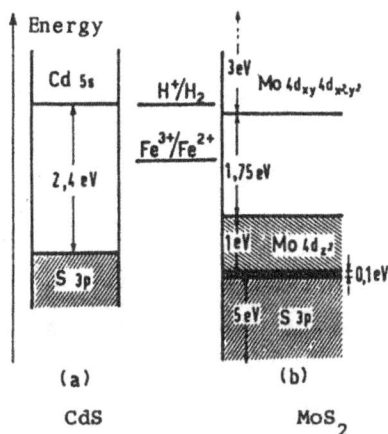

FIGURE 11- Band structures of CdS and of MoS_2.

Figure 11 compares the band scheme of a 'classical' semiconductor, i.e. CdS, with that of a layered semiconductor, i.e. MoS_2. It may be effectively noticed that while in the former the transition involves p-s bonding orbitals, in the latter only the metal d- orbitals are separated by the gap. A similar situation

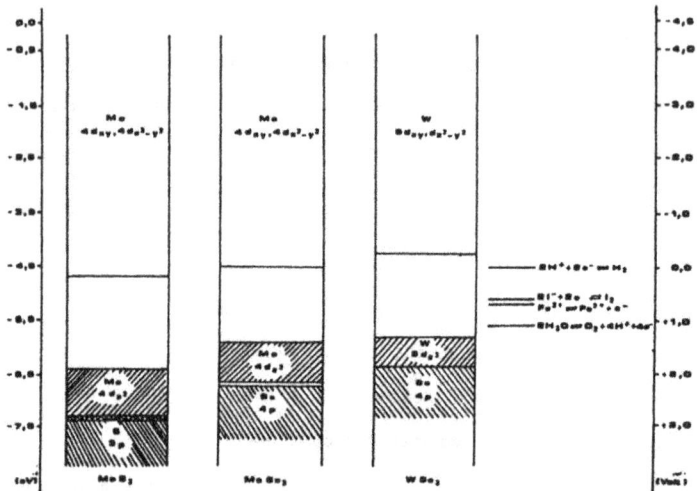

FIGURE 12- Energy levles of molybdenum and tungsten dichalcogenides and of suitable redox couples (10).

applies for $MoSe_2$ and WSe_2, where also the optical transitions concern d-d Mo or W orbitals (Figure 12). In Figure 12 the energy levels of various redox couples are also shown; among them the I^-/I_2 one appears as the most promising and indeed PEC cells using tungsten and molybdenum dichalcogenides in junction with the I^-/I_2 couple in aqueous solutions, have been effectively developed.

The promises of good stability have been experimentally confirmed in these cells. Figure 13 shows the output photocurrent trends for 3 types of Mo and W PEC: the current levels remain unchanged for a consistent length of time, this effectively indicating an excellent semiconductor electrode stability (10).

Some of the molybdenum and tungsten dichalcogenides, and $MoSe_2$ and WSe_2 in particular, have favourable band-gap values (i.e. around 1.5 eV) for the solar spectrum, so that reasonably high efficiencies may be obtained. Figure 14 shows the output characteristic of a single-crystal $MoSe_2/I^-$, I_2 PEC: fill-factor and conversion efficiency values of 0.6 and 6%, respectively have been obtained (12).

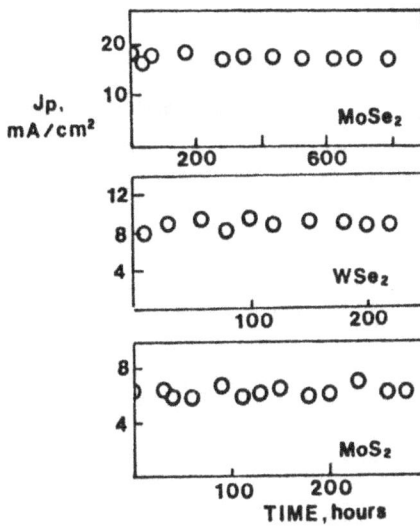

FIGURE 13- Output photocurrents versus time of illumination of transition metal dicharcogenide photoanodes.

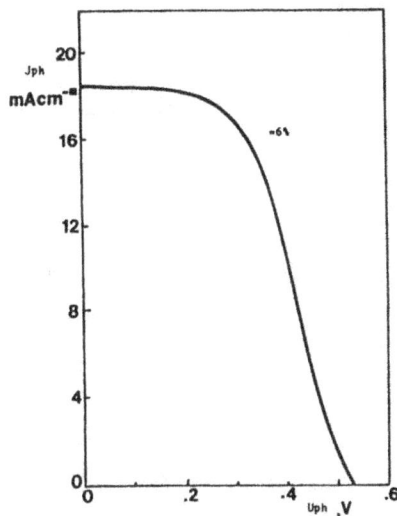

FIGURE 14- Output characteristic of a $MoSe_2/I^-/I_2$ PEC under an illumination of 100 mW/cm^2.

324

The problem of surface states

One of the major problem of PEC, and in general, of solar cells, is related of the recombination of the photogenerated carriers which may lead to severe decay in efficiency. As well known in semiconductor technology, recombination may be largely induced by surface states. Figure 15 illustrastes the effect of surface states Ess on the operation of a PEC: a fraction of the photogenerated holes, rather than oxidize the redox species in solution is trapped in ther surface state Ess where it recombines with electrons tunneling from the conduction band. Obviously, this fraction is not available for the photoelectrochemical reaction and thus the efficiency of the overall process drops.

The transition metal dichalcogenides have a layered structure as typically shown in Figure 16. The layers are loosely stack by a van der Waals bonds and

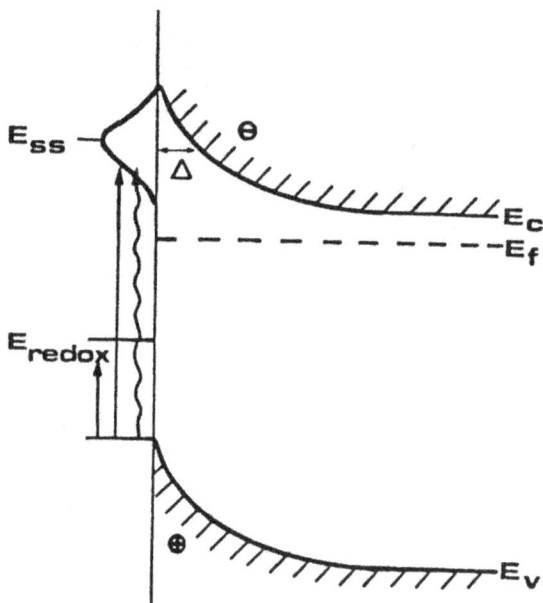

FIGURE 15- Change recombination by electron tunneling to a surface state (E_{ss}).

FIGURE 16- Van der Waals layers in transition metal dichalcogenides.

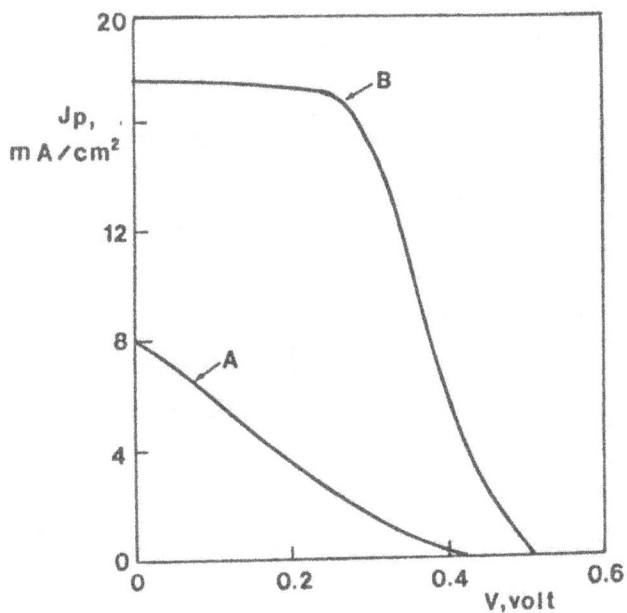

FIGURE 17- Output characteristics of 'smooth' (B) and of 'structured' (A) MoSe$_2$

photoanodes in junction with a I$^-$/I$_2$ redox couple.

thus steps or fractures may easily form on the crystal surface. These structural defects act as recombination sites and thus 'irregular' crystals have much poorer conversion efficiency than smooth crystals. This is clearly shown in Figure 17 in the case of a $MoSe_2$ crystal in junction with a I^-/I_2 a redox couple: when then smooth portion is illuminated, a good fill-factor and a high efficiency is obtained; when the structured portion is illuminated, the performance of the cell decays dramatically (11).

Furthermore, at the edge of a step on the crystal surface, there are unsaturated or 'dangling' bonds of the transition metal and/or chalcogenide atoms. These bonds and preferrred sites of attack from the solution species, so that structured crystals suffer of severe corrosion effect.

Practical applications of PEC based on the layered semiconductors would require polycrystalline electrodes which probably would have 'structured' rather than a 'smooth' surfaces. Therefore, recombination and corrosion at surface steps are serious problems and, consequently, various routes have been followed to overcome or, at least, to attenuate them.

One possible approach consists in treating the structured crystals with EDTA, which, being a selective complexing agent for transition metals, should block the atoms exposed to the solutionn at the edge of a step. Effectively, Figure 18 shows that after treatment with EDTA, the output characteristic of a MoS_2 crystal largely improved (12). Relevant is the enhancement in photocurrent and the depression is dark current, which show how the recombination and the corrosion rates have been indeed reduced. Unfortunately, this beneficial effect does not last upon illumination (Figure 19) probably because of oxidation effects induced by light.

More efficient appears to be a treatment based on a selective passivation of the defect sites. Figure 20 illustrates a transport model for a stepped layered crystal (13). Since the conductivity of the carriers is much faster along the van der Waals layers than across them, there is a drift of charges to the step edge. Therefore, by polarizing a layered semiconductor in the dark in an electrolytic cell,

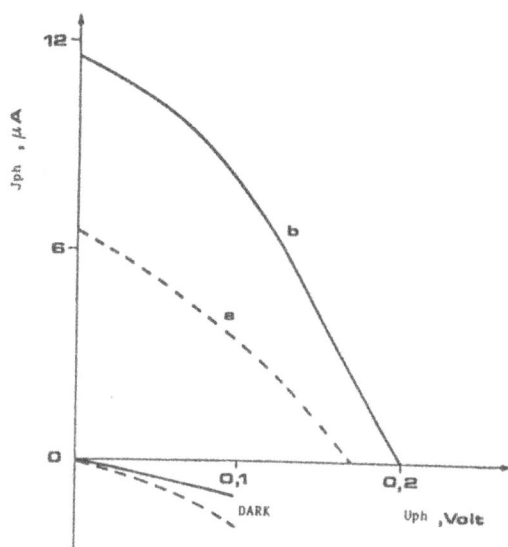

FIGURE 18- Output characteristic of a largely defective MoS_2 photoanode in junction with the I^-/I_2 redox couple. a) no treatment; b) after 12 hours of immersion in EDTA solution (12).

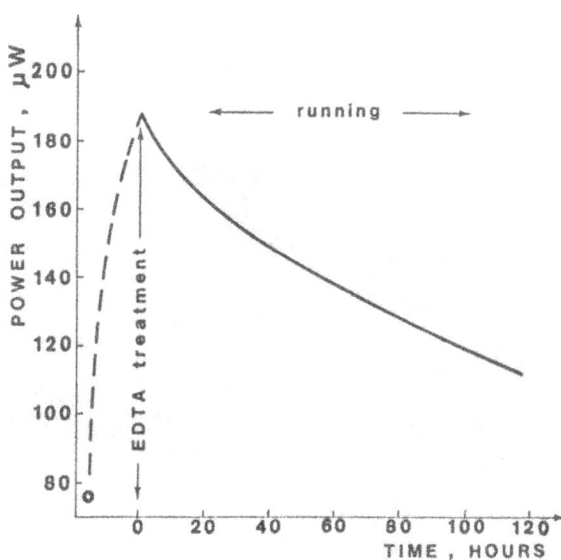

FIGURE 19- Maximum output power of a $MoSe_2/I^-$, I_2 PEC upon illumination after EDTA treatment (12).

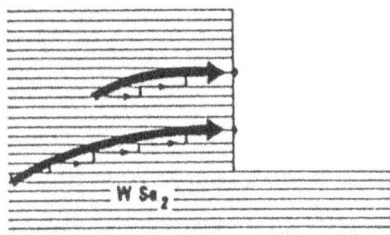

FIGURE 20- Trajectories of charges in a stepped tungsten dichalcogenide crystal
in contact with an electrolyte (13).

it is possible to run an electrodeposition seletively directed to the edges of
steps (14). Figure 21 shows the surface of a stepped $MoSe_2$ crystal after cathodic
polarization in the dark in an A_gNO_3 aqueous solution. It is clearly seen that the
silver crystals are preferentially deposited along the surface imperfections (15).

Following this principle it is then possible to selectively electrodeposit a
non-conductive material such as polymer (e.g. polyindole) and thus electrically
isolate the defective sites from the solution. Figure 22 shows that this operation
effectively improves the performance of a structured $MoSe_2$ crystal in a stable
manner (16).

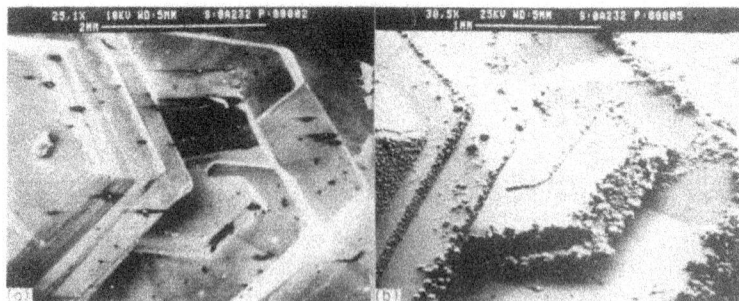

FIGURE 21- Surface of a WSe_2 semiconductor electrode having a defective surface
(a) after cathodic polarization in a silver nitrate solution (b). The
silver crystals are clearly electrodeposited along surface
imperfections (15).

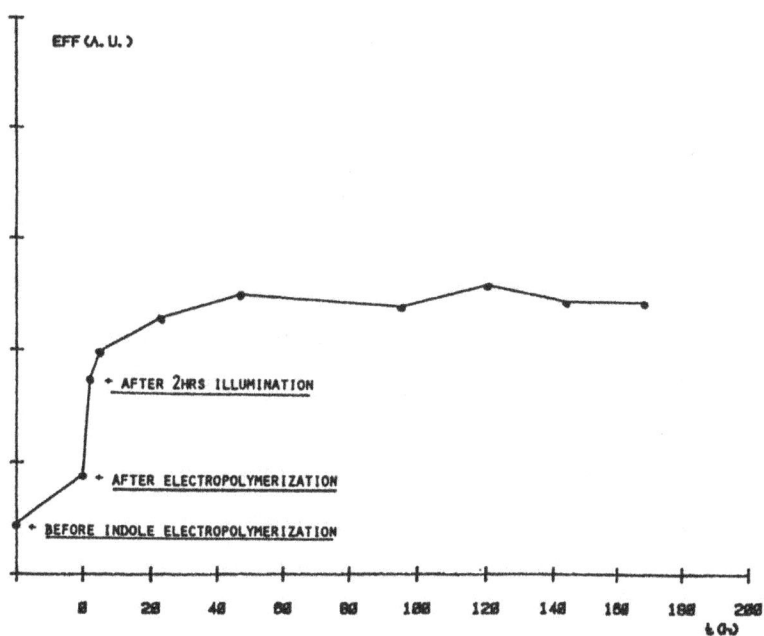

FIGURE 22- Stability under illumination of a MoSe$_2$ crystal after selective electrodeposition of polyindole (16).

However, even after this treatment, the efficiency of layered semiconductor PEC remains basically too low to attract considerable attention in the practical sense and thus, alternative, stable materials have to be considered.

The ternary semiconductors

High electrochemical stability is basically offered by ternary chalcogenides (i.e. $CuInSe_2$, $CdIn_2Se_4$....) in respect to the binary analogues (i.e. CdSe, CdS). This high stability is ascribable to the chalcopyrite structure of the ternary compounds. In fact, while in the case of CdSe and, more generally, in the case of the II – VI compounds, the upper valence band is composed entirely of states of bonding electrons (from s and p orbitals), in the case of chalcopyrites there are high density, non bounding electronic states at the top of the valence band, which allow photoelectrochemical processes without exerting a direct influence on the chemical bonds.

The most promising material of the family is copper indium selenide, $CuInSe_2$. Figure 23 shows the spectral response of a $n-CuInSe_2$ crystal: the onset at 1300 nm reveals that the band gap is around 1 e V, i.e. an optimal value for solar applications.

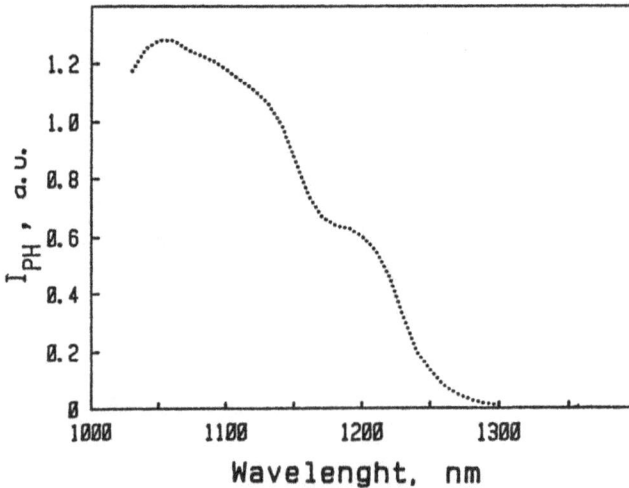

FIGURE 23– Spectral response of a $n-CuInSe_2$ semiconductor.

Indeed, by protecting the surface of this semiconductor by an indium oxide film and using a I^-/I_2 aqueous electrolyte with suitable additions of Cu^+ and In^{3+}ions, solar efficiencies around 12% have been reported for n-$CuInSe_2$ single photoelectrodes (17, 18).

Furthermore, this semiconductor has been widely prepared and characterized for solid-state applications and thus the procedures for the preparation of polycrystalline samples are well known. Effectively, by a direct synthesis from the melt using 6N pure constituents elements, homogeneous polycrystalline $CuInSe_2$ ingots, in which the average grain size was 1-2 mm have been obtained (19).

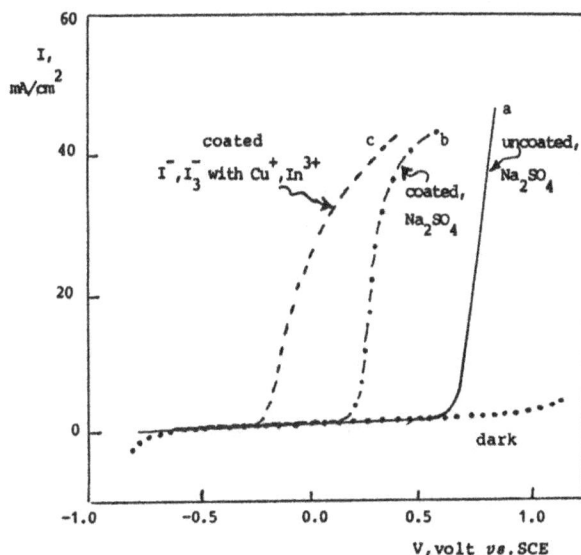

FIGURE 24- Current-voltage curves for a n-$CuInSe_2$ polycrystalline electrodes under various conditions (24).

Figure 24 shows the current-voltage curves of a n-CuInSe$_2$ polycrystalline electrode (20). Curve a refers to a the response in a Na$_2$So$_4$ aqueous solution. Since no redox couple is present, the large current flowing at about 0.6V vs SCE(SCE= standard calomel electrode) must be necessarily related to a photocorrosion process, which may be likely indicated as:

$$CuInSe_2 + 4h \rightarrow Cu^+ + In^{3+} + 2Se$$

To protect the photoanode from this corrosion, one can deposite a layer (about 50 Å thick) of indium oxide on the electrode surface. This can be achieved by electrodepositing indium on CuInSe$_2$ and thus thermally oxidize it to indium oxide. However, curve b of Figure 24, which refers to an indium oxide coated CuInSe$_2$ electrode, shows that this treatment alone is not sufficient to protect the electrode since current is still flowing. This current is probably due to the dissolution of the indium oxide layer: in fact, repeating the voltage sweeps, the onset of current progressively shifts anodically to finally match that of curve a.

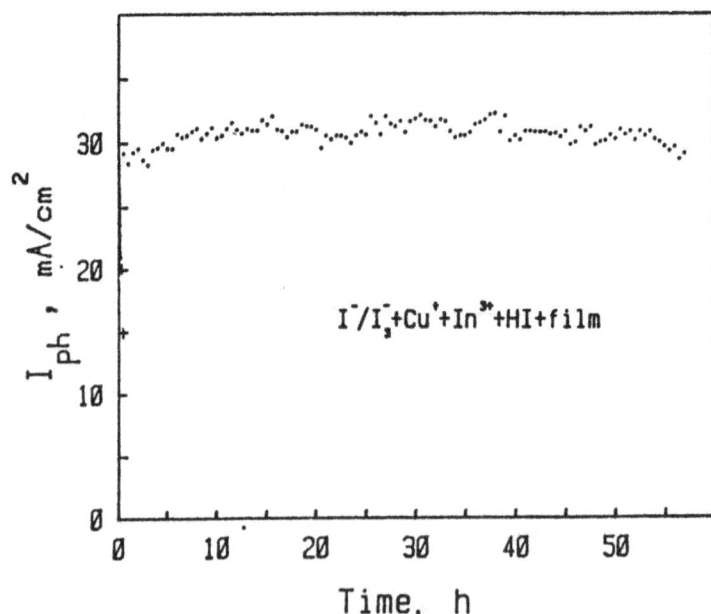

FIGURE 25- Photocurrent of a CuInSe$_2$ PEC using a modifed I$^-$/I$_2$ electrolyte.

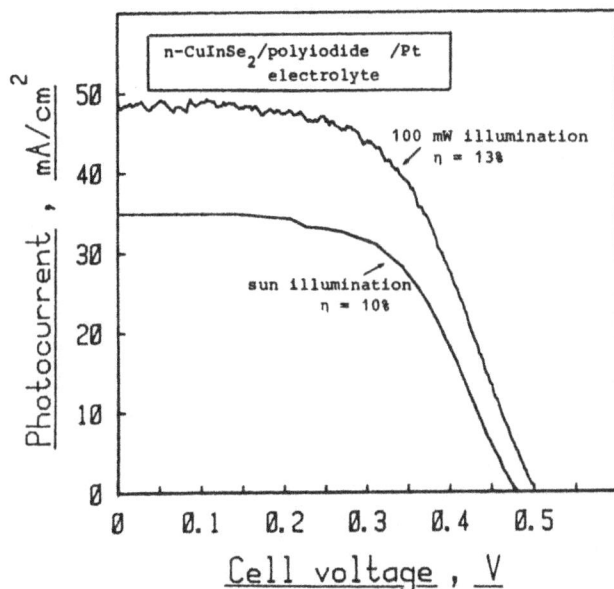

FIGURE 26— Output characteristics of CuInSe$_2$ - based photoelectrochemical cells.

Only when the oxide-coated electrode is used in a solution containing the I^+/I_2 couple plus Cu^+ and In^{3+} ions, total stabilization is achieved, as shown by the cathodic onset of the related photocurrent (curve c of figure 24).

Effectively, figure 25 shows that under these conditions, the photocurrent of a CuInSe$_2$ PEC remains constant upon prolonged illumination. Furthermore, the performance of this type of PEC are also very promising, since conversion efficiencies of 13% under simulated illumination and of 10% under direct sun illumination (figure 26) have been obtained (20).

These values are encouraging and indicated that by proper selection of the semiconductor materials and of the nature and of the composition of the electrolyte solutions, photoelectrochemical cells of practical interest may be eventually developed.

REFERENCES

1) - H. Gerischer, J. Electroanal Chem. $\underline{58}$, 263 (1975).

2) - A. J. Nozik, Ann. Rev. Phys. Chem. $\underline{29}$, 189 (1978).

3) - A. J. Bard, Science, $\underline{207}$, 139 (1980).

4) - L. Peraldo Bicelli and B. Scrosati,
    Chimica ed Industria, $\underline{63}$, 172 (1981).

5) - B. Parkinson, Acc. Chem. Research, $\underline{17}$, 431 (1984).

6) - H. Gerischer, Proceedings Symp. Electrode Materials and Processes for Energy Conversion and Storage, Electrochem. Soc. Inc., Princeton, 1977, pag.8.

7) - H. Gerischer, in 'Solar Power and Fuels', J.R. Bolton Ed. Academic Press, New York, 1977, pag.77.

8) - A. Heller, G. p. Schwartz, R.G. Vadimisky, S. Menezes, and B. Miller, D. Electrochem. Soc. $\underline{125}$, 1156 (1978).

9) - K. C. Chiang, A. Haller, B. Schawrtz, S. Menezes and B. Miller, Science, $\underline{196}$, 1098 (1977).

10) - L. Fornarini, L. Manduzio and B. Scrosati
     Chimica ed Industria, $\underline{64}$, 8 (1982).

11) - G. Razzini, M. Lazzari, L. Peraldo Bicelli, F. Levy, L. De Angelis, F. Galluzzi, L. Fornarini and B. Scrosati,
     J. Power Source, $\underline{6}$, 371 (1981).

12) - G. Razzini, L. Peraldo Bicelli, G. Pini and B. Scrosati,
     J. Electrochem. Soc. $\underline{128}$, 2134 (1981).

13) - H. J. Lewerenz, A. Heller, F. J. Di Salvo,
     J. Am. Chem. Soc., $\underline{102(6)}$, 1877 (1980).

14) - H. S. White, H.D. Abruna and A. J. Bard,
     J. Electrochem. Soc. $\underline{129}$, 265 (1982).

15) - L. Fornarini, B. Scrosati,
     Electrochim. Acta, $\underline{28}$, 667 (1983).

16) –   L. Furnarini, F. Stirpe and B. Scrosati,

J. Electrochem. Soc., 130, 2184 (1983).

17) –   D. Cahen and Y. W. Chen, Appl. PHys. Lett. 45, 746 (1983).

18) –   S. Menezes, H. J. Lewerenz, and K. J. Bauchmann, Nature, 305, 615 (1983).

19) –   G. Razzini, L. Peraldo Bicelli, B. Scrosati and L. Zanotti,

J. Electrochem. Soc., 133, 351 (1986).

20) –   G. Razzini, L. Peraldo Bicelli, M. Arfelli and B. Scrosati,

J. Electroanal Chem., in press.

Qualification and Durability Tests for Solar Thermal Collectors

G. Riesch, Joint Research Centre
of the Commission of the European Communities, ISPRA/VA, Italy

Introduction to qualification and durability tests.

We distinguish these two types of tests in the same way as we did in the lecture for photovoltaic modules:
A qualification test verifies, whether at the moment of the test a certain quality of the collector is assured. The test does not change the properties of the specimen.
In a durability test the specimen is exposed for a certain time to the influence of potentially damaging agents. It is to be checked whether these agents do change the properties of the specimen.

In the first class we count for instance a test, that verifies, that the glass of the collector resists to the impact of artificial hail stones or that the absorber resists to a certain hydraulic overpressure.

The purpose of the durability test is to provoke in a

shorter time than in reality a c h a n g e of
certain qualities. The same change would occur in real
operation in a longer time and so prediction of the lifetime
of the specimen becomes possible with tests that last only a
short testing time.

In this class we count for instance the exposure of a
thermal collector in dry conditions to the sun or to the
irradiance of a simulator. It is expected, that under these
more severe conditions, than they occur in reality during
the use of a collector, possible weaknesses of the
construction will show up.

The long term durability of thermal collectors, their
lifetime, has been considered to be a most important aspect
since the beginning of collector construction. Only the
very first generation of collectors was designed mainly on
an experimental basis and most emphasis was put on high
efficiency, while durability was then of minor priority.
However practice has shown quickly, that many of these first
generation collectors did not match to expectations, simply
because they did not function any more after two or three
seasons.

In fact economic optimisation calculations show, that the
lifetime of solar equipment must be long: of the order of
some decades of years and compareable to the lifetime of
other construction materials used in buildings. Hence the
durability of collectors and of connected systems is as
important as their high initial efficiency. And as a
consequence tests for evaluation of the durability can be
considered to be of equal economic importance as the tests
for the evaluation of the instantaneous efficiency.

Looking at the international efforts in the field of
qualification/durability testing we see that the activities
can be classified in five sectors:

A. Tests made on the properties of the materials used in
   collectors.
B. Tests made on destructive failures of whole collectors
   (qualification tests).
C. Tests made on slow efficiency degradation of collectors
   (durability tests).
D. Tests made on the atmospheric corrosion of collectors.
E. Tests made on hydraulic corrosion of the heat transfer
   systems.

More than 2o laboratories are working on different aspects of collector durability in Europe. These laboratories work in principle independently from each other, however they meet from time to time in the meetings of the "Collector Testing Group of the CEC countries" , organized by the Joint Research Centre of the EC, and exchange their knowledge.
A similar work is done on a worldwide basis in the IEA (International Energy Agency) group with its Task III, subtask E group (and the recently founded Task X group.)
In the following I am reporting on work done in these groups and especially on our own activities.

A. Tests on Materials.

Much work has been done to test the durability of the optical properties of black absorbing surfaces and of transparent covers on small samples.
In [1] results of such tests have been reported. For instance in Sweden seven selective and ten non selective absorber materials on four different substrates (steel, stainless steel, aluminium and copper) were exposed to many different treatments (elevated temperature, thermal cycling, ultraviolett radiation (with ozone and sulphurdioxyde atmosphere), moisture (including thermal cycling), salt spray, water bath. Visual examination and measurements of absorptance and emittance showed, that the most active parameters are elevated temperatures, temperature cycles, UV-radiation, moisture and atmospheric pollutants.
In the US an extensive program was conducted which contained also material tests on various coverplate and absorber samples in small mini boxes (simulating the temperature and exposure conditions in real collectors), which were exposed for several years outdoors at four different locations. At the same time real collectors were exposed at these sites and also laboratory ageing tests were conducted to provide a basis for comparison with the outdoor exposure tests. Different visible or measurable effects were produced and many detailed results are reported in [2].
If it is possible to report in a few words the conclusions from these tests one might say:
For cover materials high humidity and high solar radiation produce the most severe changes. Mini boxes and available indoor exposure tests are not capable of reproducing many of the changes observed outdoors.
For absorber materials also the high moisture in the

collectors is a most important degradation parameter. It is not well represented in the mini boxes and it is exaggerated in the laboratory tests.
Also tests on samples conducted in Switzerland on 19 different samples of absorbers show, that the combined influence of temperature and humidity produce the most pronounced degradations.

The problem with all these tests on small samples is, that visible and measureable effects (mainly changes of absorptance and transmittance, but also blistering, delamination, cracking) are observed, however it is unknown how to correlate these changes with what happens in reality in the collectors.

Tests on whole collectors.

Many tests on whole collectors have been proposed in the literature. Practice has shown, that only few of them are really important and significant.
In the following these tests which have been described in [3] are explained. Most of these tests will also be included in the guidelines for thermal collectors, to be published by the UEAtc (Union Europeenne pour l'Agrement Technique dans la Construction).

B. Qualification tests.

1.) Resistance to mechanical load.

The mechanical quality of collectors can be checked by a test with simulated wind pressure applied on the cover plate. The test can be executed with a special test chamber, producing a static air pressure on the glazing or by loading equally distributed loads (sandbags) on the glazing. This test is executed however only in the case of plastic glazings or in the case of very thin glass glazings. Normal glass of 4 mm thickness, which is rather often used on collectors, does not have to be tested because it presents no problems for pressures up to 1500 Pa, the fixed test pressure, corresponding to twice the static wind pressure at 130 km/h. The safety factor 2 (sometimes also 3) takes care of gusty winds.

For negative pressures, in suction, not only the resistance of the glazing itself has to be tested but also the resistance of the fixations of the cover plate to the case. This test is made with a number of suction cups distributed on the surface, which are drawn upwards with a strength corresponding also to 1500 Pa plus a safety factor of 1.5, i.e. 2250 Pa. The collector is lying on its back and is fixed as in usual operating conditions.

This test does not only control good quality of the collector but is also necessary for safety reasons. It is to be assured, that the whole collector or its glazing will not fall down from the roof in stormy weather.

In the case of plastic glazings the influence of the temperature may have to be studied, as the resistance of these materials is lower at higher temperature.

2.) Resistance of absorber to overpressure and leak test.

Static pressure tests for absorbers are made in two different modes and for two reasons: overpressure test and leak test.

First with a hydraulic test the resistance of the absorber to accidental overpressure is tested. This test is normally made at 1.5 times the maximum operating pressure. For safety reasons the absorber is pressurized with water, not with air, using a plumbers, hand-operated pressuring device. This test is especially recommended for plastic absorbers. (And in this case again higher temperature has eventually to be considered for the test, which however becomes not so easy, if higher temperatures are used.) This test should normally be done by the manufacturer, as pressure resistance is an essential aspect of the collector quality.

In a second step (leak test) small leaks can be detected which might be present at the joints especially in the case of roll-bonded absorbers. This test is conducted at the nominal operating pressure but with air as pressurizing agent as with this method a higher sensibility is reached, compared to pressurisation with water. Very small leaks can also be detected with a helium leak test apparatus.

3.) Hail test.

The resistance of collector covers to impact of hail (or other stones) can be tested in two ways:
The best simulation is obtained using a hail gun, with which ice balls are shot on the specimen. The hail gun available at the JRC uses ice balls of diameters between 12 and 40 mm and can apply velocities between 10 and 50 m/s, thus covering a wide range of impact energies. The ice balls are prepared in special moulds in a freezer and are propelled by pressurized air (from 0.1 to 5 bar). The velocity is measured by a double light barrier. The standard test conditions are: ice balls of 27 mm diameter (9.3 gr of weight) are shot with a velocity of 23 m/s, corresponding to a kinetic energy of 2.5 Joule.
A simpler simulation can be obtained with falling steel balls. Steel balls of 250 gr of weight, falling from 1 m of height, hence with an energy of 2.5 Joule are proposed. However these steel balls do not have the same momentum as the falling or wind driven hail stones and hence the effects for breaking glass or plastics might be different.
It has been found experimentally, that for flat plate collectors with glass glazing with more than 4 mm of thickness the impact with our standard test energy (of 2.5 Joule) is unproblematic. Even much higher impact energies do not cause any damage on window glass.
Hail damage to plastic covers has however to be checked from case to case and especially after possible embrittlement of the plastic by ultraviolet irradiation.
Also for evacuated tube collectors hail resistance has to be checked experimentally. There existed samples of tubes that already broke at 1 to 2 Joules, others resist still at 12 Joule.
It is difficult to fix a value of the impact energy to which a solar device must resist, as this depends on the meteorological conditions of the site. A rather reasonable value for European weather conditions seems to be the value of 2.5 Joule. Devices, that do not withstand to such hail impacts do not have a long service life expectation.

4.) Short term dry exposure.

Exposure in dry conditions to sun or to the irradiation of a simulator for a short time (for instance three days or three cycles in a simulator) is a valid test to assess the general useability of a collector. In these conditions the collector reaches higher temperatures than in normal operation. For some collectors, especially those which are covered with plastic glazing or those that contain plastic absorbers the exposure to full sun in dry conditions is forbidden by the manufacturer. These collectors cannot be considered to be of good quality as it is very difficult to assure that the conditions of dry exposure can always be avoided: An accidental loss of coolant from a collector which is already mounted on the roof cannot be excluded and covering of the collector may not be possible in these conditions.

The dry exposure can be made outdoors on sunny days when a minimum of 900 W/m2 of global irradiance is available for about four hours. It can also be made in a simulator. The test can be combined with the measurement of the dry exposure temperature.

If a collector shows visible damage after such a test it cannot be qualified as a "good" collector.

5.) Thermal shocks.

Two types of thermal shocks are used:
The outer thermal shock consists of a water spray applied to the hot collector in operating conditions and at thermal equilibrium. This test did not show much effect on the collectors tested at the JRC.

The inner thermal shock consists of filling the hot, empty collector with cold water. In this case the absorber is tested not only by a thermal shock, but also by a slight, transient overpressure.

This test caused in some cases permanent distortion of the absorber (roll-bonded absorber) while tube-and-fin type absorbers showed no effect. It might be that such a test is too severe, as the conditions simulated in this test can be avoided in real life by taking proper care (avoid to fill or to start circulating cold liquid in the hot collector). But a collector is certainly better if it can withstand an inner thermal shock.

6.) Freeze test.

Some collectors are claimed to be freeze resistant, either intrinsically or by drain-down in case of low ambient temperatures. These collectors have to be tested in a freeze test in an appropriate climatic chamber where the freeze temperature (typically - 20 Deg.) and the freeze duration have to be chosen according to the indications of the manufacturer.

7.) Rain water tightness test.

This test is listed here as the last test of the type "qualification" because it has already some aspects of a "durability" test. In fact, penetration of rain water into a collector is an often encountered event and it seems to be very difficult to avoid it. And the rain water entered into a collector can cause slow damage (corrosion of the outside of the absorber or damage of the optical properties, soaking of the thermal insulation, condensation on the glazing). It is very probable that the high humidity existing inside of the collector case is a major parameter for its long term destruction. Hence a good collector construction has to avoid that large amounts of rain water enter into the collector or it has to provide for good draining and ventilation so that the water which could not be prevented from entering is rapidly eliminated.
Seen under this aspect the rain water penetration test is an important qualification and durability test.
Two types of apparatus can be used for this test:
- a simple shower mounted on top of the collector which is fixed in its operating inclination or
- a grid of spray nozzles, mounted in front of the collector which is fixed in a test chamber allowing to pressurize the atmosphere around the collector. This gives the possibility to simulate the "breathing" of a collector under the influence of gusty winds.
Water is sprayed during four hours onto the collector. The water penetration into the collector is determined by weighing the collector before and after the treatment on a precision balance with an accuracy of +- 1 g (Mettler).
It has been found that most collectors are not tight to rain water. In fact it seems to be practically impossible to prevent rain from entering into a collector. It must hence be recommended that collectors should have a good ventilation, so that the humidity is eliminated from the collector atmosphere. This aspect is of course of minor importance in the case of dry climates.

C. Durability tests.

1.) Exposure to ultraviolat irradation.

Tests made with irradiations of several weeks duration on collectors susceptible to be damaged by UV have shown, that no appreciable effects are produced. Hence this type of durability test has been stopped.

2.) Outdoor exposure to sun and weather
    (with analysis of the dry exposure temperature [4]).

Outdoor exposure in dry conditions is the best known and most commonly used method to test the durability of thermal collectors.   In these conditions the collectors are exposed to daily thermal cycles, to general weathering (rain, wind, freezing etc.), and to maximum temperatures which are considerably higher than the temperatures reached in normal operation.
In particular this higher temperature of all the inner parts of the collector can cause some accelerated ageing , for instance
-by outgassing of the insulation and sealant materials and by formation of an opaque layer on the inside of the transparent cover;
-by thermal degradation of the optical quality of the absorber surface;
-by thermal degradation of the insulation material.
The collector efficiency can be measured before and after the dry exposure if an appropriate measurement facility with long term reproduceability is available. However this is a rather tedious, circumstantial and time consuming procedure, as the collector has to be demounted from its exposure rack, mounted on the efficiency measurement facility (for instance a solar simulator) and then brought back to continue the exposure.
The dry exposure lasts from a minimum of two months (depending on weather conditions) up to several years.   It is very helpful to have available during such long exposures some quantitative indication of a possible degradation of the collector efficiency.
This information can be obtained by monitoring during the exposure: the temperature of the collector absorber, the temperature of the ambient air, the global sun irradiance, and the wind speed.

The data have to be taken with a data logger for easy subsequent data analysis by a computer.In our case they were taken every 30 seconds with a PDP 11/03 computer.

Two methodes of data analysis are possible:
1.)Correlation of dry exposure temperature (DET) and global irradiance (G).

On most days the dry exposure temperature is not a steady function of time, as clouds or haze cause irregular changes. Hence the problem lies in the selection of thermally stable values of the collector temperature and the corresponding (stable) irradiance.
In a first attempt a simple search was made for the maximum value of the irradiance and for the maximum value of the difference between absorber temperature and ambient temperature reached on that day. This procedure does not give satisfactory results on days with strong fluctuations of the insolation.
Therefore another attempt was made choosing those time intervals of at least 15 minutes duration during which the global irradiance remained constant within +- 30 W/m2, with the average value $G(i)$. Then time intervals of at least 10 minutes duration were searched following that interval of stable global irradiance during which the difference of the dry exposure temperature and the ambient temperature remained constant within +- 1 degree Centigrade, with the average value $DET(i)$. In this way one obtains pairs of values of irradiance and corresponding temperature differences that can be correlated.
A plot of $DET(i)$ vs. $G(i)$ does not allow us to analyse whether any changes of the collector does occur with time, as the date of each point is not easily identified in such a plot. In order to avoid three dimensional plots it is hence convenient to characterize the collector on each day by a single figure. This figure can be the normalised dry exposure temperature difference $NDET(i)$, defined as the temperature difference (between absorber and ambient air) extrapolated to an incident irradiance of 1000 W/sqm.
The empirical formula (1) gives the relation between measured values of $DET(i)$ and $G(i)$.

$$DET = p * G^{1/n} \quad ; \tag{1}$$

where DET is the temperature difference between the absorber and the ambient temperature,

p is a constant, characteristic for each collector,

G is the irradiance and

n is a coefficient that characterizes the non linearity of the heat losses of the collector.

n and p are determined empirically by choosing some good values (same, low wind and same ambient air temperature) of data pairs DET(i) and G(i) and by fitting a curve to equation (1).

The normalized dry exposure temperature difference NDET is then obtained by the equation (2):

$$NDET(i) = DET(i) * (1000)^{1/n} / G(i)^{1/n} \; ; \qquad (2)$$

where DET(i) and G(i) are the experimental values.

The NDET values, when plotted vs. the day of the year, still show some scatter, due to influences from wind and ambient air temperature.

Wind correction.

The average wind speed at our site varies from zero to 3 m/s, but most of the data were taken for wind speeds between 0.5 and 1.5 m/s. A plot of the normalized dry exposure temperature differences vs. wind speed showed a decrease of the normalized dry exposure temperature of the order of 6 Deg. per 1 m/s of wind speed. (indicative value for the range of 0.5 to 1.5 m/s). A plot of the normalized dry exposure temperature difference (NDET) with this correction vs. day of the year has a smaller scatter than the uncorrected plot.

Temperature correction.

A plot of the wind corrected data showed an influence of the ambient temperature on the NDET values (seasonal fluctuation: higher heat losses at higher ambient and higher collector temperatures). Empirically a correction by −0.25 % per degree of ambient temperature difference allows us to eliminate this effect.

A result of such an analysis is plotted in Fig. 1 as an example with corrections of the normalized dry exposure temperature difference of + 6 Deg. per each m/s of wind speed and of −0.25 % per each degree of the air temperature below 32 Degrees. The scatter of the resulting temperature difference is of the order of +−4 Deg. (or +− 4%), which is a rather satisfactory result. A slight decrease of the NDET by about 8 degrees (over 2 years of outdoor exposure) can be noted.

2.)Analysis by Integrals.

The heat balance equation for a point model of the flat plate solar collector is:

$$G(t) * \tau_e \alpha_e = Ue*(T(t)-Ta(t)) + Ce(dT(t)/dt) ;$$ (3)

where:
G = G(t) = global irradiance (in Wm-2),
τ_e = tau = effective transmittance of the collector,
α_e = alpha = effective absorptance of the collector,
Ue = Ue(T,w) = effective overall heat transfer coefficient
     of the collector (in Wm-2K-1),
Ce = effective overall heat capacity of the collector per
     unit surface (in Jm-2K-1),
Ta = Ta(t) = ambient air temperature (K),
T = T(t) = effective surface temperature of the absorber (K),
t = time (s),
w = wind speed (m/s).
All the "effective" definitions used in the point model of eq.(3) represent some spatial average values, which are not very exactly defined. As only changes . in time of these values are of interest, their vague definitions can be accepted.
The effective overall heat transfer coefficient depends on T: for higher absorber temperatures the irradiation heat losses become more important and Ue increases. It also increases with the wind velocity w.
For the analysis, integrals of equation (3) are made over time and over such intervals that the absorber temperature is the same at the beginning and at the end. By extending the integral over that time interval the integral over the second term of the right hand side of equation (3) is zero because the collector has the same temperature and hence the same energy content when it has the temperature Tm in the morning and when it reaches the same temperature Tm again in the afternoon.   In other words all of the radiant energy absorbed by the collector during that time interval has been lost in heat losses to the environment.
These heat losses are represented by the integral over the first term of eq.(3), the temperature differences multiplied by the heat loss coefficient.

Equation (3) then becomes:

$$\frac{\tau_e \alpha_e}{\overline{Ue}} = \frac{\int_{t_1}^{t_2} (T(t) - Ta(t)) * dt}{\int_{t_1}^{t_2} G(t) * dt} \quad ; \quad \begin{array}{l} \text{integrals taken from} \\ \text{t1, when } T(t){=}Tm \text{ to} \\ \text{t2, when } T(t){=}Tm. \end{array}$$

Assuming an average value of the heat loss coefficient $\overline{Ue}$ for the range of Tm to the maximum absorber temperature T.

In reality the heat loss coefficient is not a constant, as the heat loss is higher at higher absorber temperatures (radiant energy transfer). So the ratio of the integral $IT = \int(T(t)-Ta(t))dt$ and the integral $IG = \int G(t)dt$ depends on the value of Tm chosen and is smaller for higher Tm's. It is hence preferable to use the differences of the integrals taken from Tm = 50 Deg. and from Tm = 100 Deg. as these differences correspond to the mean heat loss coefficient between 50 and 100 Degrees and not to the mean heat loss coefficient between Tm and a changing upper temperature limit, which depends on the maximum temperature reached on that day.
The ratio of the difference of the IT integrals and of the difference of the IG integrals, which is according to eq.(4) Q = (tau)e*(alpha)e/Ue is called the collector "quality" and is a measure for the efficiency of the collector. A decrease of this ratio with time would indicate a degradation of the collector.
The quality is plotted in Fig. 2 versus the day of measurement; the scatter is due to unidentified errors and limits the sensitivity of the test.A slight decrease of the quality by about 6% over the two years of dry exposure can be seen.

In order to check the sensitivity of the method a collector was covered for two months by a transparent plexiglass cover, which reduced the transmittance "tau". The data for this test are also plotted in figures 1 and 2. A sharp decrease of the NDET by about 10 degrees and of the quality by about 8 percent is to be noted.

Monitoring the Dry Exposure Temperature and analysing the data in the proper way is a valid method to detect slow degradation of the efficiency of thermal solar collectors The scatter of the data obtained by either method 1 or 2 coming from unidentified influences is such, that a decrease of the quality and hence of the efficiency of a collector of

5 % would certainly be detectable.

A similar test with measurement of the dry exposure temperature and correlation with the irradiance can be made in a simulator. At the JRC the degradation simulator AT-6 is used for that purpose with irradiations lasting 5 days.

D. Tests on atmospheric corrosion.

At the JRC (Ispra) two activities are carried out:
- Accelerated corrosion tests in climatic chambers. There exist a large corrosion chamber (AT-1) permitting treatment in high humidity atmosphere with addition of sulphurdioxyde gas according to standard IEC 68-2-42 and a smaller chamber permitting treatment in salt mist, according to standard IEC 68-2-11.

- Weathering in service conditions in typical atmospheres. Five different commercially available collectors are exposed as in real operating conditions in two different atmospheres: rural clean atmosphere as in Ispra, and industrial (SO2 containing) atmosphere as in Turbigo. The operating conditions are monitored including SO2 content and after several years of exposure the collectors will be checked for possible efficiency decrease and for corrosion and other damage [5].

A literature study and some laboratory experiments have been done on a contract basis at the University of Leuven . The results are reported in [6].

E. Tests on hydraulic corrosion.

This aspect of the durability of complete collector systems must not be neglected. A corrosion loop is in operation in Ispra to study these effects.

References:

[1] Proceedings: Workshop on Service Life of Solar Collector Components and Materials, Dec.6-8,1983 at the Technical University of Denmark (to be published as IEA report)

[2] D. Waksman et al. NBS Solar Collector Durability/Reliability Test Program: Final Report (NBS Technical Note 1196 (1984).

[3] G. Riesch, Qualification and Durability of Thermal Solar Collectors, Report EUR 8786 EN (1984).

[4] G.Riesch, Collector Degradation Evaluated by Analysis of Dry Exposure Temperature, ISES-Proceedings of Intersol 85, Montreal, Canada (1985).

[5] R. Bauch et al., Weathering of Solar Collectors, Report EUR 9311 EN (1984).

[6] W. Bogaerts et al., Atmospheric Corrosion and Degradation of Construction Materials for Solar Collectors, Report EUR 9042 EN (1983),
Corrosion Studies on Collector Materials, Report EUR 9597 EN (1984).

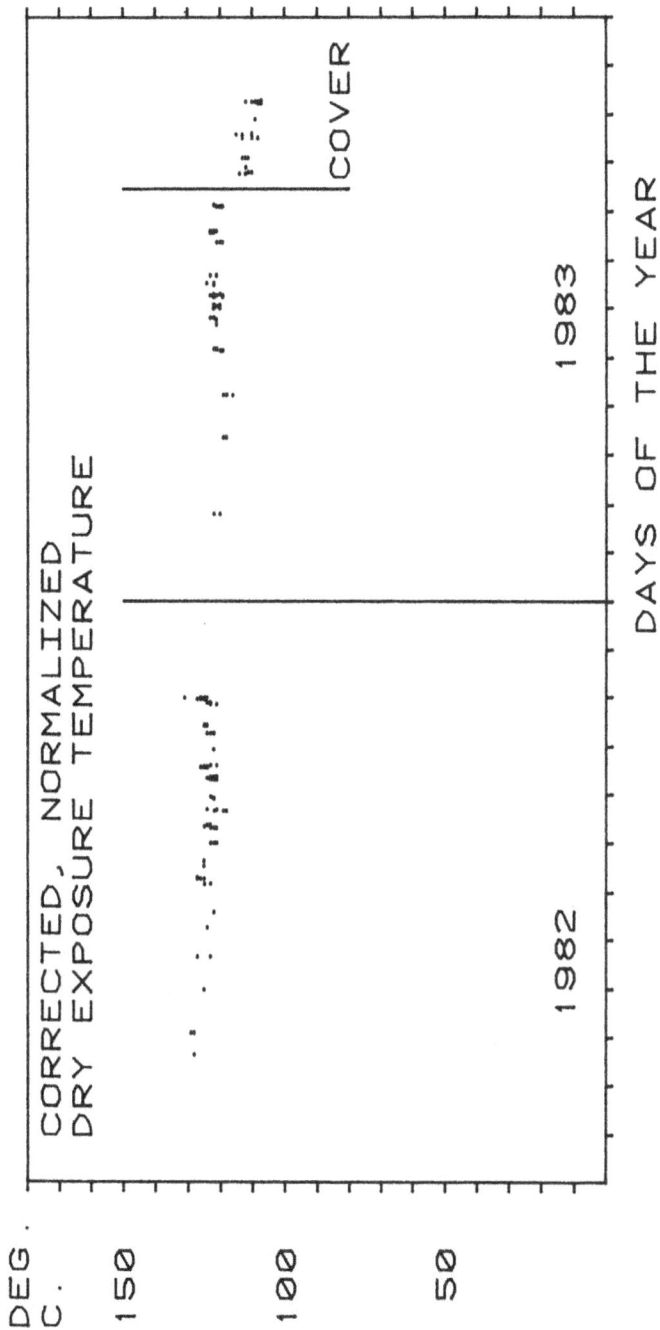

CORRECTED, NORMALIZED
DRY EXPOSURE TEMPERATURE

DEG.
C.

150

100

50

1982

1983

COVER

DAYS OF THE YEAR

FIG 1 NDET VS TIME

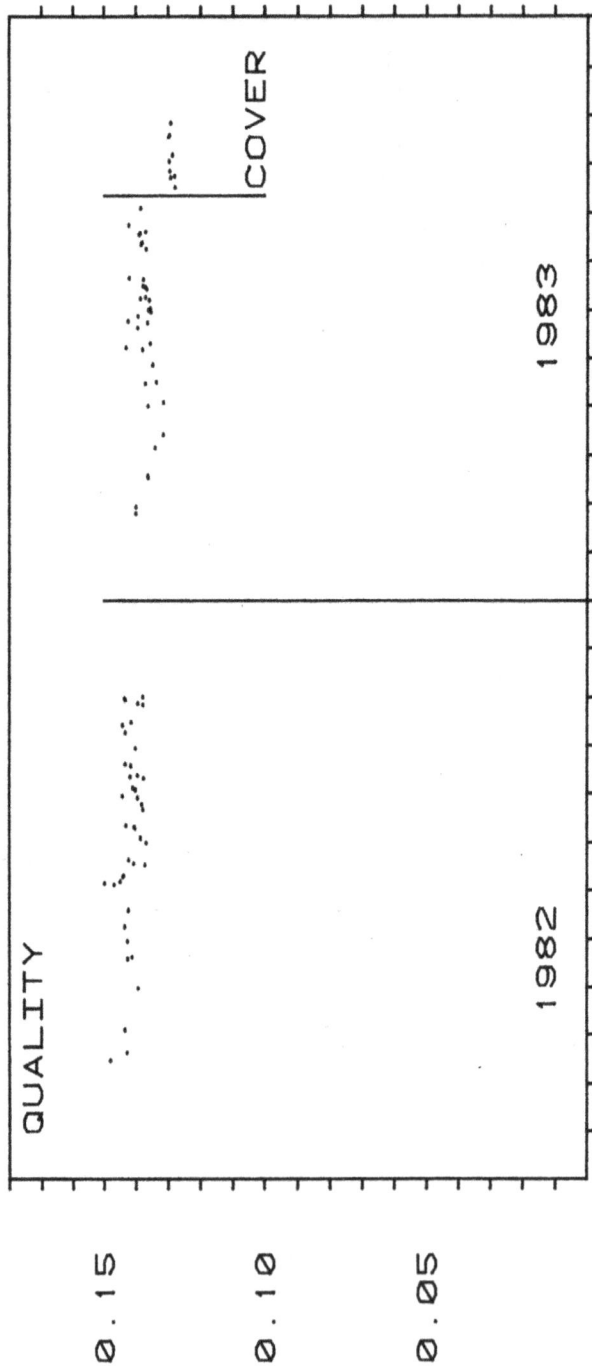

FIG. 2. QUALITY VS. TIME.

DUST EFFECT ON SOLAR FLAT SURFACES DEVICES IN KUWAIT

A.A.M. Sayigh*, S. Al-Jandal** and H. Ahmed**

*Energy Dept. Resources, OAPEC, P.O. Box 20501, Safat, Kuwait.

**Kuwait Institute for Scientific Research, Kuwait.

ABSTRACT

The dust problem is one of the major reducing agents of solar radiation intensity in Kuwait. The paper is divided into three parts: one, which deals with dust occurence situation and the solar radiation in Kuwait. Part two deals with the instrumentations for measuring dust effect which consists of two detectors and voltmeters. The set up made of four benches. Each bench has six stands for different tilt angle, namely 0^0, 15^0, 30^0, 45^0, 60^0 and 90^0. The samples are placed facing south and made of glass, plastic, mirrors and stainless steel material. Part three is made of all the data collected in three locations, KISR solar house, which is near the sea, Al-Shagaya area which is a typical desert location and in built-up area with grass. Readings are collected for different exposures ranging from one day to forty days over a period of nine months.

KEYWORDS: dust degradation, angstorm, Raleigh, Mie absorptivity, emmissivity.

INTRODUCTION

The dust effect on solar energy equipment is a major problem in Kuwait. Storms and suspended dust particles in the atmosphere occur all year round due to the situation of Kuwait which is at the edge of the Arabian Desert and to the prevailing wind which is invariably westerly or north-westerly. Fine dust particles accumulate on solar energy devices such as collectors, reflectors, concentrators, and photovoltaic panels and prevent the full amount of solar radiation from reaching their surfaces. This effect leads to the reduction of the incoming radiation by as much as 50% during one month if the solar device has not been cleaned, (1).

Little work to assess the degradation and solar radiation depletion from dust accumulation has been carried out elsewhere. In Saudi Arabia, Sayigh, investigated the dust effect on flat plate collectors extensively and found that a reduction of about 30% in the energy collected was reached after 25 days without cleaning, (2). In another study, in Saudi Arabia Sayigh, et. al, (3), found in another study about the dust effect on a photovoltaic panel at 30° tilt, it was found that a reduction of power of 2%, 14% and 30% after one, thirteen and thirty two days respectively without cleaning. In Kuwait a limited number of tests were carried out on photovoltaic panels which registered a reduction of 17% after 6 days without cleaning, (4). Work carried out by Pettit et. al, (5) on silver glass mirrors exposed for 35 days in Albuquerque, NM, USA shows a reflectivity reduction of 20% due to dust accumulation. Testing heliostats under desert conditions, Blackman and Cureija, (6), found a degradation in reflectivity for single surface glass mirrors of 9%, 17%, 19.5% and 3% corresponding to mirrors facing: up during the day and down at night; near vertical stow; permanent face up; and permanent face down respectively, after 30 days of exposure during the summer.

THEORETICAL ANALYSIS

Light impinging on glass or reflective surface normally refracts or reflects from these surface in a definite pattern. But when dust particles are present, the light will scatter, or be absorbed so very little of it can be reflected. It was believed that radiation scattered by dust particles would have a Lambertian angular distribution, (7). Young found experimentally that the scattered rays from particles contaminants on low scatter mirrors was not characterized by the Lambertian model. Particles on a mirror surface smaller than the incident radiation wave length produce less scatter than was predicted, and large particles scattered more energy at near specular angles than the Lambertian model predicted. Kerker proposed several models, (8), but very sophisticated equipment is needed to measure and validate them.

The two well known theories for solar radiation scattering due to dust particles or aerosols are the Rayleigh theory and the Mie theory. Rayleigh's theory is limited to spherical particles with a diameter smaller than the light wavelength (λ), while Mie's theory is more general and can be used for any particle size.

If D represents the particle diameter in micron (μm), and n is the index of refraction, then the following cases will result:

if $\pi D/\lambda < 0.6/n$, scattering is governed by Rayleigh's theory and if $\pi D/\lambda > 5$, then scattering is a reflection process, and if $0.6/n < \pi D/\lambda < 5$, then Mie's theory is the governing theory, (9).

Mie's theory in terms of dust scattering coefficient, (K_d) is:

$$(K_d) = 0.08128\lambda^{-0.75}$$

which is valid for a wide range of number of particles per cubic centimeter (Ma) between 1 and 800. An atmosphere with 200 particles per cubic centimeter is a clean atmosphere, while an atmosphere with 800 particles per cubic centimeter is a very dirty one. Therefore the spectral transmittance (τ_d) can be expressed as:

$$\tau_d = \exp\left[-0.08128\lambda^{-0.75} (d/800)\, ma\right]$$

Angstorm proposed the following formula: $K_d = \beta\lambda^{-\alpha}$, where β is called turbidity coefficient, and α is the wavelength exponent. Also $(\tau_d) = \exp(-\beta\lambda^{-\alpha}ma)$, (9).

EXPERIMENTAL RESULTS

Four stands were especially made with the following dimensions, 1.5m tall, 3.5m long, and 30 cm wide. Platforms at tilts of 15^0, 30^0, 45^0, 60^0 and 90^0 to the horizontal were placed on top of each stand.

Two stands were located in a built-up area, one in a relatively planted area and the fourth was placed in a desert environment.

For transmittance tests, several 4mm window glass and 2mm plexi-glass-G specimens were prepared for testing. For reflectance tests, several 0.5mm stainless steel and 6mm glass mirror specimens were used. The specimens of the same material were tested for different durations and at different inclinations to the horizontal, and vice versa, night only, day only, one whole day, 2, 3, 7, 14, 21, 28, 35 and 45 days. Figure 1 shows one of the stands while Fig. 2 shows the results for a set of specimens of one material at a given tilt. Figures 3, 4 and 5 show dust accumulation on glass specimens, P-V panel and flat plate collectors.

Fig. 1 Two stands for dust specimens.

Three different apparatus were used to read the dust effect. The first one consisted of a light source, a sensor and a wattmeter. The sensor and the light source are coaxial and the specimen test was placed between them. A rigid steel bench with spring loaded platform that slides in two perpendicular directions on a horizontal plane was used. The whole arrangement was used to measure transmissivity of glass and plexiglass specimens only.

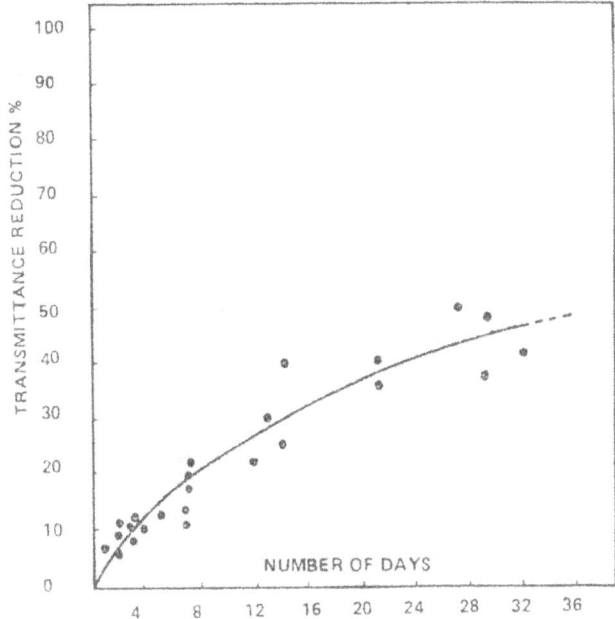

Fig. 2 Transmittance reduction in a glass pane at 30° - tilt due to dust effect.

Fig. 3 Dust accumulation on glass specimens.

Fig. 4 Dust accumulation on a PV panel.

Fig. 5 Dust effect on flat plate collectors.

The second apparatus was made of a 6V light source, a detector and a voltmeter. A special electrical arrangement was made to change the voltage from 220 volts AC to low DC voltage. The light source and the detector were placed at 45° to the vertical axis, ensuing a prop reading from surfaces. This equipment was used for reflectivity measurement of glass mirror and stainless steel samples. The third apparatus was an electron microscope. Random dusty samples were viewed by the electron microscope in order to measure the various sizes of dust particles. Figure 6 shows a photograph of different dust particles on a specimen. The magnification was 8000.

Fig. 6   Dust particles' sizes as seen through electron
microscope with 8000 magnif.

The dust pattern in Kuwait is shown in Fig. 7, (10). The effect of dust on daily radiation measurement can be seen in Fig. 8. Figures 9 to 12 show the results of more than 200 tests during a nine months period. In order to simulate the actual situation of the sun light

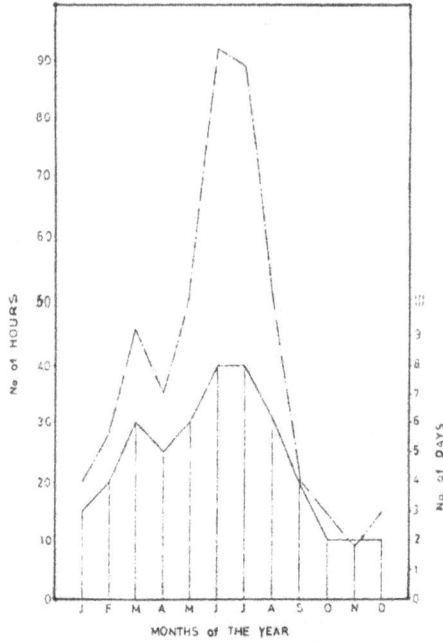

Fig. 7 Dust pattern in Kuwait 1962-1973, (10).

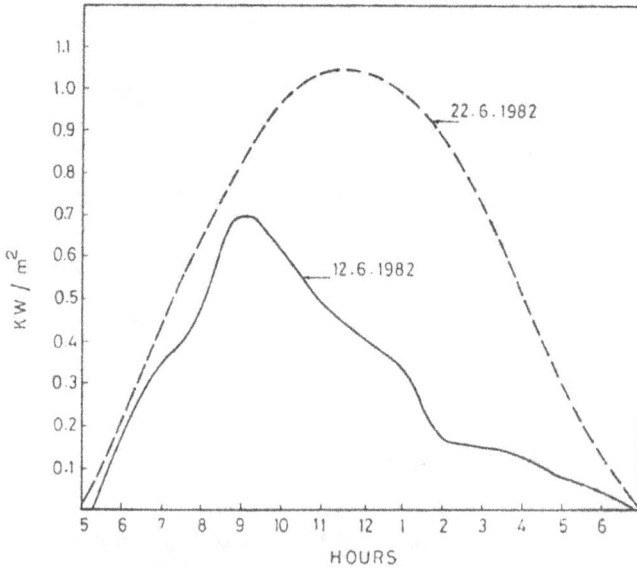

Fig. 8 A comparison of irradiance between dusty day and clear day in June 1982.

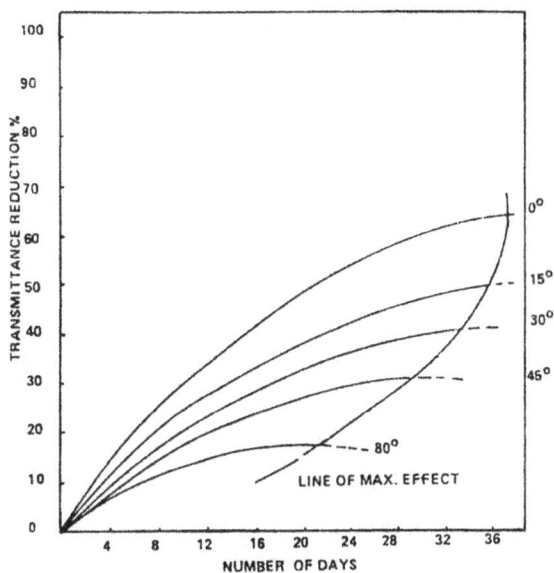

Fig. 9   Transmittance reduction in glass panels.

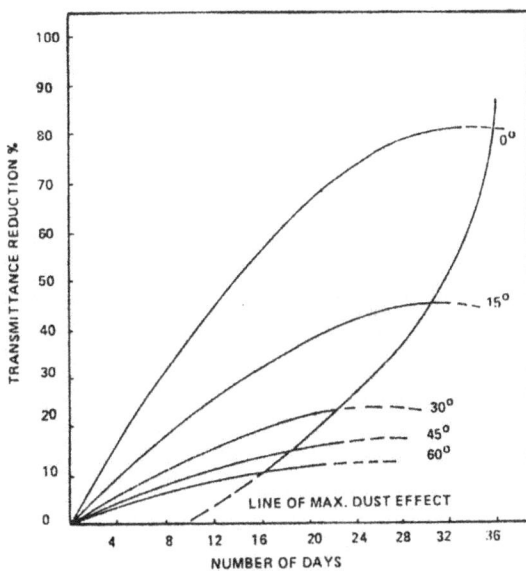

Fig. 10   Transmittance reduction in plexiglass-G.

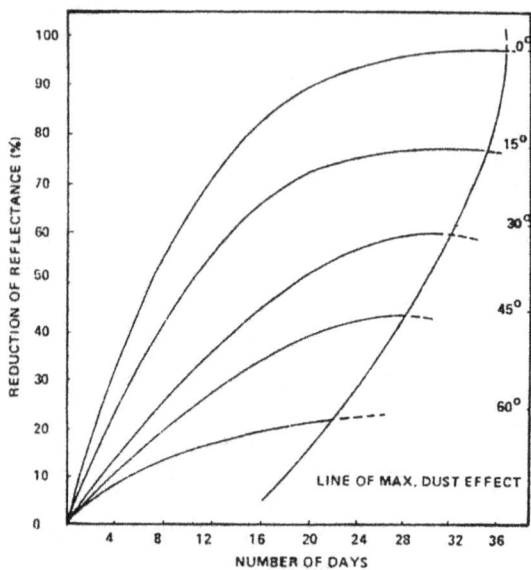

Fig. 11 Reflectance reduction of stainless steel sheets at different tilts.

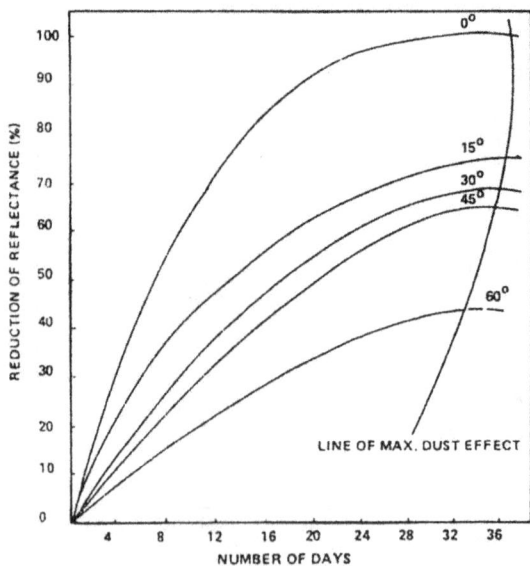

Fig. 12 Reflectance reduction due to dust effect at different tilts for flat mirrors.

going through a glass of a given collector at a fixed inclination, a special adjustable angle was made and various glass samples were tested using the second apparatus arrangement. Figure 13 shows these results. Several tests were also carried out to study the effect of dust on a heliostat. The results are shown in Fig. 14. Another set of results were carried out on flat plate collectors which shows markedly the effect of dust on the efficiency, see Fig. 15.

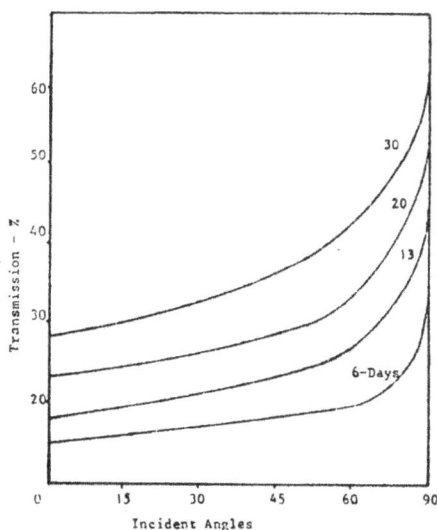

Fig. 13 Dust effect on glass transmittance when sunlight hits it at different angles.

CONCLUSIONS

In reviewing Figs. 7 to 15, one can deduce the following:

1. Dust exists all year round in Kuwait and it was found that there is little difference between various locations because of the small area of the country and the large desert which borders it on the west side. Dust is normally maximal during the months of June and July and normally minimal during the months of October, November and December. Dust storms are also frequent but do not last more than few days, especially during the spring season.

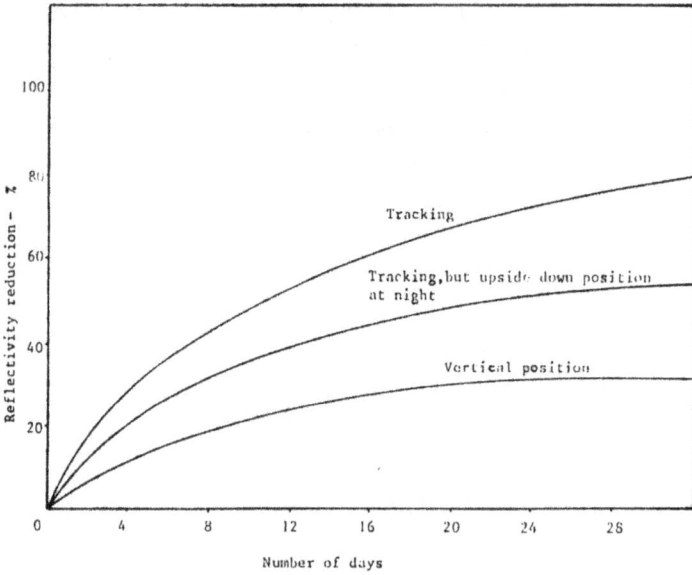

Fig. 14 Dust effect on heliostats.

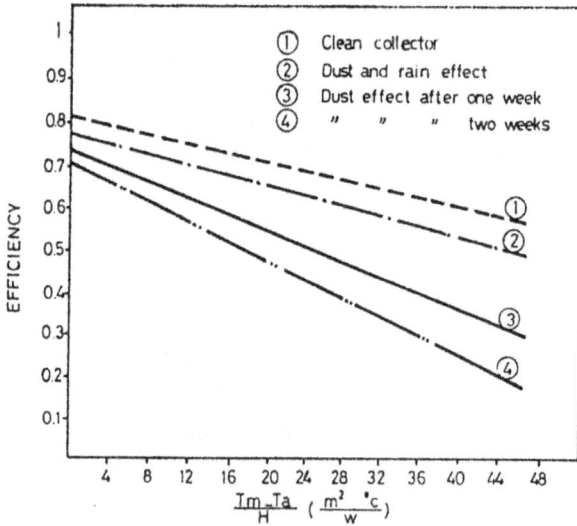

Fig. 15 Dust effect on flat plate collectors.

2. A summary of the results has been tabulated in Table 1.

Table 1 Dust effect on transmittance and reflectance pro-
         perties of various samples

Type	Angle tilt with the horizontal	No. of days for maximum dust effect	Percentage of Reduction
Glass	0	38	64
Transmittance	15	35	48
	30	33	38
	45	29	30
	60	22	17
Plexiglass-G	0	36	80
Transmittance	15	31	46
	30	23	22
	45	19	15
	60	16	10
Stainless	0	37	96
steel reflec-	15	35	77
tance	30	33	60
	45	28	42
	60	22	22
Mirrors	0	38	100
Reflectance	15	37	73
	30	36	67
	45	36	63
	60	33	42

3. The dust particles have various sizes and in general it was
   found that about 5% of the surface was covered with particles
   greater than 5um, 30% with particles between 5um and 1um, and
   65% was covered with particles less than 1um in size.

4. Several mathematical models exist to simulate the effect of
   dust on solar radiation, such as Raleigh, Mie, and Angstorm.
   It is possible to use a polynomial such as

$$Y = e^{A + Bx} - Cx^2$$

where y is the number of days, x is the percentage of

transmissivity reduction or reflectivity reduction, A, B and C are constants dependant on the material under test and the amount of solar radiation.

5. A cleaning procedure is essential. This depends on the number of days and the percentage of dust accumulation which can be tolerated. Using 10% reduction as a guide in Kuwait, a cleaning cycle is needed every three days. If the dust is dry then simple wiping will do for cleaning while if it is mixed with moisture then a water jet is required followed by wiping technique.

ACKNOWLEDGEMENT

The authors acknowledge the help and assistance of Dr. N. Kolias, Physics Department, University of Kuwait and Mr. Ali Baqer, Al-Sabah Hospital, particularly in the preparation of the dust measuring equipment. Also Mr. Muhammad Rashead and Mr. Zain Al-Abidean for their parts in collecting Al-Shaqaya samples.

REFERENCES

1. A.A.M. Sayigh, S. Al-Jandal and H.H. Ahmed, "Degradation of flat surface by dust solar applications" Final report SE-43, KISR 1487, October 1984.

2. A.A.M. Sayigh, "Effect of dust on flat plate collectors", Int. Solar Energy Congress, ISES7, Jan. 16-21, 1978, New Delhi, India.

3. A.A.M. Sayigh, S.H. Charchafchi and A.A. Al-Habali, "Experimental evaluation of solar cells in arid zones", Izmir International Symposium -11, August 6-8, 1979, Izmir.

4. F.G. Wakim, "Introduction of photovoltaic power generation to Kuwait", KISR report No. 440, 1981, Kuwait.

5. R.B. Pettit, J.M. Freese and D.E. Arvizu, "Specular reflectance loss of solar mirrors due to dust accumulation", Seminar Proceedings, Inst. of Environmental Sciences and Conference on Testing Solar Energy Materials and Systems, PP. 164-167, 1978.

6. T.B. Blackman, M. Curcija, "Heliostat reflectivity variation due to dust build-up under desert conditions", Seminar Proceedings, Inst. of Environmental Sciences and Conf. of testing Solar Energy Materials and systems, pp. 169-179, 1978.

7. R.P. Young, "Low-scatter mirror degradation by particle contamination", Optical Engineering 15(6), pp. 516-520, 1976.

8. M. Kerker, "The scattering of light and other electromagnetic radiation", Academic Press, New York, 1969.

9. M. Iqbal, "An introduction to solar radiation", Academic Press, New York, 1983.

10. Meteorological Department, Climatological Division, Directorate General of Civil Aviation, "Climatological summaries", Kuwait International Airport, 1962-1982.

SOLAR ENERGY UTILIZATION IN THE ARAB COUNTRIES
A.A.M. Sayigh
Energy Resources Dept., OAPEC,
P.O.Box 20501, Safat, Kuwait

ABSTRACT: The paper outlines the solar intensity and sunshine hours in the region and compares them with some industrialized countries. Brief descriptions of various solar energy projects are tabulated. After specifying the number of months for each when a) heating and b) cooling are required project details are given. More specifically the cooling of the solar energy research building in Baghdad (120 tons of solar absorption chillers, 80 tons of heat pumps); the heating of the King Abdul-Aziz Airborne and Physical Training School, near Tabuk, Saudi Arabia; the 350 kW PV. field of the solar energy village near Riyadh; and the 100 kW solar thermal plant in Kuwait are discussed. Five desalination plants are also mentioned plus the Yanbu plant of 2400 m^3/day, which is one of the largest in the world. The potential of wind energy utilization is considered. Obstacles hindering the progress of solar energy in the region are also outlined.

INTRODUCTION: The Arab countries enjoy some of the maximum solar radiation intensities in the world lying as they do between 30^0 parallel, the sunny belt of the earth. Compared with other regions on the earth the Arab region receives 70% of the extraterrestial radiation, (1,2). Most countries have about 3000 hours of sunshine and an average solar radiation intensity during the day of 500 W/m^2. Figures 1,2, and 3 show the minimum, maximum and average solar radiatioin intensities in the region. A further comparison can be made for the twons of the region and some industrial towns not in the region and figure 4 will result.

As early as 1907, Shuman had built his solar pumps near Cairo, Egypt which was used for irrigation. Then during the late fifties, a study was made of the Qattara depression for electricity generation, which lies between Egypt and Libya with a view to utilizing it for power generation. At the same time several French-made furnaces were installed in several Arab countries in North Africa. In the mid-sixties R and D activities were started in the national research centre of Egypt mainly in heating and desalination. These activities later were enlarged to cover most of the Arab countries where solar energy research centres were established either seperately as in Iraq, or part of national research centres like those of Saudi Arabia and Kuwait.

SOLAR ENERGY PROJECTS : In the Arab countries most of the R and D in the field of solar energy takes place in the universities and educational institutions. However, there is some progress in the private sector in fields such as water heating and in the government sector such as repeater stations, cathodic protection and telecommunication.

1.   Solar heating and cooling: Many countries have different requirements for airconditioning but the most widely used one is that of ASHRAE which specifies 75^{0}F (23.9^{0}C) and 40% relative humidity. However, in fact in many parts of the world, 68^{0}F (20^{0}C) and 50% of relative humidity are used. Using the 20^{0}C as a criterion for comfort, table 1 can be deduced for cities in the Arab countries. Nevertheless, despite the abandant oil resources, most of the Arab countries have developed the know-how and embarked on several solar energy utilization projects for heating and cooling. Table 2 shows some major cooling projects in the area. Several small projects in cooling were run in Tunis, Morocco, Sudan and Kuwait employing lithium bromide-water, amonia-water, zeolite, and desiccant cooling systems, (3,4).

2. <u>Solar heating projects</u> : As shown in table 1, heating is required in some Arab countries during the winter months. Countries such as, Jordan, Suadi Arabia, Tunis, Egypt, Iraq, Syria, Lebanon, Morocco and Libya have some manufacturing capability to produce solar water heaters, (1). In Jordan for example, 50000 water heaters are in operation in Amman alone, while in Iraq if solar water heaters were used extensively it would lead to a saving of about two million tons of oil a year. In the experimental side, hot water and solar heating have been utilized in all the solar houses which exist in Jordan, Kuwait, Libya, Tunisia, Algeria and Iraq. The Jrdanian solar house utilizes under floor heating which proved to be very effective. In Saudi Arabia, King Abdul Aziz airborne and physical training school near Tabuk, has 2592 flat plate collectors with 1.64 m^2 cross sectional area each. The collectors are used to produce 36000 gallons of hot water at 60°C. The system provides 100% of the hot water needs and 40% of the heating requirement for the school,(5). The other heating project is that of Abu Nawas street - Baghdad, where there are 273 apartments using solar water heaters for their hot water supply and partial heating demand, (4).

3. <u>Photovoltaic applications</u>: Most Arab countries have photovoltaic applications. These applications vary from electricity generation to be used as part of the utility, to a specific application such as traffic light signals or cathodic protection in the oil field. Table 3 details some of the important photovoltaic applications in the Arab world. Additionally there is experimental research on thin film cell in Tunisia, Iraq, Kuwait, Egypt, Syria, and Qatar (6,7).

4. <u>Solar thermal power generation</u>: Solar thermal power plant programmes did not get off to a good start in the Arab world.

This was mainly due to operational problems as well as dust problems. Table 4 shows some of these applications, (8)

5. Solar desalination: Most Arab countries have experimented with solar desalination and Simple Still experiments are part of each research group's programmes. Table 5 shows some of the large scale solar desalination plants in the region.

6. Solar pond applications: Solar ponds in an experimental stage has been tested in UPM-Saudi Arabia, Egypt, Kuwait and Qatar. An active programme is about to start in the Royal Scientific Societies RSS-Jordan to build a large size solar pond in the dead sea, (9).

7. Wind power generation: Wind generators of reasonable size (20-50) kW are not yet in use in the Arab countries. Both Jordan and Morocco have several (1-2) kW wind generators employed for pumping water. The RSS in Jordan has developed its own wind towers, but the rest of the region does not have any manufacturing capability. Using a reference speed of 4 m/s as a good propsect for wind power generation, the following countries have a good wind energy potential in a descending order: Yemen A.R., Yemen D.R., Jordan, Mauritania, Tunis, Bahrain, Morocco, Oman, Qatar, Somalia, Lebanon, Libya, Kuwait, United Arab Emirates, Sudan, Syria, Egypt, Algeria, Iraq and Saudi Arabia, (10).

CONCLUSIONS: Solar energy utilization in the Arab world has made reasonable progress mainly due to the research activities in its universities and research centres. Fundamental research has been carried out in solar radiation, selective coatings and photovoltaic technology to name a few topics. However, there are several obstacles hindering this progress, namely:

1. Availability of oil and its superiority to solar energy as a source of energy and its relatively low cost in Arab oil producing countries.
2. The dust effect which in some part reduce solar energy by 50%.
3. Land and material intensive nature of solar energy technology.
4. The availability of governmental subsidies for oil-electricity generation,and non availability of similar subsidies for solar energy programme which hinders the establishments of large scale installations which are usually needed to gain experience and adaptability.
5. Environmental capability and adaptability are invariably lacking. Thus technologies more suited to a cold climate, for example are used for a hot climate. The development of local technology and expertise should be of priority.

Communication and exchange of solar energy material among Arab scientists has reached a reasonable level, whether through official channels such, The Arab Union of Scientific Centres, The Organization of Arab Petroleum Exporting Countries (OAPEC), The ALECSO and The Arab Section of International Solar Energy Society.

Table 1
Heating and Cooling Periods in Some Arab Cities

City	Number of Months for		City	Number of Months for	
	Heating	Cooling		Heating	Cooling
Sanaa	0	0	Kuwait	2.5	7
Nouakchott	0	12	Baghdad	4	5
Mogadishu	0	12	Riyadh	4	6
P. Sudan	0	6	Beirut	6	1
Abu Dhabi	2	8	Damascus	6	2
Muscat	2	8	Alger	7	0
Doha	3	7	Rabat	4	0
Bahrain	2	7	Amman	8	0

Table 2
Cooling Projects in Some Arab Countries

country	Name of Project	Cooling capacity & Type	Type of collectors	Area of collectors	Cool-ing Tower Type	Hot water storage m³
Egypt - Cairo	Cold storage - for fish and vegetables	6-Absorption	Heatpipe			
Saudi Arabia	Experimental P-V-cooling unit - Riyadh	1/2 - vapour compress-ion	P-V	1.5	-	batteries
	Agricultural Lab-oratory (Ministry of Agr-iculture	7-Rankine	Flat P.	2400	Air	960
	College of Engin-eering - King Saud University	3-Absorption	Flat P.	56	water	12
	Soleras - Engin-eering field Tests-SANCST	18-Rankine	Para.Conc. Collectors	122.3	air	456
		14-Rankine	Evacuated tube-collectors	218.5	air	-
		15-Absorp-tion	Para-Conc. collectors	133.8	water	174
		10-Absorp-tion	Fresner Conc. Collectors	88.8	air	152
Kuwait	KISR-Solar House	7.5 Absorp-tion	Flat P.	96	water	7.5
	Saqr Al-Rushood Kindergarten	40(4x10) Absorption	Flat P.	360	water	30
	Offices - Ministry of Defence	40(4x10) Absorption	Flat P.	300	water	25
	Private Villa	20(2x10) Absorption	Flat P.	160	water	10
Iraq	Solar House-SERC	20(2x10) Absorption	Flat P.	243	water	20
	Abu-Nawas Project	156 units (7.5 & 10) Absorption	Evacuated tube collectors	-	water	-
	SERC-Kindergarten	12x2270 m³/hr. Aircooler	P-V	-	-	-
	SERC-building	120(2x60) Absorption and 80(2x40) Heat Pumps winter	Evacuated tube collectors	1557 units	water summer 300	30

Table 3
Photovoltaic application in some Arab countries

Country	Type of Application	Remarks
Saudi Arabia	- Solar Village	350 kW, Tracking with Fresnel lenses,Riyadh
	- Fence electrification	59 arrays - 80 Km. long P-V fence. King Khalid Military City.
	- Car park lighting	20 kW, Jedda Int. Airport car park.
	- Tunnel lighting	10 kW, Abha road.
	- Cathodic protection	several large projects of oil pumping across the desert and oil installations.
	- Short wave transmission	Radio transmission station near Riyadh
	- Telephones	Highway telephones between Riyadh and other cities, Dhahran, Jeddah and Medina
	- Desalination	10 kW., reverse Osmosis unit-Jeddah
Kuwait	- Solar House	2 kW, lighyts and pumps, KISR.
	- Battery charges	Several installation for battery charging of fire-engines and military vehicles
	- Traffic	3-sets of traffic lights in Kuwait city, 240 W each.
	- Cathodic protection	In some oil installations
	- School lighting	24 kW, Kuwait English School-Kindergarten
	- water pumping	6kW, Multi-Stage-Flash unit, KISR
Iraq	- street lighting	40-Units of 75W each to be installed near Baghadad.
	- water pumping	A unit for pumping water to 40-m height (about 10 kW)
Tunisia	- solar village	27 kW, electricity for lighting and water pumping
Jordan	- general purposes	About 3kW, electricity generation and degradation tests.
	- Battery chargers	Several stations for battery charging of police cars, civil defence repeater stations.
Bahrain	- Petrol station	3kW, operates pumps in a car filling station.
Morocco	- water pumping	Several medium size water pumping stations
Oman	- T.V. Transmission	6-stations near Muscat.
	- Cathodic protection	All oil installations
United Arab Emirates	- Street lighting	Many street lights were installed near Jabal Abu Ali, tunnel road.
Mauritania	- water pumping	3.7kW, and 0.9 kW pumps for irrigation.
Libya	- solar village	Generation of electricity and in water pumping
	- cathodic protection	All oil installations in the south of the country.
Qatar	- Desalination	9.6kW, revrse Osmosis unit - Qatar University.

Table 4
Solar Thermal Power Plants In Some Arab Countries

Country	Project Name	Remarks
Kuwait	Sulaybia project	100 kW$_E$, consists of 56 dishes – Rankine engine, KISR.
	Solar Furnace	1 kW$_E$, KISR demonstration furnace.
Saudi Arabia	Solar village	100 kW$_E$, Two dishes with sterling engine
Algeria	Solar Furnace	21 kW$_{th}$, used for chemical reactions.
Morocco	Solar Furnace	6 kW$_{th}$, used for ceramic work.
Qatar	Power Generation (Experimental station)	High temperature collector test bench, near Doha)

Table 5
Solar Desalination Plants In the Arab Countries

Country	Project type	Remarks
Kuwait	KISR – multi stage Flash unit (MSF)	10-m^3/day, uses linear$_2$concentrating collectors (Area 220 m^2)
	Sulaybia (MSF)	50-m^3/day, uses power plant cooling water
Saudi Arabia	Yanbu-Desalination Plant	240 m^3/day, uses concentrating collectors (MSF and indirect freezing system)
	Mobile Club	10-m^3/day, utilizes P-V and reverse Osmosis – Jedda.
Qatar	Qatar Univ. Project	7-m^3/day, utilizes P-V and reverse Osmosis process–Doha
United Arab Emirates	Emirates Project	100-m^3/day, uses evacuated tube collectors and MSF method.
Jordan	Aqaba Project	2$_3$m^3/day, Area of collectors is 375 m^2, heat pipes.

References

1. A.A.M. Sayigh, Solar Energy in the Arab World, the 3rd Arab Energy Conference - Energy and Arab Cooperation, Alger, 4-9 May, 1985.
2. W.M.O., Meterological Aspects of the Utilization of Solar Radiation as an Energy Source, Technical Note No. 172, 1981.
3. Tunisian National Paper, Solar Heating and Cooling Symposium, Baghdad, 16-20 Jan., 1984.
4. K. Abdul Jabbar, and S. Abdul hay, Solar Airconditioning in Iraq, Solar Heating and Cooling Symposium, Baghdad, 16-20 Jan., 1984.
5. Sverdrup & Parcel and Associates, Inc., Solar Energy for the Airborne and Physical Training School-King Abdul Aziz Military Cantonment.
6. L. Al-Houty, M.H. Omar and H. Abou-Leila, Gold Film Studies Towards Optimizing MIS Solar Cell Upper Electrodes, Solar and Wind Technology - An Int. Journal, Vol.1, No.1, 15-19, 1984.
7. M. Dacharaoui, S. Belgacem, r. Bennaceur and H. Latrous, Realization of Solar Cells Cd_{1-y} Zn_yS - Cu_2S by Airless Spray. Proceedings of First Arab Int. Solar Energy Conference, Kuwait, 231-234, 2-8 Dec. 1983.
8. M.M. Haruni, The Solar Pilot Project in Qatar, Solar and Wind Technology - An Int. Journal, Vol.1, No.2, 93-109, 1984.
9. B. Nimmo and M. Elhadidy, The Solar Pond Program of the Research Institute/UPM, proceeding of first Arab Int. Solar Energy Conference, Kuwait, 361-365, Dec. 2-8, 1983.
10. S. Quraeshi, B.M. Pederson, and A. Sayigh, Wind turbine generators: State of the - Art, Solar and wind technology - an Int. Journal - volume 1, No.1, 37-49, 1984.

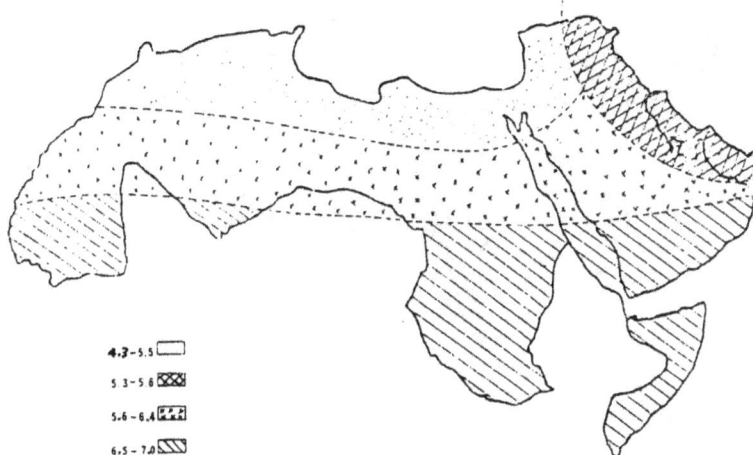

Fig. 1   Annual Average of Solar Radiation in the
Arab Countries, $KWh/m^2/d$.

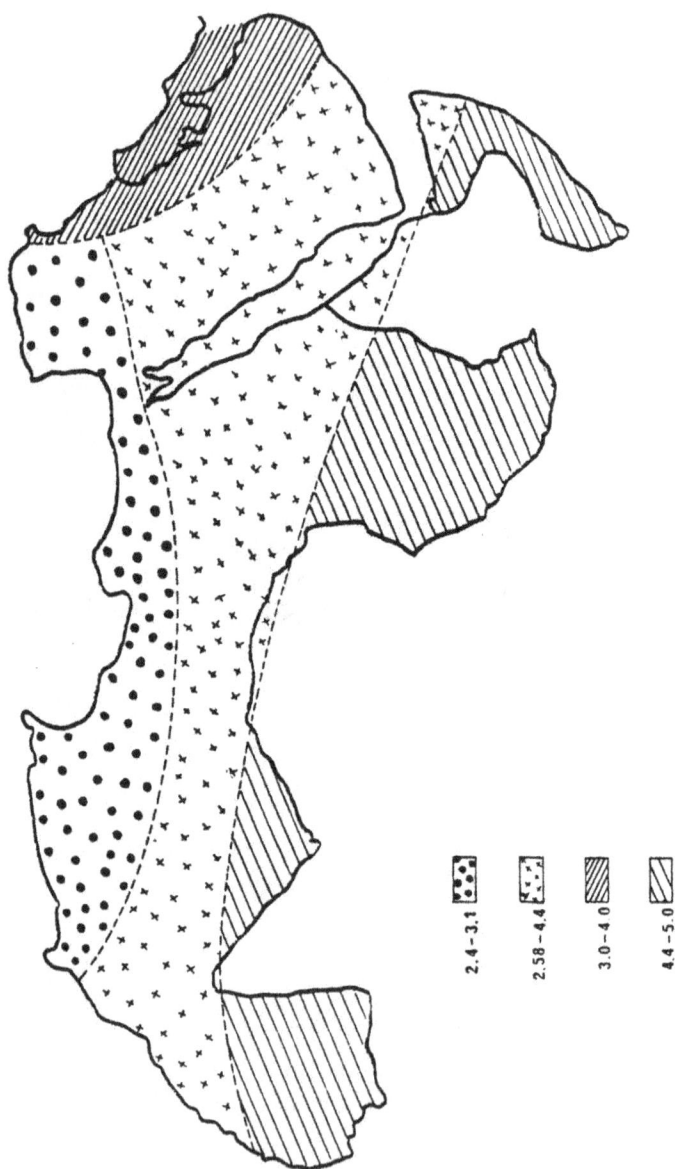

Fig. 2 Average Minimum Global Radiation in the Arab Countries, kWh/m^2/d (December and January).

2.4 - 3.1

2.58 - 4.4

3.0 - 4.0

4.4 - 5.0

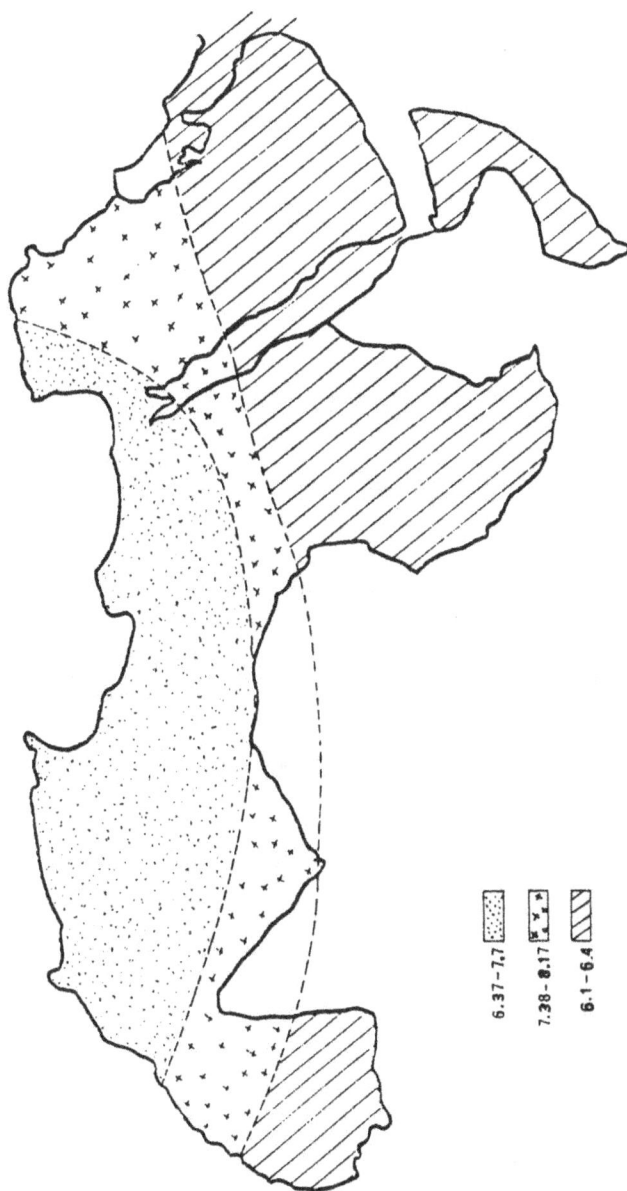

Fig. 3 Average Maximum Global Radiation in the Arab Countries, $kWh/m^2/d$ (June, July and August).

6.37 - 7,7

7.38 - 8.17

6.1 - 6.4

Fig. 4 Average Energy
        VS shunshine
        hours in some
        Arab and indust-
        rialized cities.

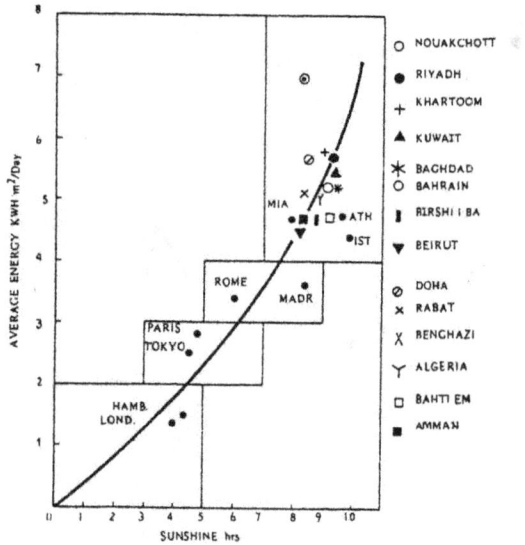

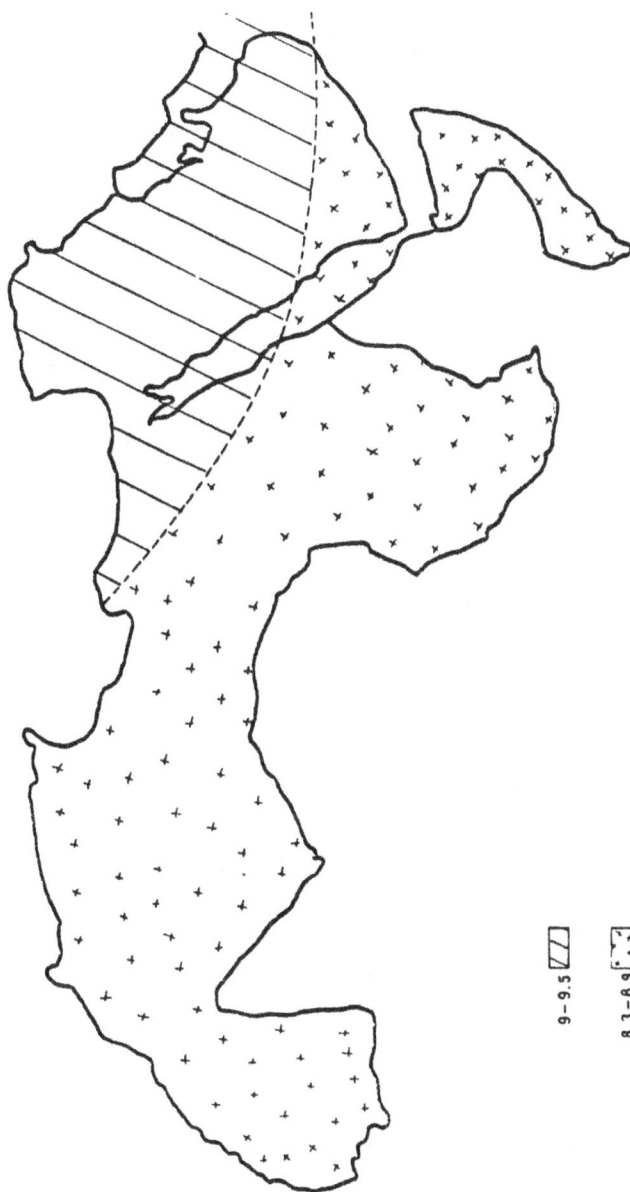

9 – 9.5

8.3 – 8.9

Fig. 1   Daily Annual Average of Sunshine Hours in the Arab Countries.

ENEL's ACTIVITY IN THE FIELD OF RENEWABLE ENERGY SOURCES

P. Peiser

ENEL - Ente Nazionale Per l'Energia Elettrica

PREFACE

Italy is a country which has only very few domestic sources of energy: hydro-electric power, geothermal energy, small reserves of low-grade coal and of natural gas.

In the past, electricity was produced mainly from hydro power: in 1963, when the private utilities were nationalized and ENEL, the Italian Electric Power Authority, was formed, hydro-power accounted for 64.6% of Italy's total electricity production, against 31.5% from fossil-fuel thermal plants, 3.4% from geothermal energy, and 0.7% from nuclear plants.

At that time all the important sites for hydro-electric power were already exploited, and therefore in the following years, due also to the low cost of oil, the main new plants were all oil-fired.

In 1984, when ENEL's installed gross capacity reached 44,459 MW and Italy's total production reached 182,700 million kWh (2.5 times the value of 1963), the subdivision among the different primary energy sources was the following one: fossil fuel 69.9%, hydro-power 24.7%, nuclear energy 3.8%, geothermal energy 1.5%.

The increasing cost of oil induced recently ENEL to plan for the future only coal-fired and nuclear power plants, and to develop as much as possible, from the technical and economic point of view, the renewable energy sources: minor remaining hydro-electric sites, geothermal energy, and also solar and wind energy.

In the same time the Italian industry devoted a huge effort to the development of technologies for the renewable energy sources.

In the following paragraphs ENEL's experience and programs in the field of power production from renewable energy sources will be described.

SMALL-SCALE HYDRO-POWER

In the past many small hydro plants were shut down, because they were no longer competitive; other plants were kept in service, but converted to completely automated operation in order to reduce costs. Whilst in 1963 only 15% of the small hydro plants were in automated operation, today this percentage reaches 77%. In the last years 15 small plants were restored to operation, with a total installed capacity of about 10,000 kW and a yearly mean production of 31 million kWh; during the next three years it is foreseen to reactivate further 50 plants, with a total installed capacity of about 23,000 kW and a mean yearly production of 120 million kWh.

In order to further reduce the cost both of restoration of old plants and of construction of new plants, ENEL and the main Italian manufacturers prepared a standardization of the equipment for the small plants (between 100 kW and 5,000 kW). The same standardization may be used in future for the plants which may be built by private firms, according to the provisions of Law No 308.

GEOTHERMAL ENERGY

The first experiments for producing electricity from geothermal energy were carried out in Italy, at Larderello, in 1904, and commercial exploitation started in 1912; presently the installed gross capacity reaches 459 MW, with a yearly production of about 2800 GWh.

The classical geothermal areas of Italy as well as the most promising new areas being investigated by ENEL and ENI are all on the pre-Appennine belt of Tuscany, Latium and Campania.

The data and information arising from many years of intensive research let us predict that the annual average production from geothermal sources could increase to an average electric capacity of 700 MW by 1993, which corresponds to about 5000 GWh/year.

Much more important may be in the future the utilization of geothermal energy for non-electric purposes, i.e. for heat supply to district-heating schemes, and to agricultural and industrial users. The main projects under way are as follows:

- 400,000 m^2 of greenhouses in the Monte Amiata area;
- 70,000 m^2 of greenhouses at Radicondoli;
- 600,000 m^3 of buildings of the Infantry School at Cesano (Rome);
- a district-heating scheme in the town of Ferrara, with an initial supply of 150,000 Gcal/year and a long-term goal of 600,000 Gcal/year.

ENEL is also very active in the transfer of geothermal technology to other countries.

We should however bear in mind that geothermal energy is not a really renewable energy source, because of the extremely long times needed by the heat from the deeper parts of the earth to reach the geothermal reservoir to replace the heat extracted by the power generation process. A geothermal system is therefore more similar to a conventional energy reserve, and the rate of exploitation is determined on the basis of the commercial life-time of the power plants (around 40 years).

Research is presently being carried out in the field of very deep geothermal drilling (around 5000 m) where it may be possible to find fluids with better thermodynamic characteristics.

As concerns the equipment for geothermal power plants, a new turbine characterized by a large flexibility as concerns steam conditions and by a reduced installation time was recently developed by ENEL together with the Italian manufacturers.

SOLAR ENERGY

In the present paper, attention is focused essentially on the conversion of renewable energy sources into electricity. A detailed examination will thus not be made of applications which enable conversion of solar energy into low-temperature thermal energy, even if such applications - hot water production for domestic uses, heat production for space heating, etc. - are the closest to economic competitivity and even if relevant efforts are devoted in Italy to the development of such applications from both a research and industrial standpoint.

As concerns solar water heaters, ENEL, the Italian Electricity Board, provided last year for a form of incentivation by anticipating to its customers 70% of the cost of the installed solar water heater; the loan will be paid back by the customer on the electricity bill at a relatively low rate of interest. This operation is financed by the European Community through a special grant.

As concerns electricity production, we should bear in mind that Italy is a country where the electricity distribution system reaches nearly all inhabitants: only about 231,000 inhabitants with permanent residence and 268,000 with seasonal residence in the whole country are not connected to the grid. The cost for connecting part of these houses to the distribution system is very high, exceeding in about 1,000 cases the amount of 15,000 dollars per house.

Taking into account this situation, the following three cases of solar electricity uses may be envisaged in Italy:

a) supply of electricity to isolated houses with an exceedingly high cost of connection to the distribution system;

b) supply of electricity to the minor islands (using also other renewable energy sources and electric storage);

c) supply of electricity to the national grid by medium-large sized plants or by small plants located at the consumers sites (in parallel with the other plants and without local storage).

Thermodynamic solar electricity generators

The operating principle of these plants is well known: the sun radiation is reflected by heliostats, which automatically track the position of the sun in such a manner as to concentrate the reflected radiation in a receiver (steam-generator) set on a tower. The receiver can be of the water-type (in which case the steam feeding the turbine is produced directly) or of the air-type (for feeding a gas turbine) or of the liquid metal-type or molten salt-type with subsequent steam production in a heat exchanger.

Italy has a very long tradition in this field: the first central receiver power plant in the world was built in 1965 at Sant'Ilario, near Genoa, by the late Prof. Giovanni Francia with a funding of the National Research Council. The plant consisted of 121 mirrors with a total area of roughly 30 m^2 and produced superheated steam at 500°C and 150 atm. The mirrors had special steering mechanisms whereby they tracked the position of the sun.

Of particular interest was the conception of the receiver which exploited the anti-irradiating structures patented by Prof. Francia at the beginning of the 60s.

Since 1965 the S. Ilario plant has underwent various modifications and improvements and today the relevant data are as follows: about 100 kW_{th}, 140 mirrors with a diameter of 1.1m, improved receiver and improved steering mechanisms. The most important function of this plant is to test solar receiver models which, as is known, represent one of the most delicate points of central receiver power plants and which can be developed only through an intense experimental activity.

As a confirmation of the vanguard position reached by Italy in this field, in 1976 the Georgia Institute of Technology of Atlanta bought from the Italian company Ansaldo, which in the meantime had taken Prof. Francia as consultant, a plant similar to that of S. Ilario, but with a capacity four times greater (about 400 kW_{th}); the receiver is designed for production of superheated steam up to 600°C and 150 atm.

The plant, that is made up of 550 mirrors with a 1.1 m diameter, came into operation in november 1977 and is used for tests on receivers and other equipment.

The experience acquired in Italy in the field of central receiver solar power plants was one of the fundamental conditions for the most important initiative that was developed in this sector, namely the 1-MW$_e$ "Eurelios" solar power plant built at Adrano, Sicily, as part of a research program of the European Economic Community (EEC).

The plant was designed and built by a consortium of European industries formed by ENEL and Ansaldo (Italy), CETHEL (France), Messerschmitt-Bölkow-Blohm (MBB) (Federal Republic of Germany).

The plant became operational in April 1981 and was connected with the Italian electric network. ENEL is the operator of the plant and co-proprietor along with the EEC.

The success of this initiative was largely determined on one hand by the know-how which Ansaldo acquired in this sector, and, on the other hand by ENEL's interest in this application of solar energy and its willingness to manage the plant operation and testing.

The choice of the plant location was made after thorough investigations which ENEL conducted in various regions of Southern Italy and which led to the choice of Adrano, in Sicily, near Catania.

The most significant technical data of the Adrano plant are given below:

- Nominal power   1 MW$_e$
- Total efficiency   16%
- Total mirror surface   6200 m^2
- Required mirror field area   35000 m^2
- Steam flow   4700 kg/h
- Steam temperature   510°C
- Steam pressure   64 atm
- Tower height   50 m
- Storage capacity   30 min

The mirror field consists of 182 heliostats, of which 70 (built by CETHEL) have an area of 52 m^2 and 112 (built by MBB) an area of 23 m^2.

An automatic computer control system positions all the heliostats so as to concentrate the sun rays in the receiver at all times.

The receiver, representing one of the technologically most advanced parts of the plant, is of the type developed by Prof. Francia and was built by Ansaldo.

A series of computer programs was developed in order to solve such delicate design problems as the determination of energy distribution in the receiver under the different operating conditions.

A reduced-scale receiver model was also built and installed in the test plant of S. Ilario where a test program was carried out that allowed to test the most important hypotheses made during the calculation phase.

The results of the operation of the plant, from april 1981 until now, are very similar to those recorded in other plants of the central receiver and mirror field type installed in other countries: the Nio plant in Japan (1 MW_p, water-steam boiler); the CESA-1 plant in Spain (1.2 MW_p, sodium receiver); the CRS plant in Spain (0.5 MW_p, sodium receiver); the Themis plant in France (2.5 MW_p, molten-salt receiver); the Barstow plant in the USA (10 MW_p, water-steam boiler).

The overall operating experience with all these plants gives some quite clear indications, despite of the relatively short period of operation. It has indeed been demonstrated that such plants can generate electricity and can operate in parallel with a large power system. Their operation, however, has been marked by a number of outages, in addition to those due to lack of sunlight, and by production levels often below those theoretically possible, because of unavailability or reduced efficiency of some of the equipment. These arise from the boilers, which suffer frequent large thermal stresses, which in a few cases have given rise to leaks; to the heliostats, the availability of which, in some cases, has been lower than expected; to the control and storage systems, which have not always succeeded in overcoming transients due to the

passage of clouds; to the startup system, which required, in some
plants, very long times (2 to 3 hours) for synchronization. In fact,
many of these troubles are due to deficiencies in plant design or
construction, and to the novelty of the system. They are likely to be
eliminated, at least partially, through improved design of the
components, of the system and of the plant as a whole, taking into
account the operating experience gained. On the other hand the operation
confirmed that these are complex and very sophisticated plants,
particularly in relation to their modest size (a few tens of MW) of the
possible commercial plants, and to the expected limited annual output
(of the order of tens of millions kWh).

The experience acquired with some of the prototypes was unsatisfactory
with regard to the annual production per kW_p. In some plants, high
auxiliary power requirements led to no net production at least in the
season of low insolation. Allowing for the improvements suggested by
operating experience, there are doubts that the future plants could give
a net yearly output exceeding 1000 kWh per installed kW_p.

Moreover, from an economic standpoint, the plant cost for these solar
demonstration plants hass been very high (of the order of 10,000-20,000
$/kW_p$). therefore, successful commercial adoption of solar thermal
stations would require a drastic reduction in plant costs. In many
countries exhaustive evaluations have been carried out, which have led
to the conclusion that whilst substantial plant cost reductions were
possible through design optimization, the use of cheaper material, the
use of advanced construction and erection techniques and of mass-
produced components, these improvements even if substantial would be far
too small to achieve an acceptable kWh production cost. Therefore, since
one cannot see firm prospects for industrial applications, at least for
the next 15-20 years, further installation of thermodynamic solar plants
is not planned in countries like Italy, France and Japan.

Power production by photovoltaic process

As is well known, the direct conversion of solar energy into electricity through the photovoltaic process - which has received an enourmous impetus from space programs - has made extremely relevant progress over recent years, so much that it can be already competitive today for some very special applications.

As regards conventional applications, however, competitivity is still far away: very strong reductions will be needed not only in the cost of solar cells but also in the cost of all the equipment and facilities which, together with the solar cells, make up a photovoltaic power plant (e.g. the structures supporting the solar cell panels, the necessary cables, the concentrators, when needed, the electric equipment converting direct current into alternating current, etc.)

The sizeable reduction in the cost of the solar cells and of the more conventional components of the photovoltaic power stations that is required to reach competitivity must be achieved not only through a considerable technological progress but also through a notable contribution of imagination and innovation to the conception of photovoltaic power plants.

In Italy, ENEL is active in the photovoltaic sector both through a promotional effort of the research and development activities of the Italian industries and through a program of installation and testing of photovoltaic generators with capacities ranging from a few kW to some tens of kW.

As concerns demonstration programs, ENEL has undertaken, in cooperation with the Italian industries, a coordinated multi-annual program.

Moving from small to large-size systems, we may mention:

- tests on a 1 kW silicon unit without concentration, capable of supplying the electricity to an isolated home for lighting, refrigerator and television set;

- the development of small silicon cell systems, with and without concentration, totalling some tens of kW;

the construction of a 80 kWp photovoltaic plant without concentration but with storage for the island of Vulcano in the Aeolians; this plant was commissioned in 1984.

This plant is the first step in the development of the "Eolie Project", aimed at making a group of small islands completely independent from imported energy.

On one of these islands geothermal energy will generate electricity and supply also two other islands through submarine cables and pipelines.

The remaining islands, which are too far away, will rely on electricity generated by solar and wind energy.

As concerns very large plants, ENEL has already carried out the preliminary study for a 10 MW silicon cell plant without concentration and for a 1 MW low-concentration silicon cell plant; a 1 MW plant to be connected to the grid ("Delphos Project"), but able to supply also an isolated load (and therefore provided with storage means) was studied in cooperation with ENEA.

WIND ENERGY

In our Country there has been, and still is even today, a modest application of small-scale wind plants manufactured in Italy, mainly those built to meet local needs, such as those for farmhouses and roadman's houses located along national highways.

Today, the main research and development activities in Italy are related to ENEL's VELE Project.

The VELE (Wind for Electricity) Project was launched in 1980, and is scheduled to last seven years.

The aims of the Project are:

- identification of suitable wind energy production areas;

- definition of the technical characteristics and use of wind-turbine generators for wind power plants;

- acquisition of experience in the design, construction and operation of such power plants.

Some years ago research has begun to assess the wind energy resources that could, in practice, be utilized in Italy for the production of mechanical and electrical energy.

The annual average wind speed map, drawn up by the Italian National Research Council (CNR), show that, from the point of view of wind resources, Italy is not one of the more favoured countries. Nevertheless, it may be considered that Italy, too, has a limited number of areas that would lend themselves to the exploitation of wind as an energy source.

In order to pinpoint these areas, however, the isovent maps of a country or region, such as those at present available, are not sufficient, inasmuch as the wind characteristics of a site are strongly influenced bu the local orography.

ENEL started operating in Sardinia, but recently wind research was extended to other Italian regions.

In the localities being researched, 15 m-high meteorological towers have been installed that record the speed and direction of the wind over a period of 10 minutes.

Each met tower has instruments for measuring the aformentioned quantities, plus equipment for digitalization and recording on magnetic tape.

The quantities thus digitalized on the spot are subsequently collected, checked and processed. The frequency distributions of the intensity and direction of the wind, both overall and for each wind-rose sector, are obtained and plotted, thus determining the chief parameters.

As concerns wind turbine generators, experiments on prototypes are conducted in a test area set up by ENEL at Santa Caterina near Cagliari in Sardinia. It is also planned to instal prototypes in other parts of Italy with different environmental characteristics from those of Santa

Caterina, for example in mountainous parts, where it would be possible
to observe the performance of machines under particularly severe
conditions (ice formation, ghusts of wind, etc.).

The test area can take up to six small and medium wind turbine
generators, and is equipped with the measuring and data-acquisition
apparatus required for monitoring the performance of wind-turbine
generators over a period of time.

ENEL has under test, at the present time, at the Santa Caterina Test
Center, a down wind horizontal-axis machine built by "Centro Ricerche
FIAT", and two Aeritalia/Grumman 15 kW wind turbine generators.

Another part of the VELE Project concerns the design and construction of
a 500 kW wind power plant, consisting of 10 wind turbine generators of
50 kW each. The subsequent operation of the machines will make it
possible to acquire knowledge and experience of the running and
maintenance of such power plant.

Some of the wind turbine generators are fitted with the instruments
required for continuous checking of the mechanical, electrical, and
thermal stresses acting on the various components during operation.

Special care is being devoted to problems concerning with the hook-up of
the power plant to the electricity grid, and to study possible problems
of integration with the environment. As regards the latter, the
following steps are planned:

- evaluation of any effects on the micro-climate in the immediate
  machine vicinity and of the size of the area affected;

- evaluation of any effects on vegetation and wild-life;

- evaluation of the possibility of utilizing, for other purposes, the
  land surrounding the wind-turbine generation sites, while observing
  all the necessary safety regulations;

- definition of the noise made by the plant, and study of ways to
  install the machines in order to minimize such noise;

- checking any interference with TV, telecommunications and radar.

In order to ascertain the effect of grouping wind turbine generators together in large array on total plant efficiency, the power station was set up in such a way as to make it possible to experiment with different layouts of the production area.

The power station is located at Alta Nurra, near Porto Torres, Sassari and was commissioned in 1984.

The wind turbine generators for the power station have been built by Sistemi Energia Sud S.p.A. (SES), a company of FIAT Group, on the basis of a joint ENEL-FIAT design.

The activities involved in this sub-programme are conducted by ENEL with the co-operation of SES, in order to assess the performance and reliability of the wind turbine generators, and with the University of Cagliari, for evaluation of the impact of the power plant on local micro-climate and vegetation.

As concerns future programmes, ENEL's interest in wind energy is aimed mainly at the development and testing of large wind turbine generators for hooking-up to the electricity grid, because it is such units that offer the best prospects from the economic point of view.

To this end, ENEL has promoted an important programme involving various Italian manufacturers (FIAT Aviazione, Aeritalia), in order to develop in co-operation with ENEA, a detailed research plan for the construction of a multi-megawatt wind turbine generator.

POWER PRODUCTION FROM URBAN WASTE AND AGRICULTURAL RESIDUES

The energy crisis and the problems related to resource conservation and environment protection brought up in the last years the need for a correct management of waste, which should not be considered only as a polluting agent but also as a source of recoverable materials and of energy.

Particularly important is the problem of the disposal of the large amount of domestic waste: a recent Italian survey indicates a yearly per capita production of 260 kg of domestic waste, having a mean humidity content of 20-30% and a mean heat value of 1,500 kcal/kg.

By extrapolating those values to the whole of the Italian population, we obtain about 14 millions of tons per year; if we could use 20% of this amount for generating electricity, we would be able to cover 1% of our today's demand. It is however clear that, as combustion of waste with electricity production is better than a simple incineration, so other options may be preferable to the combustion. In particular we could envisage:

- recovery of energy-intensive materials like paper and metals;
- heat production for process use or for space-heating and air-conditioning purposes;
- production of a pulverized fuel with a heat value of about 5,000 kcal/kg;
- utilization, after pretreatment, in agriculture as a fertilizer.

In the case of combustion for electricity production, the highest efficiency will be obtained if we add the pre-treated waste to the coal in conventional power plants, because we are able in this way to benefit of the high efficiency of large power plants, which may be twice the efficiency of an incinerator.

Furthermore we avoid in this way the capital costs associated with the construction of an incinerator.

Experiments and pilot tests are being carried out in Italy by ENEL since 1977, and a first large programme involving the combustion of 300,000 tons of waste per year (corresponding to the production of waste from an area with one million inhabitants) is in the design phase.

As concerns agricultural residues, some of them (e.g. olive husk) may be utilized in the same above-mentioned way, and ENEL will burn within short about 120,000 t/year of olive husk, mixed with lignite, in one of its power plants.

RECENT SITUATION OF SOLAR HEATING AND COOLING IN JAPAN

Ken-ichi Kimura
Professor, Department of Architecture
Waseda University
Okubo 3, Shinjuku, Tokyo 160, Japan

INTRODUCTION

Solar heating and cooling were regarded as most promising areas of solar energy utilization just after the oil crisis in 1973 and many efforts have been made to develop viable systems supported by fundamental and applied studies since then. As Japan is not blessed with natural energy resources, a lot of governmental and private funds have been paid to encourage and stimulate the people's mind towards energy conservation and natural energy utilization.

After the decade, however, it has become clear that the use of solar energy for domestic hot water supply is much more predominant and space heating and cooling less publicized than expected. The followings are general descriptions about the recent trends on solar heating and cooling in Japan with some considerations.

SOLAR WATER HEATERS

It is estimated more than 10% of the total households in Japan have solar water heaters on the roof top, which can be seen everywhere in Japan and especially more in the south Pacific coast. There are two popular types of the solar water heater. One consists of a flat-plate collector of $2 - 4$ m^2 coupled with an insulated tank of $200 - 300$ litres in which thermosyphon action takes place. The other is made of a row of black cylinders that act as collector and water storage. The total amount of energy contribution by these solar water heaters might amount to 0.2 % of the national total energy consumption. This can be roughly calculated as follows.

Number of solar water heaters currently used: 4×10^6 units

Average collector area per unit: 3 m²

Average daily collection of solar energy per unit area of collector: 4.5MJ/m²

Annual total solar contribution by solar water heaters:

$4 \times 10^6 \times 3 \times 4.5 \times 365 = 20 \times 10^9$ MJ

Annual total solar contribution of oil equivalent taking account of

boiler efficiency: $20 \times 10^9 / 25 \times 10^3 = 0.8 \times 10^6$ kl

Annual national total energy consumption of oil equivalent in Japan: 4×10^6 kl

Percentage of contribution by solar water heaters:

$0.8 \times 10^6 / 4 \times 10^8 = 0.2 \times 10^{-2}$ (=0.2%)

Although solar water heaters may not be aethetically attractive from the

viewpoint of urban landscape, the modern society may have to accept them as the

symbols of energy crisis.

STATISTICS OF ACTIVE SOLAR SYSTEMS

The Solar System Development Association was established in 1978 as a

government subsiduary non-profit organization to promote mainly active solar

systems. They are in charge of arranging governmental support funds and low

rate loans for solar systems, formulating standards of solar equipment and

planning various promotion of solar systems.

The number of solar installations has increased up to over 220,000 at

the end of 1984 and most of them are applications to domestic hot water supply

as shown in Table 1 and Table 2. It must be noted that the solar water heaters

a described in the previous section are not included in these tables and only

forced circulation system of solar domestic hot water supply is included in

these tables.

As shown in Table 3 the total collector area installed amounts to over

1.7×10^6 square meters. It can be estimated that half of them belongs to solar

— 2 —

Table 1 Number of Solar System Installtion

Year / Type	1975 -78	1979	1980	1981	1982	1983	1984	Total
Single Family	1841	2305	24420	30525	50428	63165	48006	220690
Apartment	19	11	60	37	62	126	62	377
Commercial	192	251	981	1462	1043	736	524	5189
Industrial	16	13	69	55	32	54	21	260
Total	2068	2580	25530	32079	51565	64081	48613	226516

(Solar System Development Association)

Table 2 Number of Solar System Installation at the End of 1984

System / Type	DHW	DHW + Space Heating	DHW + Space Heating & Cooling	Swimming Pool, etc.	Total
Single Family	219669	943	72	6	220690
Apartment	362	7	7	1	377
Commercial	4612	187	325	65	5189
Industrial	159	18	16	67	260
Total	224802	1155	420	139	226516

(Solar System Development Association)

Table 3 Total Installation of Collector in Effective Area (m^2)

Year	Effective Collector Area (m^2)
1975 - 80	350182
1981	306696
1982	385607
1983	392998
1984	290388
Total	1725871

(Solar System Development Association)

domestic hot water and another half the rest of solar systems which usually require larger area of collector.

Application to solar space heating has not been so well publicized as expected. This may be because of higher initial cost for shorter period of heating requirement in the year compared to domestic hot water supply.

Larger number of space cooling application may be attributed to the fact that financial support of government to public buildings gave incentives in the case of solar systems with solar cooling. In fact MITI's Sun Shine Project has put an emphasis on development of solar cooling from its strat in 1974.

PASSIVE SOLAR HEATING

All of the Japanese vernacular houses may be called as passive solar houses in the sense that they usually have large openings on their south facade, inviting a lot of solar radiation into the living space during winter and cutting strong solar beam by deep caves in summer. It may not be properly stated that their system is so called direct gain system, because they do not have large thermal mass to store solar energy and are not well insulated.

Because of the above tradition it seems that the passive solar houses as explained in the texts published in the United States have not well popularized in Japan. As wooden construction is still predominant in Japanese single family residences, it is quite difficult to provide large thermal mass within the interior spaces. On the other hand provision of insulation for house structure has been widely spread throughout the country and people have get realized money savings with insulation as well as energy conservation.

There are many interesting examples of passive solar houses actually built, but measured results of energy are very scarce. Experimental passive solar houses are being built by various companies and organizations and extensive data are being analysed and reported. Passive solar system may be still under developing stage, because clients never know as to how effective they will be and rather prefer traditional conservative type of houses with a little reinforcement of insulation and with no more unfamiliar details of passive systems.

It may take a little more time for the passive solar systems to be accepted by people at large in Japan until many good reputation prevails.

A low interest loan for Trombe wall of more than $7m^2$ or equivalent has been implemented since fiscal year 1983, but none for the direct gain system which may be regarded too commonplace for Japanese.

No one has estimated solar contribution of direct gain system to the national annual total of energy consumption, but it is undoubtedly very large as the annual energy consumption for space heating in dwellings is far smaller than those in the European countries.

PROBLEMS WITH SOLAR ABSORPTION COOLING

Solar absorption cooling system has been adopted in large scale build-ings as seen in Table 2. In many cases it is reported that solar contribution turned out less than expected. Causes of unsuccessful behaviour have been made clear after several years of operation experience.

Heat loss from pipings from collector to tank and tank to absorption chiller is quite large because high temperature liquid flows through these pipings. Heat loss occurs from valves, pumps, control devices and measuring sensors provided in pipings. In many cases insulated pipings are exposed to the sun and rain, which may cause cracks in the covering and water may get soaked in insulation after long period of exposure. All of these are making collection less than expected.

The length of piping is an very important factor. If the storage tank and the solar absorption chiller were installed quite apart, the temperature drop from the tank to the chiller because of heat loss naturally makes the operable time of the chiller shorter and overall COP lower.

Combination with auxiliary equipment is another point of interest. If a boiler was installed to feed hot water to the regenerator of the absorption chiller when solar energy is not sufficient as shown in Fig. 1, precious fuel must be used for driving the absorption chiller at such a low value of COP. In this case the solar cooling system would not be energy conserving compared with a conventional centrifugal chiller system unless more than 72% of tatal cooling load was shared by solar absorption system. Otherwise double effect type of absorption chiller may be used. This machine operates in double effect with auxiliary at a COP of 1.0 and in single effect with solar at a COP of 0.6.

If the solar absorption chiller and a heat pump were installed in parallel as shown in Fig. 2, energy saviing can be achieved at all times. In this case, however, the initial cost tends to be higher, but it would be realistic for a large installation where plural chillers would be required. Most economical system with solar absorption chiller would be the system without auxiliary. This works only when solar energy is available and the period when solar system is idle may be the period when cooling load is less.

OPEN CYCLE SOLAR COOLING

For most of residential buildings, perfect cooling is not necessarily required. Under hot and humid regions dehumidification would be much valuable. Various types of desiccant may be used for open cycle solar cooling system where solar energy is used to regenerate wet desiccant which absorbed moisture from the room air.

In order to make the system work effectively, it is desirable to try to find the desiccant which can be regenerated at lower temperature. In this respect LiCl solution is considered suitable for open cycle solar imperfect cooling, though zeolite, silica gel, calcium chloride and other desiccants are being used for experiments.

Followings are the open cycle solar absorption cooling system using LiCl solution proposed by the author's laboratory.

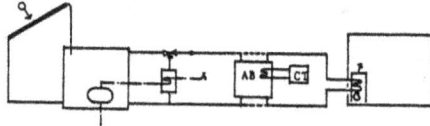

Fig. 1 Solar Absorption cooling system with auxiliary
heat directly fed to regenerator

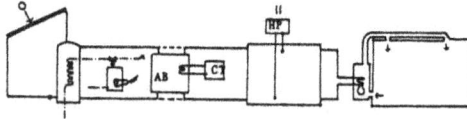

Fig. 2 Solar absorption cooling system with heat pump
provided in parallel

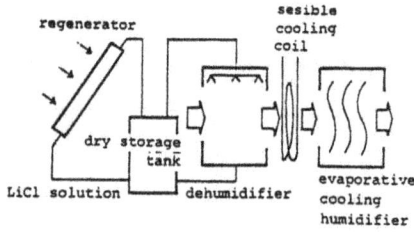

Fig. 3 The fundamental scheme of open cycle solar
absorption cooling system using LiCl solution

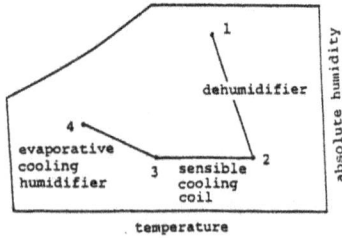

Fig. 4 The dehumidification and cooling process

Fig. 3 shows the fundamental scheme. The system consists of a dehumidifier, a sensible cooling coil, an evaporative cooling humidifier, a dry storage tank, and a regenerator. LiCl solution is supplied to dehumidifier from dry storage tank through heat exchanger and absorb moisture from inlet air. Weak LiCl solution after dehumidification is returned to the tank. Regeneration of the LiCl solution is made by natural convection of air stream within the open air space in the collector exposed to the sun. Strong solution is stored in the dry strage tank. Other possibilities to utilize forced circulation type of regenerator may be found. City water pre-heated by air and LiCl solution is used for domestic hot water supply.

Fig. 4 shows the dehumidification and cooling process in the psychrometric chart. From 1 to 2 inlet air is dehumidified by LiCl solution and from 2 to 3 the air is cooled by sensible cooling coil which in turn functions to preheat city water for domestic hot water supply and from 3 to 4 the air is cooled by evaporative unit if needed.

SUMMARY

Recent trends of solar heating and cooling in Japan are outlined. It is recognized that roof top solar water heaters share a considerable fraction of 0.2 % of the annual national total energy consumption. Other solar systems share only 0.03 % of it, a half of which is attributed to solar hot water heating and another half to solar heating, cooling, industrial and other processes.

Passive solar systems are gradually popularized in Japan as regarded one of the most promissing version of solar energy utilization.

Problems with solar absorption cooling are discussed and further consideration is necessary to make it viable. Open cycle solar cooling system as an alternative is introduced to realize imperfect cooling by simpler means, with which domestic hot water preheating system is intergrated.

Although public interest in solar systems seems to be a little faded away in reflecting the current situation of oil prices. Nevertheless research and development on solar energy utilization should be continuously made in preparation for another energy crisis to come. It is because oil supply is limeted and environmental disruption from burning oil must be avoided for future generations.

REFERENCES

1. Kimura, K., Solar Houses in Japan, (Sayigh ed.), Solar Energy Application in Buildings, Academic Press (1979) 287-315

2. Kimura, K., Utilization of Solar Energy — The Japanese Experience, (Lim ed.) Solar Energy Application in the Tropics, D. Reidel Pub. Co. (1983) 187-216

3. Kimura, K. and Tamura, F., Experimental Study and Energy Analysis on Open Cycle Solar Absorption System Combined with Domestic Hot Water Supply, Proceedings of Solar World Congress Perth 1983, Pergamon (1983) 130-134

4. Kimura, K., Tanabe, S. and Monose, T., Experimental Study and Energy Analysis on the Open Cycle Solar Absorption Cooling System Using LiCl Solution, Proceedings of International Symposium on Thermal Application of Solar Energy, Japan Solar Energy Society (1984) 571-576

SOLAR PONDS—AS AN ENERGY STORAGE DEVICE

Prof. H.P. Garg
Centre of Energy Studies
Indian Institute of Technology
Hauz Khas, New Delhi-110016
India

1.0   INTRODUCTION

Solar pond is an innovative system which simultaneously collects and stores the solar energy for some useful applications like power production, space and greenhouse heating, process heat for industries, space cooling, desalination, agricultural crop drying, and production of renewable liquid fuels such as ethanol for gasohol. The depth of the non-convective or salt-gradient solar pond varies from 1 to 3 metres and in which a portion of the pond's depth is stabilized against convective heat motion by dissolved salt with more salt toward bottom than the top. This salt gradient counteracts the thermal density gradient caused by solar radiation absorbed at the pond's bottom and prevents fluid motion within the gradient layer. This layer serves as both insulation and storage, and permits the lower layer of the pond, which does convect, to reach temperature close to the boiling point, while the surface convective layer is near ambient temperature. Such a situation indeed occurs in nature in a few lakes (e.g. Madve Lagoon lake in Transylvania, $42^{\circ} 44' N$, $28^{\circ} 45'E$) which is described by Kalecsinsky[1]. In such lakes, typical temperatures of around $70^{\circ}C$ at a depth of $1.32 m$ at the end of summer have been observed. Artificial salt gradient stabilized solar ponds, operating on the same principal as natural lakes, have been developed in several countries including Israel, USA, Australia, and India, and temperatures

as high as 90-95° C have been attained in the bottom of the
pond after about 1 year of operation. Energy from the bottom
layer of the pond is extracted by pumping the hot brine through
an external heat exchanger or through an inpond heat exchanger
i.e. by cycling external water through a heat exchanger
immersed in the lower layer of the pond. Large area solar
ponds appear to be one of the most attractive and economical
solar system, especially since it looks natural and fits
easily into the landscape.

2.0   CONSPECTUS

The phenomenon of temperature inversion i.e. increase
in temperature with depth, in natural salt lakes was observed
only recently and reported by Kalecsinsky in the year 1902.
Anderson[2] also reported a similar lake in Oroville
(Washington state) where a temperature of 50°C at a depth of
2 m was observed in a summer season. In 1948, R. Block of
Israel for the first time suggested the use of salt gradient
solar pond as a solar energy collector cum storage device,
but solar pond research in Israel began only 10 years later.
Solar pond research in Israel continued until 1967 under the
dynamic leadership of Prof. Tabor[3,4], and resulted in very
important contributions to the understanding of the problems
involved in operating solar ponds and in developing their
possible applications. Research on solar ponds in Israel
was resumed again in 1975. Also, theoretical and experimental
observations on laboratory solar ponds for understanding the
physics of solar pond were carried out by Weinberger[5], Elate
and Levin[6], Tabor and Matz[7], and Hirschman[8].
Weinberger[5] in his classical paper has derived expressions
relating the key parameters involved in the operation of the
salt gradient solar pond collection efficiency, pond depth,
optimum temperature, rate of energy withdrawal, etc., and
discussed in detail the stability of the pond. In a paper

by Tabor and Matz[7], the results on studies conducted during
the year 1959-64 are summarized and concluded that the best
application of solar ponds was salt production. Hirschman[8]
explored the possibility of making solar ponds on natural salt
flats to collect solar energy for electric power, industrial
heat, and desalination. Hirschman and Gaete[9] made some
detailed studies of some physical characteristics associated
with the operation of solar ponds. Stolzenbach et al[10], at
MIT developed numerical methods to predict temperature distribution within the solar pond. Some theoretical investigations
on solar ponds have been carried out by Usmanov et al[11] in
the USSR. No significant work on solar ponds was carried out
during the year 1964-1974. The oil embargo of 1973 revived
the interest in solar ponds throughout the world. Since that
time there has been a steady and phenomenal increase in the
amount of work and the number of publications on solar ponds.

Several review articles[12-17] on solar ponds have
appeared recently in periodicals and books. Research and
development on solar ponds is in progress in several countries
including USA, Israel, India, U.K., USSR, Australia, Chile,
Argentina, Zambia, Canada, France, etc. The details of some
of the major solar pond projects are given in Table 1.

Table 1 Major Salt-gradient Solar Ponds

Contractor/ location	Area (m^2)	Depth (m)	Main objective	Achievements/Problems
DOE Mound Lab. Miamisburg Ohio (USA)	2020	3.0	To provide heat to city recrea-tional building and swimming pool	Construction of this pond was completed in 1978 and is the largest solar pond operating in the USA (1981). It worked

Contractor/ location	Area (m^2)	Depth (m)	Main objective	Achievements/Problems
				well for one year and then liner started leaking. The salt concentration varies from zero to 18.5 percent NaCl at 1.5m. The lower convective zone is 1.5m deep, composed of 18.5 per-cent NaCl. The cost of heat is approx. $ 8.95 per GJ.
Ohio/State Univ.Columbus (USA)	200(old) 408(New)	2.5 3.0	To conduct rese-arch, understand the physics of solar ponds and behaviour of materials, con-duct heat ext-raction experi-ments and possi-bly demonstrate the grain drying.	First pond was built in 1975 and the sec-ond in 1980. Success-ful heat extraction experiments were conducted which es-tablished the exis-tence of finite thickness gradient zone and correlation between salt and thermal gradient at stable zone boundary. The clarity got re-duced due to wind blown debris.

Contractor/ location	Area (m^2)	Depth (m)	Main objective	Achievements/Problems
University of New Mexico, Albuquerque, (USA)	167	2.5	To conduct experiments on gradient maintenance, heat extraction, stability with NaCl	Built in 1975, boiling temperature was reached and successful experiments of heat extractions were conducted for one complete year. Thermal efficiency of about 8 percent was observed. The layer stability analysis compared favourably with observed pond behaviour.
Ohio Agricultural Res. & Development Centre, Wooster, Ohio (USA)	155	3.0	Green House heating, heat pump source, test cover and reflector, studies for possible improvements.	The pond built in 1975 mainly for providing greenhouse heating started leaking after one year due to poor design and materials. Cover and reflector were found ineffective and a maximum efficiency of 12percent was observed. Chemical treatments were developed to maintain clarity.

Contractor/ location	Area (m^2)	Depth (m)	Main objective	Achievements/Problems
Ein Bokek, Dead Sea (Israel)	7000 1,00000 10,00000		Power generation, multistage desalination and possibly aircon- ditioning	This is the world's lar- gest solar pond project. The first pond was com- pleted in the late summer of 1978 and has reached a temperature of 88°C. The first pond is pro- viding 150 KWe of peak power. Future work inc- ludes the development of 5000 KW and 10000 KW power station.
Ormat Turbines Company, Yavne(Israel)	1500		Power generation and operating experience and power production	The pond is supplying hot brine at a temperature of 80-90°C to the boiler of an organic Rankine turbine. The pond is working since 1977 and supplying elec- trical power on a 24 hour basis.
Dead Sea Potash Works, Near Sadom (Israel)	1100		Demonstration, operating expe- rience and pow- er production	The pond reached a maximum temperature of 103°C. The pond operated at an annual collection effici- ency of 15 percent and successful heat extraction experiments were conduc- ted.

Collector/location	Area (m^2)	Depth (m)	Main objective	Achievements/Problems
Aspendale (Australia)	100	0.86	Operating experience and materials behaviour with NaCl	This pond was built in 1964 and closed in 1966. The maximum temperature recorded was 65°C which may be due to poor clarity, algae growth and high upward heat loss due to low thickness i.e. 37-75cm of non-convecting (insulating) layer.
Alice Springs (Australia)	2000		Power production and operating experience on large solar pond	The pond is fully instrumented and is presently being heated and monitored. This pond will supply heat to a 2 KW Rankine cycle turbine.
Bhavnagar (India)	1210	1.2	Operating experience and behaviour of materials	The pond recorded a maximum temperature around 80°C at the bottom in 1972. The pond worked only for two years.
Bhavnagar (India)	1600	2.3	Operating experience on large ponds and possible applications like desalination and power production.	This pond is presently getting heated. It is designed to supply heat to a 20 KW Rankine cycle turbine.

Contractor/ location	Area (m^2)	Depth (m)	Main objective	Achievements/Problems
Pondicherry (India)	100	2.0	Experience in solar pond operation and maintenance in India's difficult climate, evolve criteria for the materials to be used, monitor thermal performance, study the physical behaviour of the gradient zone	The pond was built in Jan. 1980. Various problems like leaking through linings, algae growth, and mineral impurities were observed. The pond is fully instrumented. Average net rise of bottom water temperature was 0.6°C per day for 45 days and 0.25°C per day for the next 42 days. Gradient Stabilization by using hot brine injection, and algal control by using a submerged chlorinator have been successfully tried.

3.0 THE SOLAR POND

The solar pond combines solar energy collection and sensible heat storage. A salt-gradient solar pond is schematically shown in Fig. 1. Part of the solar radiation incident on the pond surface gets reflected back and part of it gets transmitted and absorbed within water. Since water is almost opaque to the far infrared, only the short wavelengths (1.2 to 0.2 μm) of the solar spectrum reach

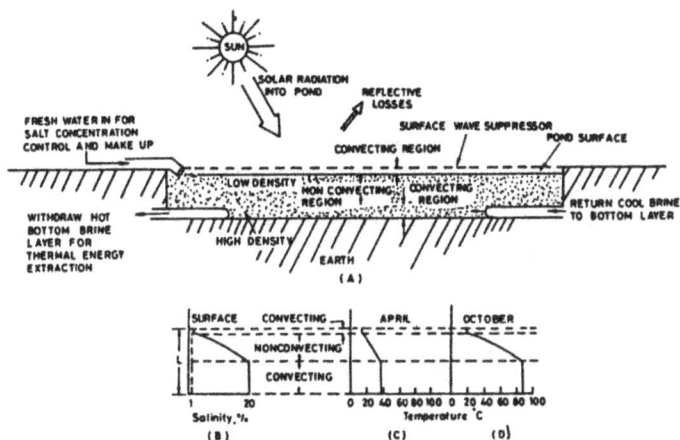

FIG.1 (A)   CROSS SECTION OF SOLAR POND-SCHEMATIC-SHOWING THREE
             ZONE CONFIGURATION.

     (B)   SALINITY PROFILE THICKNESS OF SURFACE CONVECTIVE
             ZONE VARIES SEASONALLY AND THIS PROFILE IS AN
             APPROXIMATE MEAN.

     (C)(D) TEMPERATURE PROFILE (FOR TWO TYPICAL MONTHS) IDEALIZED-
             ANTICIPATED IN SPACE HEATING APPLICATIONS.

the bottom of the deep pond. The density gradient in a pond
is obtained by using a high concentration of suitable salts,
such as sodium chloride or magnesium chloride, at the bottom of
the pond and a negligible concentration at the top. The thermal
conductivity of salt solution, which is even less than that of
stagnant water, decreases with the increase of salinity and thus
acts as an insulating layer also. As seen the Fig. 1, a salt
gradient solar pond consists of three zones, a relatively thin
(0.1 to 0.4 m) homogeneous and convective surface layer of
constant temperature (isothermal zone) and salinity formed due to
upward salt transport, surface heating and cooling, and wave

action, and serving no useful purpose; a non-convective gradient
layer acting as an insulating layer in which there are gradients
in salinity and temperature and the thickness of which depends
on the desired storage temperature, salt transmission properties
of water and thermal conductance of water and typically varies
from 0.6 to 1.0 m, and a homogeneous and convective bottom layer
acting as a high temperature (isothermal zone), constant salinity
and acting as a thermal storage layer. Useful heat is extracted
from this bottom convecting layer, the thickness of which depends
on the temperature and the amount of thermal energy to be stored.
The non-convective gradient layer transfers heat by conduction
only and allows solar radiation to penetrate and reach the bottom
of the pond and thereby trapping the solar radiation. Solar
ponds have been studied widely and the literature which covers,
amongst other things, experiments.[18,19,20] absorption of
solar radiation[15,17,21,23], physics,[15,24,25], hydrodynamics[26-28]
thermal analysis[21,22,29-31] heat extraction[32,33],
engineering[13,15,17] and economics[34,35].

4.0   EXPERIMENTS

In different parts of the world solar ponds are being used
to supply large amount of heat to industrial applications. Short
term experimental data on laboratory small size, field medium
size and field very large size solar ponds is available.
Therefore, lifetimes of ponds are not known and there is no
data at all available on medium sized or large scale ponds
concerning stability, lifetimes and operation. Experiments on
laboratory solar ponds, medium capacity (100-500m^2) solar ponds
and large size solar ponds are conducted in countries like USA,
Israel, Australia, India etc. Following is a very brief review
of work done in India as well as abroad.

4.1   Work Conducted (Abroad)

Experiments on laboratory solar ponds were conducted by several workers but the work of Wilkins and Pinder[18] will be described here. The laboratory solar ponds as shown in Fig.2.

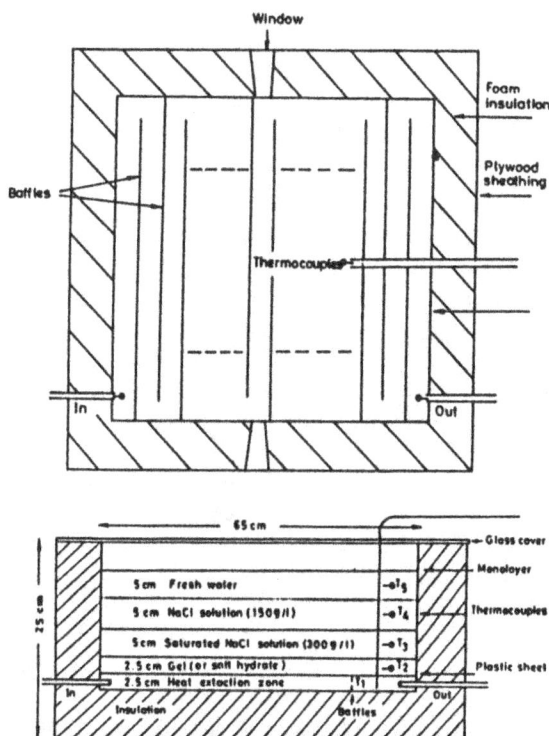

FIG.2   SCHEMATIC OF THE VANCOUVER MODEL SOLAR POND
        (FROM WILKINS AND PINDER)

is a square one with a side 0.65m and depth 0.25m.   The interior
walls and bottom is painted black with a vertical slot left
unpainted for observations of the stratification. Insulation
is used in the bottom as well as on all sides. A partition is
used to separate the lower convective zone (LCZ) from the
Non-Convective Zone (NCZ). Better heat transfer for heat
removal is obtained by using baffles in LCZ. In this pond
commercial NaCl with the concentration as shown in the figure
has been used. To minimize evaporation a monolayer of 10
percent of 1:5 stearic acid-Cetanol in N-Butanol is employed.
The pond surface was covered with a glass sheet. In this model
pond, three different horizontal partitions are tested:

   - Horizontal clear plastic sheet with no gel
   - Clear plastic sheet supporting gel
   - Black plastic sheet supporting gel

The plastic sheet used, in all cases, was of 0.2mm
polyethylene. The gel was prepared from commercial food clear
gelatin (100 gm/1). From the actual experiments conducted in
the open sun it is concluded that the pond with gel in the
bottom shows higher temperature than the pond with salt in the
bottom. Highest efficiency of 77 percent obtained in case of
Gel/Clear plastic. The efficiency was about 60 percent for
the clear plastic and 56 percent for the gel-black plastic
case.

Experiments on medium capacity (100-500m^2) solar ponds
are conducted in countries like USA, Israel, Australia, etc.
Perhaps the most systematically built, instrumented, and
studied medium capacity solar pond is at the University of New
Mexico,[19] USA, which was built in the fall of 1975. The
New Mexico solar ponds as schematically shown in Fig. 3 is having
the dimensions: diameter, 15m; depth, 2.5m; bank angle, 35°;
average collecting areas, 105m^2. The total capacity of the

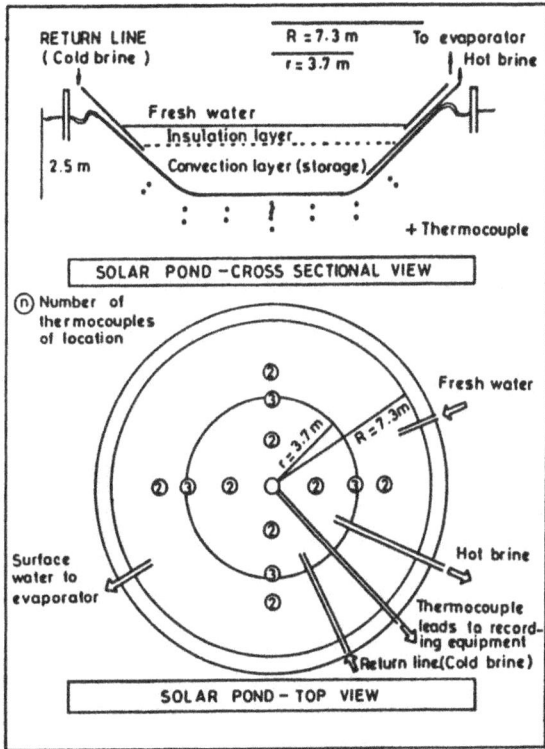

FIG.3 SCHEMATIC DIAGRAM OF THE EXPERIMENTAL SOLAR POND
(FROM ZANGRANDO AND BRYANT)

pond is $230m^2$ with an average capacity of $130m^2$, and it
contains 40 US tons of NaCl; it was built and lined with
commercially available materials and was designed to supply
33MWh per year of useful heat, which is the total heating

requirement of a 185.8m^2 house in Albuquerque. The circular shape facilitates mathematical modelling of the pond's behaviour and also maximizes volume per wall area. This pond was originally filled in November 1975; the insulating layer (70 cm thick) and a gradient of 0.20 percent per cm and the bottom convective layer (20 cm thick) had a uniform concentration of 15 percent NaCl by weight. Continuous daily observations of salt concentration and temperature in the pond was taken for the year 1975-1978. For the two consecutive years 1975 and 1976 when no heat was withdrawn, this pond has shown the maximum yearly temperature within the same week: 71°C on August 7, 1976 and 93°C on August,1977. The recorded lag between convective layer temperature and insolation was about 46 days for the two summers and 30 days for one winter. Heat extraction is done by circulating the brine from the bottom of the pond through an external heat exchanger and then returned to the convective layer through the diffuser.

Experimental studies on large size solar ponds are conducted in Israel and USA. The largest solar pond which is still in operation having an area of 7500m^2 and is used for electrical power production is in Israel. Some typical data of this pond is given in Table 2.

Table 2: Performance data of solar pond of Ein Bokek, Dead Sea(Israel)

Pond Size	7500m^2
Pond depth	2.6m
Storage zone depth	1.0m

Parameters	Summer	Winter
Hot brine temperature (°C)	92	72
Cooling water temperature (°C)	37	25
Hot brine flow rate (m^3/hr)	549	549
Cooling water flow rate (m^3/hr)	379	379
Working fluid boiling (°C)	86	86
Working fluid condensation temperature(°C)	45	45
Electric gross power (KW)	170	150
Auxiliaries hydraulic power (KW)	33	28

The pond yield and utilization varies from week to week
and typical results are given below:

Thermal yield   14263 Kwh(t)
Storage zone temp.   93°C
Cooling water temp. 27°C
Pond efficiency   19.4 percent

The largest working solar pond in USA is at Miamisburg
with an area of 2000 m^2 used for swimming pool heating of
nearby recreation building in winter. This pond is 54.5m x36.4m
at the top with sides tapered at an angle of 45° (a 1:1
slope) to a depth of approximately 3.0m. The pond uses a good
linear and partially filled with 18.3 percent sodium chloride
solution upto a depth of 2.1m and was acidified to a pH of 6.1
by addition of 0.44m^3 of concentrated hydrochloric acid.
Sufficient copper sulphate was added to achieve a copper ion
concentration of 2 ppm. This mixing of acid, copper sulphate,
and water brought the depth of the pond to 2.3m. The salt
gradient was formed by a special circular water distributor
(1.2m dia) which was placed 460mm below the surface of the
salt water. Fresh water was pumped at high pressure through
a 3.2mm slit at the edge of the distributor at the rate of
7.4 x 10^{-3} m^3/s. The velocity of this water, 0.6m/s was
sufficient to inject a horizontal layer of fresh water
across the width of the pond at the same depth as the distributor.
This water because of its low density will rise and dilute
the salt solution. Successive injections of water, 25mm thick,
were added in this fashion, and then the distributor was
raised 50mm for the next injection of fresh water. The experimentally determined salt profile[20] is shown in Fig. 4. An
inbuilt heat exchanger is used to remove the heat from the
LCZ. The acidity of the water is maintained by adding
concentrated hydrochloric acid at a rate of 0.36m^3/year. Measured
temperature profile[20] in the pond with depth and in the ground
below for two typical days, one in October and another in
February is shown in Fig. 5. From this figure a three layer

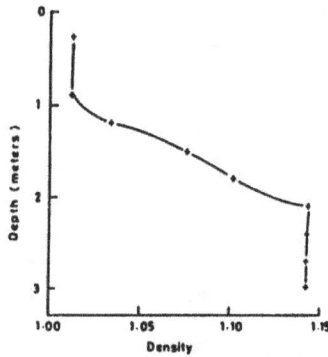

FIG.4: DENSITY DEPTH PROFILE IN THE MIAMISBURG SOLAR
POND DURING EXPERIMENTATION (FROM WITTENBERG AND HARRIS)

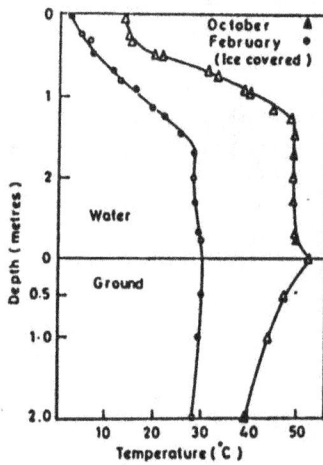

FIG.5: TEMPERATURE DEPTH PROFILE IN THE MIAMISBURG POND
AND GROUND BELOW THE POND FOR TWO DIFFERENT SEASONS.
(FROM WITTENBERG AND HARRIS)

structure in the pond is clearly seen. The convective zone
caused by sunlight and wind has a constant temperature and is
of approximately 0.4m thick. In the non-convective zone the
temperature increases with depth and goes upto 1.5m depth.
The bottom layer of the pond or the storage zone is approximately
1.5m thick and it stores the high temperature water. It is
seen from the temperature profile in the ground that during
winter some of the heat does return to the pond from the ground.
The monthly variation of temperature in the pond for two
successive years is shown in Fig.6. The operational temperature

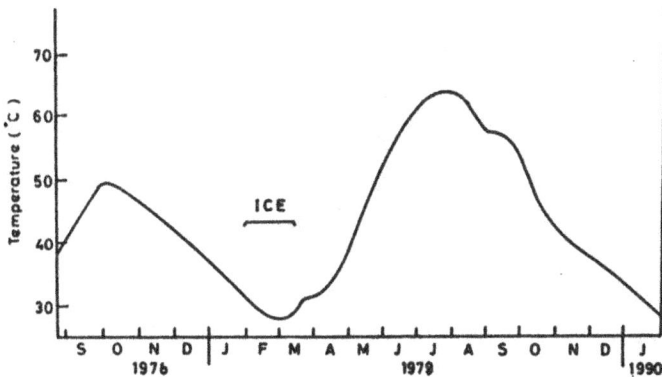

FIG.6: SEASONAL VARIATION OF TEMPERATURE OF POND STORAGE
WATER. (FROM WITTENBERG AND HARRIS).

in pond was reached only in June 1979 and between June 26 and
September 5, 1979, about 143.5 GJ of heat was extracted for the
use in the swimming pool with an average heat extraction
of 23.06KW. The efficiency of this pond under ideal conditions
is predicted to be nearly 20 percent, which would produce
1372 GJ of useful heat.

There are numerous practical problems in a large size
solar pond. The pond water does not remain transparent for
longer duration due to dirt, leaves and other foreign materials
falling in the pond and due to biological growth (algae,
bacterial). Further, there can be a reaction between the
salt and chemicals present in the water which may cause precipitation and non-transparency. There can be a serious environ —
mental hazard due to salt leakage through the brine or from the
surface of the pond. In an unsaturated pond there is a continuous
upward diffusion of salt which is about 10 Kg/m^2 yr in NaCl
pond. This salt is to be flushed and disposed of at some
safe place without any environmental hazard. Soil can be made
impervious and there are certain natural areas (such as salt
lakes) where ponds can safely be made without any leakage.
Generally in an artificially made solar pond a linear is used
which should be stable at high temperatures (~110°C), resistant
to strong solutions, resistant to UV degradation, ease of sealing
in situ, and a minimum mechanical strength. Some of the
technical problems that we encountered and approaches used in
USA are listed by Sargent and Neeper[38] and the same are
reported in Table 3.

Table 3: Technical Problems and approaches

Problems	Approaches
Prevent convection	1. NaCl salt gradient
	2. Other salts where cheap (like bittern)
	3. Saturated solutions
	4. Gels
	5. Membranes
Water Clarity	1. Copper sulphate (for algae)
	2. Chlorine (for bacteria)
	3. Selective precipitation for minerals
	4. Fences and surface flushing for debries

Table 3 /- contd..

Table 3 contd.

Problems	Approaches
Heat extraction	1. Inpond heat exchanger for small ponds
	2. Optimize hot brine withdrawal for large ponds
Slow migration of layer boundaries	1. Model pond and full scale experiments
	2. Theoretical and numerical hydro-dynamic studies.
Surface layer growth	1. Model pond and full scale experiments
	2. Underground effect of diurnal heating and cooling plus surface evaporation
	3. Theoretical and numerical hydro-dynamic studies.
Wind driven instabilities	1. Wave break may prove adequate, problem needs theoretical hydro-dynamic study
Scale upto many hectares pond for industrial process heat or electricity	1. Field experiments including design studies
	2. Establish maintenance requirements and cost.
Salt pollution	1. Liners for small ponds
	2. Natural saline environments or imprevious soil for large ponds
	3. Recycle diffused salt
Pond life times	1. Test and develop improved materials

4.2   Work Conducted (India)

In India research work on solar ponds was started in 1973.
Experiments on medium capacity solar ponds were conducted by
G.C. Jain[36], C.L. Gupta[37], etc. An experimental solar pond
was constructed at Bhavnagar in 1973. The pond recorded a
maximum temperature around 80°C at the bottom. This pond was
experimented for a period of two years. However, the renewed
response has been shown from 1980 onwards. In 1980 a larger
pond of 1800 m^2 area was constructed adjacent to the site of the
earlier pond at Bhavnagar and was experimented throughly.
Another experimental solar pond was put into operation at
Pondicherry in 1980 under the direction of Dr. C.L. Gupta of
the Tata Energy Research Institute, Field Research Unit. This
solar pond was of an area of 100 m^2. It was operated for a
period of two years and provided valuable data on pond operation,
ground losses, and efficiency. At present the construction of
a 250 m^2 research pond supported through the Karnataka Energy
Board is in progress at the Indian Institute of Science under
the direction of J. Srinivasan. In 1980, computer simulation
programme was undertaken by M.S. Sodha and associates on the
performance of salt gradient solar ponds at Indian Institute of
Technology, Delhi. The studies include an investigation of the
thermal behaviour of a salt gradient solar ponds with diffusely
reflecting bottom. Simulation as well as experimental studies
on the performance of solar ponds are also carried out by IIT,
Kanpur, University of Rajasthan, Jyoti Ltd. of Baroda etc.

Work is underway to establish a thicker gradient zone
and improve the clarity of the water so that adequate solar
input and higher temperature can be achieved.

5.0   ABSORPTION OF RADIATION IN POND

It is recognised that transmission and absorption of
solar radiation in water depends on many factors, including angle
of incidence of rays, the spectral composition of radiation,

multiple scattering by water molecules and impurities, reflection
and diffusion from the water surface and the bottom, etc. The
biological organisms and other debries present in the pond also
effect the transmission. In addition, the presence of transition
metal ions (e.g. Fe, Cu) produces a weak absorption in the red
part of the spectrum. The salts present in the pond do not
appear to produce much of attenuation. The effect of wavelength
and thickness of water on the absorption of solar radiation[17]
is shown in Table 4. It is seen from this table that the short
wavelength portion of the sun's spectrum (0.2-0.9μm) penetrates
metres and tens of metres, while the near infrared is absorbed
within the first few cm or mm. For the infrared radiations the
water is practically opaque.

Table 4: Percent absorption of solar radiation in different
          spectral ranges through water layers of different
          thickness

Wavelength			Thickness of water		
(μm)	0	1cm	10 cm	1m	10m
0.2-0.6	23.7	23.7	23.6	22.9	17.2
0.6-0.9	36.0	35.3	30.5	12.9	0.9
0.9-1.2	17.9	12.3	0.8	-	-
1.2	22.4	1.7	-	-	-
Total	100.0	73.0	54.9	35.8	18.1

The fraction of the direct radiation, penetrating the
pond surface, Tm is given by Snell's law as:

$$T = 1 - \frac{1}{2} \left[\frac{\tan^2 (\theta_z - \theta_r)}{\tan^2 (\theta_z + \theta_r)} + \frac{\sin^2 (\theta_z - \theta_r)}{\sin^2 (\theta_z + \theta_r)} \right] \qquad (1)$$

where Q_i and Q_r are the angle of incidence and angle of reflection respectively. Here $O_i = Q_z =$ Sun's Zenith angle $= (90-$altitude of Sun α). The daily values of radiation penetrating the pond surface can be determined by integrating the product hourly direct radiation and T from sunrise to sunset. If only daily values of direct radiation are available then the radiation penetrating the pond surface can be determined by taking the product of daily direct radiation and daily transmission coefficient T. The daily transmission coefficient T can be determined as

$$
T = \frac{\int_{-\omega s}^{+\omega s} Ta^{cosec\alpha} \sin \alpha\, dw}{\int_{-\omega s}^{+\omega s} a^{cosec\alpha} \sin\alpha\, dw}
\tag{2}
$$

where ω_s is the sunset hour angle and is given as

$$
\omega_s = \cos^{-1} [-\tan L \tan\delta]
\tag{3}
$$

where L is the latitude of place, δ declination of sun, cosec α the air·mass and, a, the atmospheric transmission coefficient. The altitude of sun α is given as

$$
\sin \alpha = \cos L \cos \delta \cos \omega + \sin L \sin \delta
\tag{4}
$$

Where ω is the hour angle from solar noon.

Tabor and Weinberger[15] have calculated the values of fraction of solar radiation that penetrates the pond surface using the above expressions. The values of daily transmission coefficient T at air/water interface as calculated by them are shown in Table 5. The inclusion of diffuse component will change the figure slightly but it all depends on the percentage of

Table 5: Daily values of transmission coefficient (T) for direct
radiation at air/water interface for various latitudes

Latitude (deg)	Summer solstice	Winter solstice	Equinox
0	0.97	0.97	0.97
10	0.97	0.96	0.97
20	0.97	0.95	0.97
30	0.97	0.93	0.96
40	0.97	0.89	0.96
50	0.96	0.78	0.94

diffuse radiation. Exponential relations are recommended for
determining the attenuation of solar radiation in pond. In the
exponential model suggested by Rabl and Nielson[17] known as
R-N model, the solar spectrum is divided into 4 parts between
the wavelengths of 0.2 to 1.2μm with known attenuation coefficients.
Recently Hull[39] has used the newer absorption data and divided
the spectrum into 40 equal parts between the wavelengths of
0.31μm and 1.31μm and observed that this modification has
given a 0.10 higher transmission at most depths compared to that
of the R-N model. Recently Kaushik and Bansal[23] have shown
that radiation intensity H(x,t) at a depth x in the pond at time
t can be obtained more closely to the observations by the
superimposition of five exponentials as compared to four
proposed in R-N model and suggested the following expression

$$H(x,t) = T H(x = 0,1) \sum_{n-1}^{5} \eta_n - e^{-\mu_n x} \qquad (5)$$

where H(x=0,t) is the total radiation at the pond surface, η is
the fraction of solar radiation having absorption coefficient
μ_n , μ_n is the absorption coefficient for the nth portion of

solar spectrum, and T is the transmission coefficient($T = 1 -$ reflection loss). The values of these constants for 5 spectral bends are given in Table 6.

Table 6:   Exponential absorption coefficients of water

n	η	μ (m^{-1})
1	0.237	0.032
2	0.193	0.450
3	0.167	3.000
4	0.179	435.000
5	0.224	255.000

The fraction of radiation $h(x)$ reaching a depth x in the pond is thus given as

$$h(x) = \frac{H(x,t)}{TH(x=0, t)} = \sum_{n=1}^{5} \eta_n e^{-\mu_n x} \qquad (6)$$

A simple empirical logarithmic transmission model as suggested by Bryant and Colbeck [40] which very well fits down to a depth of 5 m is as follows:

$$h(x) = 0.36 - 0.08 \ln x \qquad (7)$$

where x is the depth in metres.

The angle of incidence of direct sun rays at the pond surface varies with the time of the day and year at a particular place, and this variation in angle of incidence not only effects the reflective losses but also the light has to travel a factor of $1/\cos \theta_r$ farther to penetrate to a given depth of the pond and which is calculated as

$$\frac{1}{\cos\theta_r} = [\ 1- (\frac{1}{n^2})\ \sin^2\theta_1]^{-\frac{1}{2}} \tag{8}$$

Table 7 lists the angle of incidence θ_i, angle of reflection θ_r, the reflective losses, and the factor $1/\cos\theta_r$ water with a reflective index n=1.33.

Table 7: Angle of refraction θ_r, reflective loss, and $1/\cos\theta_r$ for various angles of incidence

Angle of incidence θ_i	Angle of refraction θ_r	Reflective loss	$1/\cos\theta_r$
0	0	0.02	1.00
10	7.50	0.02	1.01
20	14.90	0.02	1.04
30	22.08	0.021	1.08
40	28.90	0.024	1.15
50	35.17	0.033	1.23
60	40.63	0.059	1.32
70	44.95	0.133	1.41
80	47.77	0.247	1.29
90	48.75	1.00	1.52

It is very difficult and time consuming to calculate the daily and monthly reflection loss and the radiation attenuation with depth to find out the solar pond performance. Rabl and Nielson (17) have suggested a reasonably simple but accurate method to calculate the attenuation of solar radiation in the pond by the use of an effective absorption coefficient $\bar{\mu}_n$ given as

$$\bar{\mu}_n = \mu_n/\cos\theta_r \tag{9}$$

where θ_r is the effective angle of refraction at pond surface corresponding to the position of the sun at 2 pm at equinox. This approximation does not also consider the direct and diffuse radiation separately.

The transmittance τ and absorptance α product i.e. $\tau\alpha$ is required for determining the performance of solar pond. A simple expression of transmittance absorptance product ($\tau\alpha$) is given by Kooi (21) with a simple analogy of flat-plate collector and is

$$\tau\alpha = \int_{x_1}^{x_2} h(x) \, dX/(x_2-x_1) \qquad (10)$$

or

$$\tau\alpha = a+b-b \left[\frac{x_2 \ln(x_2-x_1)\ln x_1}{(x_2-x_1)} \right] \qquad (11)$$

where a and b are constants and as given by Bryant and collbeck(40) are a = 0.36 and b = 0.08 where x is measured in meters. Here the pond is assumed to be consisting of three zones, the upper convective zone (UCZ) starting from $x = 0$ to $x = x_1$. This is followed by non-convective zone (NCZ) starting from $x = x_1$ to $x = x_2$ and then followed by the lower convective zone(LCZ) from $x = x_2$ to $x = x_3$.

6.0   MAINTENANCE OF SALT GRADIENT

The salt used in a solar pond for creating density gradient should need the following characteristics:

1.   It must have a high value of solubility to allow high solution densities.
2.   The solubility should not vary appreciably with temperature.
3.   Its solution must be adequately transparent to solar radiation.

4.   It must be environmentally benign, safe to handle, and ground water.

5.   It must be available in abundance near site so that its total delivered cost is low, and

6.   It must be inexpensive.

Since salt constitute the major cost of the pond, therefore the economic feasibility of a solar pond will entirely depend on the suitable identification of salt. Several salts such as NaCl, Na_2SO_4, $MgCl_2$, KNO_3, $CaCl_2 2H_2O$, NH_4NO_3 etc. are suggested for use in solar pond. The solubility of these salts in water as a function of temperature is shown in Fig.7.

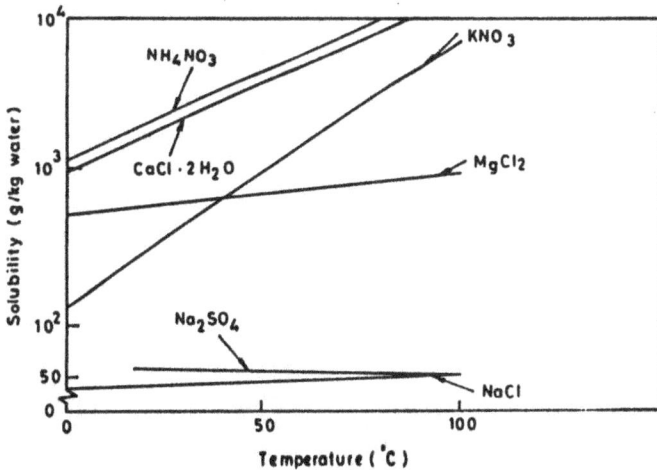

FIG.7:   SOLUBILITY OF SOME COMMON INORGANIC SALTS   USABLE IN SOLAR POND AS A FUNCTION OF TEMPERATURE

From this figure it is seen that the solubility of most of these salts increases with temperature except for Na_2SO_4. One of the main requirement of salt is that it should be transparent for solar radiation and the solubility should increase with

temperature which will assist in offsetting the expansivity of
the solution as temperature increases. Based upon this criteria
alone, a salt such as Na_2SO_4 would not be a good candidate. $MgCl_2$
and NaCl fulfill most of the requirements and are therefore most
widely used candidate materials for prototype solar ponds. $MgCl_2$
is sometimes preferred since this is residue salt in the end
brines of common salt manufacturing processes. Moreover, $MgCl_2$ has
higher solubility than NaCl thereby giving the higher densities
required at temperatures greater than 100°C. The maximum density
of $MgCl_2$ solution at nearly saturated conditions is about 1330 kg/m^3
and that of NaCl is 1220 kg/m^3 which means that pond with $MgCl_2$
will show higher maximum attainable temperature than with a
NaCl pond. Tables 8 and 9 show the effect of temperature (°C)
and concentration (kg/m^3) on the density of solution $\rho \times 10^{-3}$
kg/m^3 for NaCl and $MgCl_2$ respectively.

Table 8: Effect of temperature (°C) and concentration (kg/m^3)
on density of salt solutions ($\rho \times 10^{-3}$ kg/m^3) for NaCl

Concentration kg/m^3	Temperature (°C)					
	0	20	40	60	80	100
20	1.01509	1.01246	1.00593	0.9967	0.9852	0.9719
60	1.04575	1.04126	1.03378	1.0241	1.0125	0.9994
100	1.07677	1.07068	1.06238	1.0523	1.0405	1.0276
140	1.10824	1.10085	1.09182	1.0813	1.0694	1.0565
180	1.14031	1.13190	1.12218	1.1113	1.0993	1.0864
220	1.17318	1.16395	1.15358	1.1425	1.1303	1.1172
260	1.20709	1.19747	1.18614	1.1747	1.1626	1.1492

Table 9: Effect of temperature (°C) and concentration (kg/m^3) on density of salt solution ($\rho \times 10$ kg/m^3) for MgCl$_2$

Concentration kg/m^3	Temperature (°C)					
	0	20	40	60	80	100
20	1.0168	1.0146	1.0084	0.9995	0.9883	0.9753
60	1.0510	1.0478	1.0413	1.0325	1.0218	1.0095
100	1.0858	1.0816	1.0749	1.0663	1.0560	1.0544
140	1.1214	1.1161	1.1095	1.1011	1.0912	1.0803
180	1.1578	1.1523	1.1452	1.1412	1.1322	1.1170
250	1.2246	1.2184	1.2111	1.2031	1.1942	1.1847
300	1.2754	1.2688	1.2614	1.2575	1.2493	1.2360

Weinberger[5] in his classical paper entitled. 'Physics of solar pond' has developed a one dimensional thermal model and explained the stability of the solar pond. Weinberger asserts that there will be no vertical thermal convection or the pond will be stable when the density gradient on account of the salt concentration gradient is greater than the negative density gradient produced by the temperature gradient or the total derivative of density with respect to depth is greater than or equal to zero. At any point in the pond, the required salt concentration gradient should satisfy the equation:

$$\frac{\partial c}{\partial x} \geqslant - \frac{\frac{\partial P}{\partial T} \frac{\partial T}{\partial X}}{\frac{\partial P}{\partial C}} \qquad (12)$$

$$\text{or} \quad \frac{\partial c}{\partial x} \geqslant \frac{\beta_T}{\beta_S} \frac{\partial P}{\partial T} \frac{\Delta T}{\Delta x} \qquad (13)$$

where β_T is the thermal expansion coefficient and is given as:

$$\beta_T = \frac{1}{P} \frac{\partial P}{\partial T} \tag{14}$$

and β_s is the salt expansion coefficient and is given as

$$\beta_s = \frac{1}{P} \frac{\partial P}{\partial C} \tag{15}$$

Here X is the depth measured from surface, T is temperature, ρ is density, and C is salt concentration.

Thus for a given temperature difference ΔT between two layers of a liquid, the minimum difference concentration $(\Delta C)_{min}$ required for stability is

$(\Delta C)_{min}$ required for stability is

$$(\Delta C)_{min} = \frac{\beta_T}{\beta_s} \Delta T \tag{16}$$

The above equation represents a simple basic static stability criterian. However, in actual practice, even if a pond is statically stable against vertical thermal convection, the oscillatory disturbances can propagate vertically and grow with time, resulting in hydrodynamic instabilities. To prevent oscillatory motion or for dynamic stability, the salt concentration gradient should fulfill the following condition

$$\frac{\partial C}{\partial X} \geq - \left(\frac{\nu + \alpha_T}{\nu + \alpha_s}\right) \frac{\partial P}{\partial T} \frac{\partial T}{\partial X} / \frac{\partial P}{\partial C} \tag{17}$$

where ν is the viscosity, α_T is the coefficient of temperature diffusivity $(K/\rho C)$, and α_s is the coefficient of salt diffusivity. The above equation can also be expressed in terms of Prandtl Number

$$Pr = (K/\rho C = \frac{\nu}{T}\ )\ \text{as:}$$

$$\frac{\partial c}{\partial x} \geqslant (\frac{P_r + 1}{P_r + T_s}) \quad (\frac{\beta_T}{\beta_S}\ .) \quad \frac{\partial T}{\partial X} \tag{18}$$

where T_s is the ratio of diffusivities $(\ = \alpha_S/\alpha_T)$ or for a given temperature difference ΔT the minimum salt concentration $(\Delta C)_{min}$ can be expressed as:

$$(\Delta C)_{min} = (\frac{P_r + 1}{P_r + T_s}) \quad (\frac{\beta_T}{\beta_S}\ ) \quad \Delta T \tag{19}$$

The difference of salt concentration ΔC required between top and bottom of the solar pond can be found from equation 17 as:

$$\Delta C > (\frac{\nu + \alpha_1}{\nu + \alpha_s}) \quad \frac{\frac{\partial P}{\partial T} \frac{\partial T}{\partial X}}{\frac{\partial P}{\partial C}} \quad \alpha_s \quad \int_{o}^{x_3} \frac{dx}{\alpha_s (T,C)} \tag{20}$$

Weinberger found from an analysis of solar ponds that for the maximum value of $\Delta T/\partial X$ under meteorological conditions in Israel (maximum insolation is about 1200 W/m^2) a difference of concentration of about 340 kg/m^3 is required to maintain in a 1m deep $MgCl_2$ pond having a 0.2m mixing layer at the bottom and that a similar NaCl pond will go instable during periods of intense radiation; temperature derivative of NaCl density is relatively small.

Tabor and Weinberger[15] have derived the necessary parameters for NaCl and $MgCl_2$ for evaluating the stability criteria in solar ponds. The representative values are given in Table 10.

Table 10: Representative values of $\frac{\partial P}{\partial T}$ and $\frac{\partial P}{\partial C}$ at the pond extremities

Salt	Parameter	Pond Surface	Pond bottom 20°C	Pond bottom 90°C
MgCl$_2$	C	20	300	300 (kg/m^3)
	$\frac{\partial P}{\partial C}$	0.75	0.65	0.68
	$-\frac{\partial P}{\partial T}$	0.3	0.25	0.45 (kg/m^3°C)
	$\frac{\partial c}{\partial x}$	$>0.46 \frac{\partial T}{\partial X}$	$>0.44 \frac{\partial T}{\partial X}$	$>0.75 \frac{\partial T}{\partial X}$
		i.e.>230	(same as when hot)	300 (kg/m^4)
		for $\frac{\partial T}{\partial X}$ ~500		for $\frac{\partial t}{\partial x}$ ~400
NaCl	C	20	260	260 (kg/m^3)
	$\frac{\partial p}{\partial c}$	0.8	0.62	0.52
	$-\frac{\partial p}{\partial T}$	0.3	0.5	0.51 (kg/m^3°C)
	$\frac{\partial c}{\partial x}$	$>0.44 \frac{\partial T}{\partial X}$	$>0.92 \frac{\partial T}{\partial x}$	$>1.18 \frac{\partial T}{\partial X}$
		is >220	(same as when hot)	>230
		for $\frac{\partial t}{\partial x}$ ~500		

The non-convection stability in a MgCl$_2$ solar pond can easily be obtained while it is difficult to get in case of NaCl solar pond. The density gradient originally obtained in a solar pond is maintained against diffusion effects by controlling concentrations at the top and the bottom regions.

In the most accepted method of filling the pond, the pond is
successively filled with a number of layers of graded salinity.
Thus there is a layered or stair case like structure and the
method is known as stacking. In this case the pond is filled
with several layers of salt solution with successive layers
stepwise in concentration from near saturation at the bottom
to fresh water at the top. There can be six to eight layers
in a typical pond of 1m deep. Generally the more dense layers
are put first at the bottom and then less dense layers can
be floated successively upto the top. Alternatively, successive
dense layers can be introduced in the bottom and thus lifting
the previous less dense layer upto the top as is done in a
solar pond in Australia[41]. After the stepwise filling the
pond gradient smooths itself. This smoothing is partly due
to diffusion and partly due to the kinetic energy of liquid
flow injected into the pond during the filling process. It
is observed that for a 10 layered pond the smoothing will take
place within one week if only diffusion is involved and the
kinetic effect will further reduce the time. Because of this
mixing due to kinetic effect there is a critical flow velocity
for filling the pond which is about 0.12 m/s. The theoretical
development of a linear concentration profile[24] from an
initial six layer staircase profile is shown in Fig. 8. Table 11.

Table 11: Time (days) for the top surface to reach one half of
the concentration at the mid-level

No. of layers	Depth (cm)				
	20	40	60	80	100
2	26	104	235	417	652
3	22	88	199	353	552
6	18	72	161	287	448
10	16	65	146	260	406
20	15	60	134	238	372

shows the diffusion time in days (i.e. the time for the top
surface to reach one-half the concentration at the mid-level)
as a function of number of layers and for solar ponds of
depths 20,40,60,80 and 100 cm for a NaCl pond. From this
table[24] it is seen that for a 20 layered pond of one metre
deep, the diffusion time required is about one year.

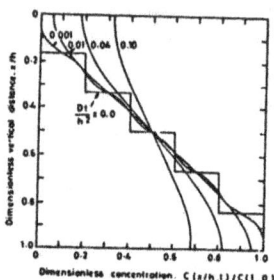

FIG.8: DEVELOPMENT OF CONCENTRATION PROFILE FROM AN INITIAL
SIX LAYER STAIRCASE PROFILE (D IS THE MOLECULAR
DIFFUSION COEFFICIENT, t IS THE TIME, h IS THE TOTAL
DEPTH AND $C(x,h,t)$ THE CONCENTRATION) (FROM CHEPURNITY
AND SAVAGE).

Another successful method of filing the pond is the
redistribution method and is based on the principal that when
fresh water is injected at some level into homogeneous brine
it stirs and uniformly dilutes the brine from a few centimetres
below the injection level to the surface. Thus a series of
layers of decreasing salinity can be produced by a series of
fresh water injections at successively higher levels in the
pond. This method is supposed to be more convenient and expedient.
In this process to produce a gradient zone of certain thickness,
the pond is filled with high salinity brine to half of its total
depth and then fresh water is added through a diffuser. Initially
the diffuser is placed at the bottom and the water is added in
the pond flowing in as under current and the level in the pond
increases; the diffuser is moved upward continuously or in steps.

It is seen that if fresh water is injected at a uniform rate
and the injection level is moved upward, stepwise or continuously
at twice the rate the surface rises, then an approximately
uniform concentration gradient is produced between the initial
injection level and the final surface.

In a pond there is a continuous diffusion of salt upwards
from higher to lower concentration. Due to density gradient,
the mass transfer rate $\dot{m}$ from bottom to top is given as:

$$\dot{m} = -\alpha_s \ A \ \frac{\partial c}{\partial x} \qquad (21)$$

where A is the area and α_s in the coefficient of salt diffusion.
If α_s is $2.16 \times 10^{-4} \ m^2/day$ then the rate of mass transfer will
be $0.0734 \ kg/m^2$ day which comes to 26,800 tons/year for a pond
of $1 \ km^2$. The salt diffusion coefficient α_s can be nearly
linearly related with temperature. For NaCl it is written as:

$$\alpha_s(T) = 1.39 \ [1+0.029 \ (T-20°C)] \times 10^{-9} \ m^2/s \qquad (22)$$

For $MgCl_2$ salt the diffusion coefficient α_s is given as:

$$\alpha_s(T) = 1.12 \ [1+0.027 \ (T-20°C)] \times 10^{-9} \ m^2/s \qquad (23)$$

Thus to maintain the stability in the pond either
this salt diffusing upwards is to be replaced periodically
or the diffusions is avoided. One of the most successful
method tried in Israel on large pond is by periodically
introducing concentrated salt solution at the bottom and
washing the surface with fresh water. A novel method known
as the 'falling pond' concept in which the salt gradient and
hence stability in the pond is maintained without addition
of salt is suggested by Tabor[42] and Shahar[43]. The salt
concentration gradient is governed by the diffusion equation

$$\frac{\partial \left(\alpha_s \ \frac{\partial c}{\partial x} \right)}{\partial x} = \frac{\partial c}{\partial t}$$

For steady state conditions $\partial c / \partial X = 0$, and on integration the salt flux q is given as:

$$q = \alpha_s \frac{\partial c}{\partial x}$$

If v is the vertical velocity (+ downwards) of the mass of fluid in pond, then we get

$$\frac{\partial \left(\alpha_s \frac{\partial c}{\partial x} \right)}{\partial x} = v \frac{\partial c}{\partial x} + \frac{\partial c}{\partial t}$$

Thus under steady state conditions

$$q = vc - \alpha_s \frac{\partial c}{\partial x} \tag{24}$$

Therefore if water is withdrawn from the bottom of the pond to give v a value

$$v = \frac{1}{c} \left(\alpha_s \frac{\partial c}{\partial x} \right) \tag{25}$$

then the net upward flux of salt is zero, i.e. the salt remains stationary in space. Thus in the 'falling pond' concept the brine is withdrawn from the hot lower layer and passed through a flash evaporation as shown in Fig. 9 where the heat is extracted. The concentrated brine is now returned to the bottom. The difference between the rates at which brine is removed and returned causes a steady fall of the pond, the rates being chosen so as to compensate for the molecular diffusion of salt upwards, and fresh water being added at the free surface to keep both the pond depth and surface concentration constant. This operation takes place without the addition of fresh salt and with no surface washing. The falling velocity has been estimated at about 0.2-0.5 mm/day. This method performs the dual function, it maintains the salt density and provides an efficient method of heat removal from the solar pond.

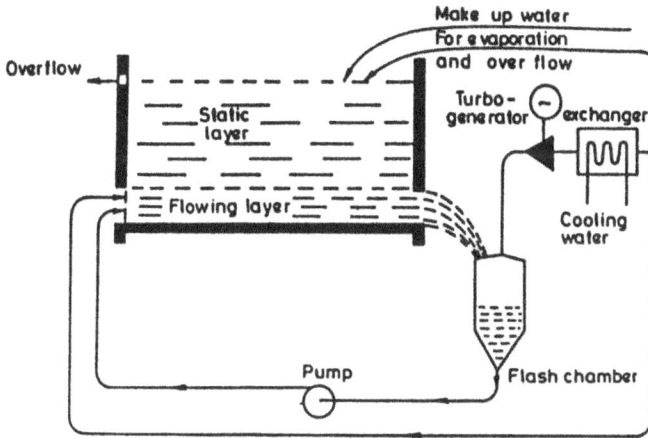

FIG.9: SCHEMATIC REPRESENTATION OF THE FALLING METHOD OF
EXTRACTING HEAT FROM THE BOTTOM OF A POND.

7.0   SIZING THE SOLAR POND

There is no method available for sizing the solar pond
depending on the heating requirements. Recently Edesess et
al[44] have discussed a simple method to calculate the required
pond surface area and depth which will enable a potential user
to determine the approximate size solar pond needed for the
contemplated application and location. According to Edesess[44]
et al, the radius r of a circular solar pond to meet the
requirement is qiven as:

$$ r = \frac{U_e T_d + [(U_e T_d)^2 + \bar{Q}_L \{\bar{H}_p - (U_s - U_b) T_d\}/\pi]^{\frac{1}{2}}}{\bar{H}_p - (U_s + U_b) T_d} \tag{26} $$

where Ue, Us, & Ub =   average heat loss coefficient from the edges of the pond, surface of the pond and bottom of the pond respectively.

T_d = $(\bar{T}-\bar{T}a)$. Here $\bar{T}$ is the annual average pond temperature desired in °C and $\bar{T}a$ is annual average ambient temperature in °C.

$\bar{Q}L$ =   annual average load in watts.

$\bar{H}_p$ = $\bar{H}$ f $(\tau\alpha)$,

f   =   reflection loss adjustment factor which depends on the latitude L and its values are given in Table 12.

$\tau\alpha$ =   the average optical transmission, i.e. the fraction of incoming insolation that reaches the pond's storage area.

Table 12: Reflection loss adjustment factors (f)

Latitude (L) degrees	f
0-29	0.98
30-43	0.97
44-49	0.96
50-53	0.95
54-56	0.94
57 -58	0.93

The determination of depth of storage layer D of pond is difficult and is determined by trial and error. The expression for the resulting minimum pond temperature t_{min} for a depth D is given as

$$A_{min} = \bar{T} - \frac{[(a+aD)^2 + (b+cD)^2]^{\frac{1}{2}}}{5.2327\ D^2 + 7.5445(U_s+U_b)^2} \rule{2em}{0pt} \tag{27}$$

A trial of D is selected and t_{min} is determined. Another trial value of D is used until a value of D is found such that $t_{min} = T_{min}$ the desired minimum pond temperature is obtained. The value of D thus obtained is the depth of the pond's storage layer. Here a,b,c and d are given as

$a = (1.4138\ \bar{H} - 2.3313\ U_s\bar{T}a - 7.5445\ \bar{Q}L\ Cos\ 2\pi\alpha)(U_s+U_b)$

$b = (-7.4110\ \bar{H} - 7.1756\ U_s\bar{T}a - 7.5445\ \bar{Q}_L\ Sin\ 2\pi\alpha)(U_s+U_b)$

$c = -1.1775\ \bar{H} + 1.9415\ U_s\bar{T}a + 6.2832\ QL\ Cos\ 2\pi\alpha$

$d = -6.1720\ \bar{H} - 5.9759\ U_s\bar{T}a + 6.2832\ \bar{Q}_L\ sin\ 2\pi\alpha$

where

$\alpha = (M-0.5)/12 - 0.25$

M = the number of the month with the highest demand. In northern hemisphere, M = 1 for January, M = 2 for February, etc. In southern hemisphere, M = 7 for January, M = 8 for February with M = 6 for December

$\bar{T}_a = \bar{T}a - Ta.min$

$\bar{H} = \bar{H}p - H_p.min$

$\bar{Q}L = (Q_{L,max} - \bar{Q}_L)/A$

A = the pond surface area (m^2)

Tmin = minimum desired pond temperature (°C)

Ta,min = average ambient temperature in the coldest month of the year (°C)

Hp.min = average insolation in the least sunny month
of the year (W/m^2)

$Q_{L'max}$ = average load in the month with the highest
demand (Watts)

The values of Cos 2πα and Sin 2πα for different values of
M are given in Table 13.

Table 13: Values of Cos 2πα and Sin 2παfor various values of M

M	Cos 2 π α	Sin 2 π α
1	0.2588	-0.9659
2	0.7071	-0.7071
3	0.9659	-0.2588
4	0.9656	0.2588
5	0.7071	0.7071
6	0.2588	0.9659
7	-0.2588	0.9659
8	-0.7071	0.7071
9	-0.9659	0.2588
10	-0.9659	-0.2588
11	-0.7071	-0.7071
12	-0.2588	-0.9659

Having calculated the depth D of the storage layer i.e. LCZ,
the total depth is calculated by adding to this the depth of
UCZ i.e. 0.3m and NCZ i.e. 1.2m.

8.0   THERMAL ANALYSIS OF SOLAR POND

The mathematical formulation of the behaviour of a
totally nonconvecting solar ponds was first given by Weinberger[5]
whose analytical solution of the partial differential equation
for the transient temperature distribution was obtained by
superimposing the effects of radiation absorption at the surface,

in the body of water, and at the bottom the effect of each being
considered separately. This technique and by using the similar
boundary conditions was adopted by Dake[51] and Akbarzadeh and
Ahmedi[30]. An iterative computer model of a solar pond which
accurately considers the external driving functions and load
demands and ground loss has been suggested by Tybout[52] and
developed by Hull[29]. Tybout pointed out some difficulties
with the Weinberger's model and suggested that a numerical approach
would be useful. Stolzenbach, Dake, and Harleman[53] have
described a finite difference method to predict the temperature
distribution in the pond in terms of solar insolation, surface
losses, radiation absorption, and heat withdrawal from the
bottom. Kooi[21] examined the three-zone salt gradient
stabilized solar pond as a steady state flat-plate collector
and assumed the atmospheric air temperature equal to the pond
surface temperature. In all these models, the approach has been
in solving the one dimensional heat conduction equation including
heat source function with appropriate boundary conditions.
Solutions to the one dimensional heat conduction equation have
been obtained by various authors both in terms of time dependent
closed form analytical solutions (Sodha et al[54]), as well as
numerical finite difference schemes (Chepunity and Savage[25],
Hawlader and Brinkworth;[22] Meyer et al[55])corresponding to
periodic and arbitrary transient boundary conditions. In some
analysis, the density ρ , the specific heat C, and thermal
conductivity K, have been approximated as being constant and in
others as linearly dependent upon concentration, but the
convective-nonconvective zone boundaries are considered constant
in time in all the models. Here the numerical modelling of
solar pond as developed by Hawlader and Brinkworth[22], which
gives greater freedom to incorporate appropriate initial and
boundary conditions, and more realistic representation of
climatic conditions, load variations, etc. is described. The
pond configuration assumed in the one dimensional numerical
model is a three region system (Fig. 10) with boundary layers

separating the convecting regions from the nonconvecting regions. The temperature distribution $T(x,t)$ in the pond can be determined from the following heat balance equation:

$$pc \frac{\partial T(x,t)}{\partial t} = \frac{\partial}{\partial x} \left(\frac{K \partial T(x,t)}{\partial x} \right) + H(x,t)_{\mu} \sec \theta_r \qquad (28)$$

where the second terms on the right hand side is the source term and $H(x,t)$ which is the solar radiation at depth x and time t is expressed as:

$$H(x,t) = TH(x=o,t) \ (1-F)e^{-\mu(x-\delta)\cos \theta_r}$$

$$= H'(x=o,t) \ (1-F)e^{-\mu(x-\delta)\cos \theta_r} \qquad (29)$$

where μ is the effective absorption coefficients and the factor F may be interpreted as a measure of the part of the radiation which is absorbed within a short distance of the surface.

Equation (28) can now be written in a dimensionless form as follows:

$$\frac{\partial \theta(x,t)}{\partial T} = \frac{\partial^2 \theta(x,t)}{\partial z^2} + \frac{x_2^2 H(x,t)\mu \sec \theta_r}{KT_o} \qquad (30)$$

where $\theta(x,t) = \dfrac{T(x,t)}{T_o(x,t)}$, $Z = \dfrac{x}{x_2}$, and $T = \dfrac{\alpha t}{x_2^2}$

Here T_o is the initial temperature.

The solution of equation (30) requires two boundary conditions and an initial condition. The initial condition can be obtained from the physical conditions of the pond. The surface boundary condition can be determined by writing the heat balance equation on the upper convective zone.

$$Pcx_1 \frac{\partial T_1(x,t)}{\partial t} = H'(x=0,t) \ [1-(1-F)$$

$$\exp\{-\mu(x-\delta)\sec \theta_r \ \}]$$

$$+k \left. \frac{\partial T_1(x,t)}{\partial x} \right|_{x=x_1} - Q_{LS} \qquad (31)$$

where T_1 is the temperature of the liquid in the upper convective zone.

The first, second and third terms of the right hand side of equation 31 represent the effect of fraction of radiation absorbed in the upper convective layer, the heat transfer by conduction between the upper convective zone and non-convective zone, and the heat loss by convection, radiation and evaporation by the pond surface to the outside respectively. Equation (31) can also be written in the dimensionless form as follows:

$$\frac{\partial \theta_1(x,t)}{\partial T} = \frac{x_2^2 H'(x=0,t)}{K T_o \, x_1} \times$$

$$x \; [1-(1-F) \, \exp \, \{-\mu(x-\delta) \, \sec \, \theta_r\}]$$

$$+ \frac{x_2}{x_1} \left(\frac{\partial \theta_1}{\partial z} \right)_{z=z_1} - \frac{U_{LS} \, x_2^2}{K x_1} (\theta_1 - \theta_a) \tag{32}$$

The second boundary condition can be obtained by writing the energy balance equation on the lower convective zone:

$$P_\omega C_\omega (x_3 - x_2) \frac{\partial T_\omega(x,t)}{\partial t} =$$

$$x'(x=0,t) \; (1-F) \exp \, [-\mu(x_2-\delta) \sec \theta_r]$$

$$-K \left. \frac{\partial T_w(x,t)}{\partial x} \right|_{x=x_2} - Q_{LG} - Q_D \tag{33}$$

where tw is the temperature of the lower convective zone and Cw are the density and specific heat of water in the lower convective zone respectively. Here the terms on the right hand side represent respectively the effects of solar radiation reaching the lower convective zone, conduction to the nonconvective zone, heat loss to the ground, and heat retrieved (delivered) from the pond.

The above equation in the dimensionless form can be written as:

$$\frac{\partial \dot{\theta}_w(x,t)}{\partial T} = \frac{x_2^2}{KT_o(x_3-x_2)}(1-F)H'(x=0,t)$$

$$\exp[-\mu(x_2-\delta)\sec\theta_r] - \frac{x_2}{(x_3-x_2)}\left(\frac{\partial\theta_\omega(x,t)}{\partial z}\right)\bigg|_{z=1}$$

$$-\frac{U_{LG}x_2^2}{K(x_3-x_2)}[\theta_w(x,t)-Q_{CW}] - \frac{\theta_D x_2^2}{KT_o(x_3-x_2)} \tag{34}$$

Equation (30) is solved numerically by Hawlader and Brinksworth[22] using the boundary conditions discussed above by finite difference technique using the Crank-Nickolson approximation.

Using the above numerical model and the climatic data of London, Hawlader and Brinksworth[22] have studied the effect of extinction coefficient, lower convective zone depth, non-convective zone thickness, heat retrieved continuously and intermittently, bottom losses, etc. on the lower convective zone temperature. Fig. 11 shows the temperature of LCZ with number of days for two values of absorption coefficients without any heat removal. Fig. 12 shows the effect of LCZ depth on the temperature of LCZ for a number of days without any heat removal. Fig. 13 and 14 respectively, show the effect of insulation layer (UCZ + NCZ) thickness on the LCZ temperature for $\mu = 0.32$ m^{-1}

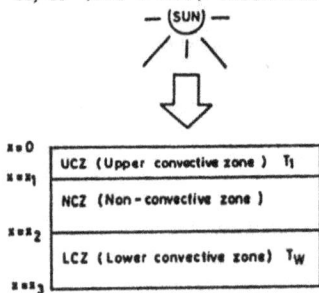

FIG. 10   POND CROSS SECTION DIAGRAM

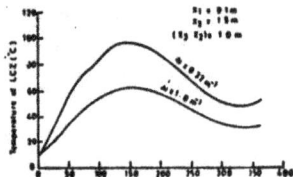

FIG.11   TEMPERATURE OF LOWER CONVEC-
TIVE ZONE WITH NO HEAT
RETRIEVAL (FROM HAWLADER
AND BRINKSWORTH)

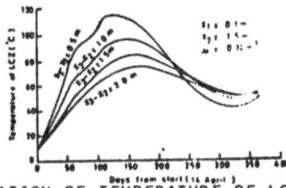

FIG. 12: VARIATION OF TEMPERATURE OF LOWER CONVECTIVE ZONE WITH
LCZ DEPTH (FROM HOWLADER AND BRINKSWORTH)

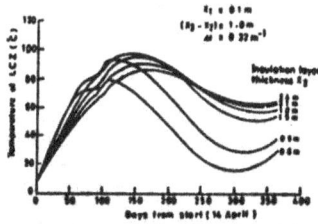

FIG. 13: VARIATION OF LCZ TEMPERATURE WITH INSULATION LAYER
THICKNESS (HAWLADER AND BRINSWORTH)

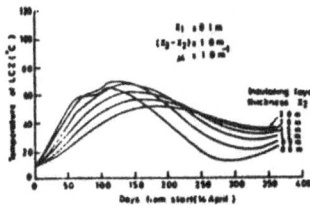

FIG. 14: VARIATION OF LCZ TEMPERATURE WITH INSULATION LAYER
THICKNESS (HAWLADER AND BRINKSWORTH)

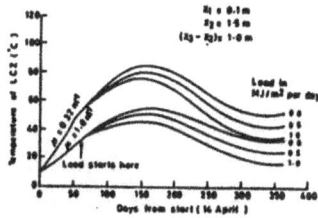

FIG.15: VARIATION OF LCZ TEMPERATURE WITH DIFFERENT LOAD CONDITIONS
AFTER TWO MONTHS (FROM HAWLADER AND BRINKSWORTH)

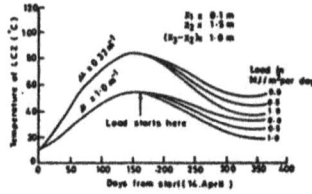

FIG.16: VARIATION OF LCZ TEMPERATURE WITH DIFFERENT LOAD
CONDITIONS AFTER FIVE MONTHS (FROM HAWLADER AND
BRINKSWORTH)

and $\mu = 1.0m^{-1}$ respectively. Figs. 15 and 16 respectively show
the effect of heat removal after 2 months and 5 months respectively
on the temperature of LCZ. The effect of continuously heat
removal and intermittently heat removal for a period of two
years on the temperature of LCZ is shown in Figs. 17 and 18
respectively. The heat loss through the ground also plays an
important role and its effect is shown in Fig. 19.

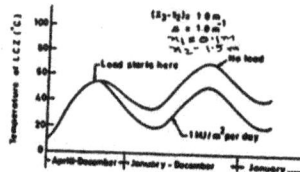

FIG. 17: VARIATION OF LCZ TEMPERATURE WITH CONTINUOUS LOAD FOR
TWO YEARS (FROM HAWLADER AND BRINKSWORTH)

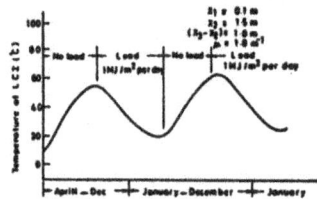

FIG.18: VARIATION OF LCZ TEMPERATURE WITH INTERMITTENT LOAD
FOR TWO YEAR (FROM HAWLADER AND BRINKSWORTH)

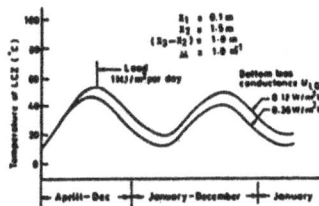

FIG. 19: VARIATION OF LCZ TEMPERATURE FOR TWO BOTTOM LOSS
CONDUCTANCES FOR TWO YEARS. (FROM HAWLADER AND
BRINKSWORTH)

The performance of the solar pond for Indian climatic conditions
has also been reported by R.R. Isaac and C.L. Gupta[58]. A
steady-state model, taking into considerations all the heat
losses that take place from the sides and the bottom of the pond
has been developed. The effect of various parameters - sizing
parameters, operating parameters and co-climatic parameters on
the performance of the solar pond have been derived by a computer
model and discussed. Studies indicate that there is an optimum
depth and storage volume of the pond for each application in
terms of temperature and heat load desired.

The schematic of the solar ponds is given in Fig. 20.
The various heat loss paths from the bottom convective zone
are shown. Assuming isothermal planes within the pond normal
to the depth, the steady-state temperature profile in the
insulating zone is given by equation.

$$\frac{d^2 T}{dx^2} - \frac{K_s}{K_w} , \frac{2}{R} (T-T_{SG}) = \frac{1}{K_w} \frac{dI(x)}{dx} \tag{35}$$

where at $x = 0$ $T = T_e$

and at $x = D_2 - D_1$ $T = T_2$

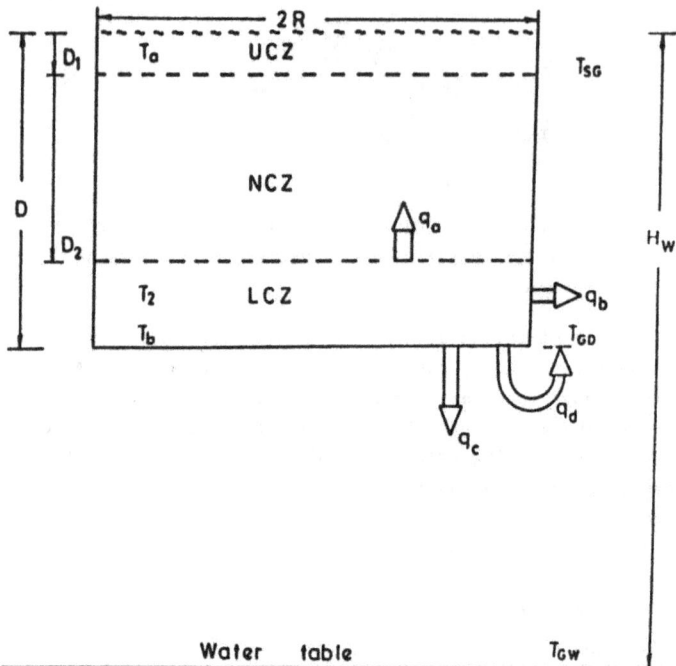

FIG. 20: SCHEMATIC OF THE SOLAR POND (FROM ISSAC AND GUPTA)

The solution to this equation in terms of T_2 and T_a was given by the author earlier[59]. To determine T_2, heat losses are considered from the sides (q_b), from the bottom to the ground water(q_c) and also to the earth at the level of the pond (q_d). The heat balance of the bottom convective zone is given by eqns. (36) and (37) at the interfaces with the gradient zone and the ground, respectively:

$$\frac{K_w}{\delta_1} [T_2 - T_{D_2} - D_1 - S_1] + U + \frac{2}{R} K_s l_n \frac{D - D_1}{D_2 - D_1} (T_2 - T_{SG})$$

$$= \tau I \lambda [\exp (-\mu D_2) - \exp (-\mu D)] + h_2 (T_b - T_2) +$$

$$(-\alpha + 1) \tau I\lambda (\exp -\mu D) \tag{36}$$

$$\alpha \tau I \lambda (\exp - \mu D) = h_2 (T_b - T_2) +$$

$$h_{eq} (T_b - T_{GD}) + \frac{K_2}{H_w - D} (T_b - T_{GW}) \tag{37}$$

For case of analysis, only icircular ponds were considered but the design data generated is valid for all ponds with aspect ratios equal to, or less than five.

Using above numerical model and the climatic data of India. R.R. Isaac and C.L. Gupta have studied the effect of sizing parameters - pond surface area and depth of the pond; operating parameters - storage volume and the heat extraction fraction; and geo-climatic parameters - solar radiation, water table depth, and upper convective zone thickness.

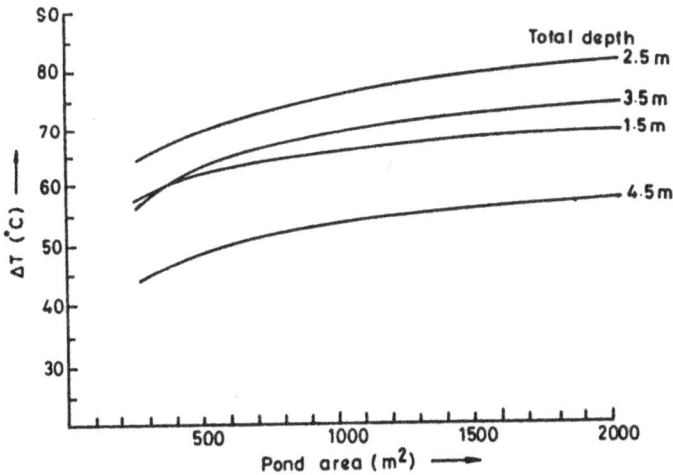

FIG.21: VARIATION OF ΔT WITH SURFACE AREA OF THE POND FOR NO
HEAT EXTRACTION. STORAGE=20% DEPTH; HEAT EXTRACTION=0%;
RADIATION =192 W/m^2;UCZ=0.2m; Hw=6.5m; Ta=27°C
(FROM ISAAC AND GUPTA)

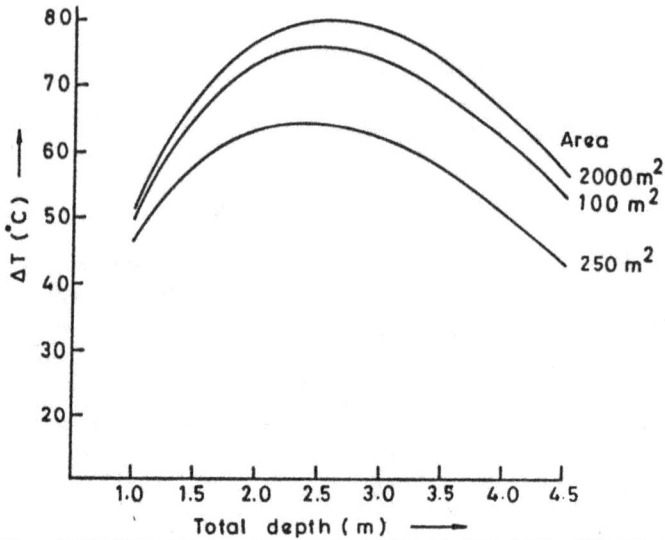

FIG.22:   VARIATION OF ΔT WITH TOTAL DEPTH OF THE POND.STORAGE=20%;
HEAT EXTRACTION = 0%; RADIATION=192 W/m^2; H$_w$=6.5 m;
T$_a$ =27°C; UC Z=0.2m. (FROM ISAAC AND GUPTA)

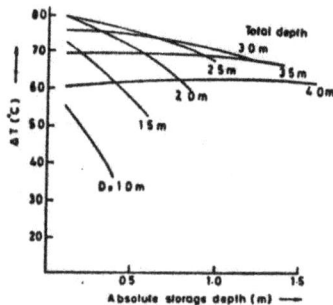

FIG.23:   CHANGE IN VALUE OF ΔT FOR VARIOUS DEPTHS OF THE STORAGE
ZONE.   POND AREA = 1000 m^2; HEAT EXTRACTION = 0%;
RADIATION =192 W/m^2 UCZ = 0.2m; Hw =6.5m; Ta=27°C
(FROM ISAAC AND GUPTA)

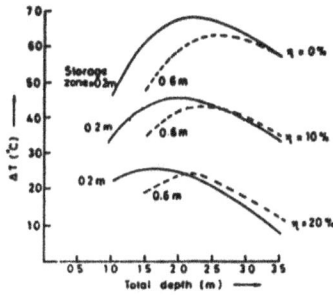

FIG.24: CHANGE OF ΔT WITH TOTAL DEPTH FOR DIFFERENT HEAT EXTRACTION RATES.POND AREA =250m^2; RADIATION = 192 W/m^2; UCZ =0.2m; Hw =6.5 m; Ta=27°C (FROM ISAAC AND GUPTA)

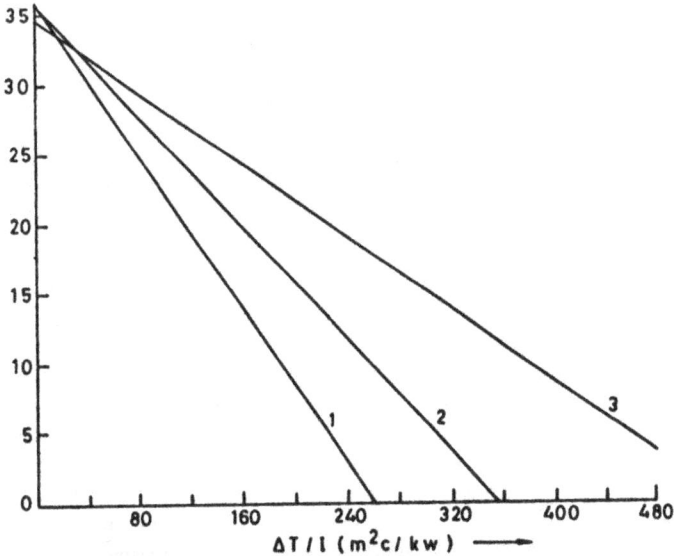

FIG.25: EFFICIENCY VERSUS (ΔT/I) CURVES.(1)100m^2 SOLAR POND.WITH LOSSES;(2) 2000m^2 SOLAR POND WITH LOSSES; (3)100m^2 SOLAR POND WITHOUT LOSSES. TOTAL DEPTH = 1.5m; STORAGE ZONE = 0.3m; UCZ = 0.2m Hw = 6.5 m; (FROM ISAAC AND GUPTA)

Figure 21 shows the effect of increase in the area on the ΔT of the pond. It is interesting to note that the elevation of temperature of the storage layer above ambient increases with increasing area and reaches an asymptotic value at areas greater than 2000m^2. Figure 22 shows the ΔT versus total depth plot for 20 percent storage. This graph shows that there exists an optimum depth of the solar pond for any given geo-climatic parameter. For a storage volume of 20 percent of the total depth, this optimum value turns out to be around 2.5m for the Pondicherry climate (12° N-hot, humid coastal location). Figure 23 shows the plot of ΔT versus the absolute thickness of storage zone total depth of the pond as the parameter for a pond of 1000m^2 areas. An increase of the storage volume upto 50 percent for any depth of the pond of less than 2.5m results in a sharp fall in the temperature of the storage layer. Figure 24 shows the plot of ΔT versus the total depth of the pond for different heat extraction rate for a fixed area of the pond (250 m^2) and for different values of storage depth (0.2 and 0.6m). The optimum depth decreases as the extraction rate increases. As can be seen from Fig. 25 solar pond have greater efficiencies than flat plate collectors at higher temperature due to greater insulation of properly designed ponds.

Figure 26 shows the change in the value of ΔT with an increase in the mean radiation for different heat extraction efficiencies and the relationship is linear. As shown in Fig. 27, the existence of the water table at depths greater than 10 m below the bottom of the pond does not seem to effect the temperature of the pond. Figure 28 shows the temperature profile of a pond with top convective zone of 20 cm and 40 cm thickness for different extraction rates. The thickness of the upper convective zone of the pond should be minimal. Figure 29 shows the maximum and minimum temperatures for different rates of heat extraction for the Pondicherry climate.

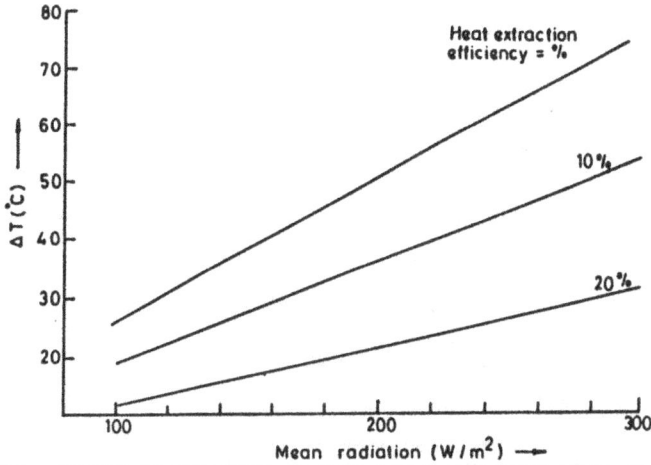

FIG.26: EFFECT OF VARIATION IN MEAN DAILY RADIATION LEVEL ON
TEMPERATURE.POND AREA = 100 m^2; TOTAL DEPTH =1.5m;
STORAGE ZONE = 0.3m;UCZ = 0.2; Hw=6.5m; Ta=27°C
(FROM ISSAC AND GUPTA)

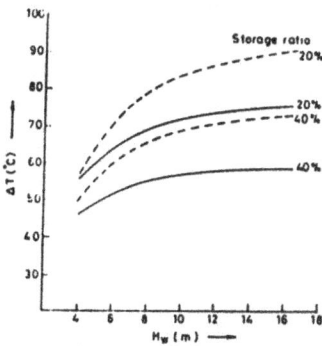

FIG.27: EFFECT OF WATER TABLE
ON STEADY-STATE TEMPE-
RATURE.POND AREA=1000m^2

HEAT EXTRACTION=0%;
RADIATION = 192 W/m^2;
UCZ=0.2; Ta=27°C.TOTAL
DEPTH=1.5m; TOTAL
DEPTH = 2.0m,(FROM
ISAAC AND GUPTA)

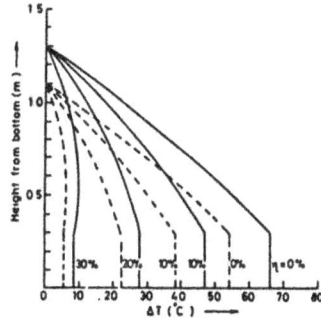

FIG.28: TEMPERATURE PROFILE FOR
DIFFERENT TOP CONVECTIVE
ZONE THICKNESS.POND
AREA=1000 m^2; TOTAL
DEPTH=1.5m;STORAGE ZONE=0.3m
RADIATION =192 W/m^2;
Hw=6.5m; Ta=27°C ,UCZ=
0.2 m: UCA=0.4m ((FROM
ISAAC AND GUPTA).

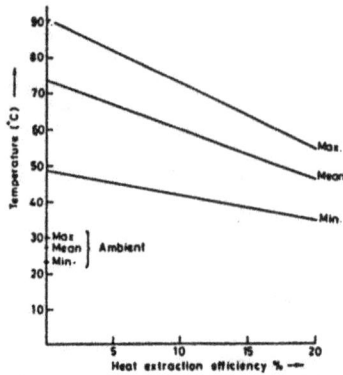

FIG.29: YEARLY TEMPERATURES MAXIMA, MINIMA AND MEAN VALUES.
POND AREA = 1000 m^2; TOTAL DEPTH = 2.0m; STORAGE =30%
WATER TABLE 5m BELOW POND BASE; UCZ=6.4 m; RADIATION=
260 W/m^2; MINIMUM RADIATION = 102 m^2 (FROM ISAAC AND
GUPTA).

The vertical temperature distribution T (x) in the solar
pond is calculated by Kooi[21] by solving the Fourier heat
conduction equations:

$$K\frac{\partial^2 T(x,t)}{\partial x^2} = \rho c \frac{\partial T(x,t)}{\partial t} + \frac{\partial H(x,t)}{\partial x} \qquad (38)$$

The necessary boundary conditions can be written as at $x=x_1$

$$-K\frac{\partial T(x,t)}{\partial x}\bigg|_{x=x_1} = h_1[T_1(t) - T(x=x_1,t)] \qquad (39)$$

and at $x = x_2$

$$- K\frac{\partial T(x,t)}{\partial x}\bigg|_{x=x} = h_2[T(x=x_2,t)-T_w(t)] \qquad (40)$$

where h1 and h2 are the heat transfer coefficient between the
upper convective zone and non-convective zone, and between the
non-convective zone and lower convective zone water of the
pond respectively.

The variation of vertical temperature distribution for a pond with $x_1 = 0.1m$, $(x_2-x_1) = 1.4m$ and $(x_3-x_2)=0.3m$, $H(x=0)250W/m^2$ W/m^2, $T(x_1) = 20°C$ is shown in Fig. 30. It is seen that the optimum collection temperature 12 is 55.3°C. Curves are plotted for temperatures above and below the optimum collection temperature. It is also seen from this figure that for collection temperature higher than the optimum, conductive heat flows out of the collection zone while for those less than optimum it flows into the collection zone. At the optimum collection temperatures, there is not conductive heat flowing across the NCZ-LCZ boundary .

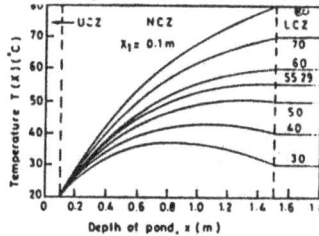

FIG.30: VARIATION OF TEMPERATURE ALONG THE DEPTH IN A SOLAR POND (FROM KOOI)

Weinberger[5] has also done detailed analysis of a solar pond for the conditions prevailing in Israel solving the one dimensional heat conduction equation (Equation 38). He observed that for a given heat removal temperature T2, there is a particular value of the pond depth x_3 at which the retrieved heat flux $\dot{Q}$ is a maximum. He calculated the optimum collection temperature T2, and the collection efficiency η as functions of pond depth x_3 and the results are shown in Fig. 31. Here it has been assumed that the ambient air temperature Ta is 26°C, and mean annual radiation of 5.87 KWh/m^2 day.. From this figure it is seen that heat can be extracted from the pond at higher temperature without sacrificing much on efficiency and thus making the system economically advantageous.

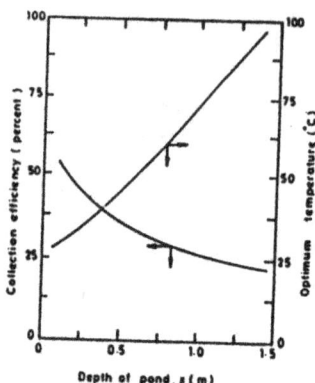

Depth of pond, x (m)

FIG.31:   OPTIMUM COLLECTION EFFICIENCY AND COLLECTION TEMPERATURE
AS A FUNCTION OF POND DEPTH (FROM WEINBERGER)

9.0   TYPES OF SOLAR PONDS

Solar ponds are of several types like shallow solar pond,
viscosity stabilized solar pond, membrane stratified solar pond
and saturated solar pond. A brief discussion of each are given
below.

9.1   Shallow Solar Ponds

A shallow solar pond is body of water with shallow depth
acting as large collector and a storage of solar radiation. It
is directly exposed to solar radiation and enclosed in a thermal
insulating base material and one or two sheets of glazing. To
keep costs low, polymeric materials are used whereever possible
instead of metal and glass. A review of shallow solar pond (SSP)
has recently been extensively studied by a group of scientists[46]
at Lawrence Livermore Laboratory, Livermore, USA. There is a
large number of design options available.   In the most popular
design of Lawrence Livermore Laboratory, a black pond linear of
a tough material such as butyl rubber, hypalon, or chlorinated
polyethylene is stretched over the insulation base and attached
to the top of the concrete curbings. Two layers of clear plastic
film are then placed over the black linear and attached to the
curbs. The space between the liner and the lower film is filled
with water and the top film is inflated by use of a small blower.

In another design recommended for large applications a two-layer
plastic bag is fabricated with a black bottom and clear top.
It rests on a insulation pad and is filled with water. Arched
over the top of the bag are semirigid, corrugated clear plastic
sheets secured along the curb edges and also over the top by
steel tie-down straps where the sheet overlaps. Several such
SSP can be connected together and hot water from them can be
pumped and stored in a large insulated storage reservoir to
reduce thermal losses during night and bad weather conditions.
Water temperatures in the range of 50-75°C can be obtained
which can be used as industrial process heat or for electricity
generation by employing a secondary fluid (such as freon)which
will derive a turbine coupled to an electrical generator. Some
of the current and planned SSP projects are listed in Table 14.

Table 14: Shallow solar ponds projects

Project/location	Area	Application	Status/Problems
Sohio Petroleum Co. Grants, New Mexico (USA)	6 Acres (Projected)	Uranium ore processing	Construction costs made system uneconomic; project currently on hold.
Sweet Sue Kitchens Athens, AL(USA)	$1600m^2$	Chicken packing	Potable water required HEX; Small size made cost/unit area high; project Cancelled by DOE.
Ft. Benning GA(USA)	$256000m^2$	Hot water for Barracks and Laundary	2 Million litres/day estimated cost $95/m^2$; Detailed design by A/E inprogress start construction June 1980. Finish December.
Ft. Gordon August, GA(USA)	$10000m^2$	Barracks Hotwater	Roof top panels, preliminary design phase.

9.2   Partitioned Solar Ponds

In a partitioned solar pond the lower convective zone (LCZ) and non convective zone (NCZ) is seperated by a transparent partition and the process of operation remains the same as the conventional salt gradient solar pond. The idea of partitioned solar pond was given by Rabl and Nielson[17] so that lower convective zone can be used for seasonal storage of heat for house heating. The partition also helps in maintaining the stability in the pond and heat can be extracted from LCZ without disturbing the NCZ. Generally the use of a flexible membrane is recommended but in this case the overall loading of the partition must be small to prevent rupture. Which means that either the lower layer density must be such that the convection zone supports the nonconvective layer, or the convective zone must be given a pressure head to balance gravitational force on the partition. In the later case fresh water can be used in the convective zone and thereby eliminating the corrosion problems which is associated with energy extraction from the brine. In the partitioned solar pond, the membrane allows for use of considerable less salt because the salt content is proportional to the square of the depth, But in this case the membranee should be fixed to the pond walls by a leak-tight seam. Rabl and Nielson[17] have shown thatthe temperature in the convecting zone decreases with increasing thickness of the convecting layer, but the mean temperature is independent of the thickness. For an infinite pond, the mean temperature is given as:

$$T_w \Big|_{A \to x} = \bar{T}_a + \frac{TH(x=0)}{k} \sum_{n=1}^{4} {}_n e^{-\mu_n x_1} \tag{41}$$

where T_a is the average ambient air temperature, x_1 is the thickness of nonconvective ozone, and K is the thermal conductivity of pond.

Some experiments on small partitioned solar pond have been conducted by Russian scientists who concluded that the efficiency of partitioned solar pond is higher than the conventional non-convecting solar pond. Recently Kaushika et al[47] have conducted analytical studies on the behaviour of partitioned solar pond having two membranes, one separating the LCZ and NCZ and the second on the top of NCZ which prevents evaporative losses. The optimal efficiency and optimal depth of nonconvective zone[47] at various collection temperature are shown in Table 15. It is seen from this table that optimal efficiency of 37 percent and 26.9 percent are obtained at collection temperature of 50°C and 100°C respectively.

Table 15: Optimal efficiency and depth for a partitioned solar pond

Collection temp (°C)	Optimal depth of Non-convecting zone(m)	Optimal efficiency (percent)
40	0.25	42.2
50	0.75	37.0
60	1.00	33.7
70	1.25	31.2
80	1.50	29.5
90	1.75	27.7
100	2.25	26.5

9.3   Viscosity Stabilized Solar Ponds

In the viscosity stabilized solar pond a kind of gel is used in water, making it nonconvective. The idea of viscosity stabilized solar pond was first given by Shafer[48] and the phenomenon can be described as static rather than stable. It is known that the Rayleigh number which relates buoyant forces and viscous drag is responsible for the circulation and its critical value for the onset of natural convection   for a layer

of fluid bounded top and bottom, and heated from the bottom is
1707. The Rayleigh number is given as:

$$R_a = \frac{g\beta \Delta T d^3}{\nu \alpha} \qquad (42)$$

From this equation it is seen that by increasing the viscosity,
the Rayleigh number can be reduced below the critical number and
thereby suppressing the natural convection. It has been shown
that even simple water with soluble gums can produce syrups that
have viscosities in the range of 36.3 to 36.3 x 10^3 kg/ms suitable
in the pond for nonconvective operation. Materials suitable
for viscosity stabilized ponds should have high transmittance
for solar radiation, high thickening efficiency and should be
capable of performing at temperatures upto 70°C. Natural
polymers such as gum arabic, locust beam gum, agin, starch, and
gelatin are all potentially useful materials. Both the synthetic
polymers like polyacrylic acid (salts), polyacrylamide, a
carboxyvinyl polymer, polymers of ethylene oxide etc. and
semisynthetic polymers like carboxymethylcellulose, hydroxyethyl
cellulose, methylcellulose, hydroxypropyl methylcellulose etc.
can also be used for stabilizing the pond. Several crosslinked
polymer gels and detergent/oil/water gels can also be prepared
which can also be suitably used in ponds. Shaffer has recommended
the use of commercial carboxy vinyl polymer as a thickner which
is found stable even at 70°C and with a proper inhibitor has
shown outstanding photochemical characteristics. The idea of
viscosity stabilized solar pond appears to be promising but
requires studies in depth and presently is not economically
competitive with salt gradient solar ponds.

9.4   Membrane Stratified Saltless Solar Ponds

A possible alternative of the conventional salt gradient
solar pond (SGSP) is the membrane stratified solar pond (MSSP)
as suggested first time by Hull[49]. The idea of MSSP appears
to be taken from flat-plate collectors where transparent honeycomb

is used for natural convection suppression. In a conventional
SGSP there are three zones UCZ, NCZ and LCZ while in a MSSP there
are only two zones, the upper nonconvective zone (NCZ at the top
serving as insulating layer) and lower convective zone (LCZ at
the bottom serving as a heat storage layer). The basic difference
between the SGSP and MSSP is in the mechanism for maintaining
nonconvection in the NCZ. In the MSSP the convection is suppressed
by using transparent membranes in the NCZ having spacings with
each other small enough to suppress convection. A few advantages
of MSSP are listed below:

1.   Since no salt is used in a MSSP, this pond can be made
        maintenance free and low cost.
2.   There is no environmental or geological hazard with the
        MSSP.
3.   There is no UCZ in a MSSP and thus making the same more
        efficient when compared to SGSP.
4.   A larger depth of LCZ can be maintained in a MSSP resulting
        in seasonal storage, less diurnal temperature variation,
        and higher collection efficiency.

Three types of membranes are suggested by Hull[49] for
MSSP (i) Horizontal sheets; (ii) Vertical tubes; and (iii) Vertical
sheets. He has also suggested the use of a thick horizontal
membrane at the top of the pond to keep out dust and debris
and to prevent optical fouling of the membranes. Teflon is
suggested to be the suitable membranes material because of its
long life, high transparency, inert to virtually all chemicals,
commercial availability in all sizes and thicknesses etc. Apart
from water as liquid in MSSP other liquids like concentrated
sugar solution, ethanol and combination of water and ethanol
are suggested as liquids.

9.5   Saturated Solar Ponds

Saturated solar ponds are non-convecting bodies of water, relying upon a density gradient brought out by differential solubility of salt with temperature. In a saturated solar pond the naturally developed temperature gradient between top and bottom develops and maintains a density gradient by ion migration. In such a pond, salt for which the solubility increases quickly with temperature is used. The pond water is kept saturated with such a salt at all levels and since the pond is hotter in the bottom than the top, more salt is dissolved in the bottom. In such a pond, vertical diffusion of salt is prevented and the density gradient is stable thus making the pond maintenance free. Generally, Na_2SO_4, $MgCl_2$ and Borax are recommended in saturated solar pond. Na_2SO_4 shows an increase in density with temperature upto transition at 30.3°C. At this point the density of the solution starts decreasing with increasing temperature which shows that Na_2SO_4 is not an acceptable working salt. Ochs et al[50] have made analytical studies of several salts and based on this study they have concluded that both $MgCl_2$ and Borax can form a stable saturated solar pond.

10.0   SOLAR POND APPLICATIONS

Because of large storage of heat and negligible diurnal fluctuation in pond temperature, solar pond has a variety of applications like heating and cooling of buildings, swimming pool and greenhouse heating, industrial process heat, desalination, power production, agricultural crop drying, the production of renewable liquid fuels such as ethanol for gasohol. Some of the applications are discussed below;

10.1   Heating of Buildings

Because of the large heat storage capability in the LCZ of the solar pond, it has ideal use for house heating even at high latitude stations and for several cloudy days,. Many scientists have attempted and sized the solar pond for a particular

required heating load for house heating. Calculations have shown
that a solar pond with a 100m diameter and a 1m deep LCZ is
sufficient to drive either an absorption or on RC/VC Chiller
capable of meeting 100 percent of the typical cooling load of
a 50-house community in Fortworth (USA). Even single storey
buildings can be heated economically with solar pond in which
case the area of solar pond can be approximately equal to the
floor area of the house.

A small exercise of cost comparisons for heating a
hypothetical 180m^2 house where the heating requirement is 2.2.5x10^7
J/°F day is done by Styris et al[56] and the results are given
in Table 16.

In the calculations a saturated potasium nitrate solar
pond which require less maintenance of an area of 145m^2 and
costing about $6,500 is assumed. The life time of the pond is
assumed to be 20 years, rate of interest 10 percent, and
maintenance cost $100 per year.

Table 16: Cost comparisons (US $ for heating 2.5x10^7 J/°F day house)

Energy Source	Unit Cost ($)	Annual Cost* ($)	Equal cost point ($)
Oil	0.30 gal	466**	0.52/gal
Electricity	0.01/KWhr	366	0.021/KWhr
	0.02/KWhr	732	
Solar pond (saturated)	0.021/KWhr	766	

*   All calculations based on 5000° F day/hr.
** Assumed 30 X 10^6 Btu/ton and 70 percent conversion efficiency

10.2   Power   Production

The concept of solar pond for power production holds great
promise in those areas where there is sufficient insolation and
terrain and soil conditions allow for construction and operation
of large area solar ponds necessary to generate meaningful
quantities of electrical energy. Even low temperatures heat
that is obtained from solar pond can be converted into electrical
power. The conversion efficiency is limited due to its low
operating temperatures (70-100° C). Because of low temperature,
the solar pond power plan (SPPP) requires organic working fluids
which have low boiling points such as halocarbons (like Freons)
or hydrocarbons (such as Propane). A sketch of a typical SPPP
is shown in Fig. 32. The hot brine pumped from the storage
layer of the solar pond delivers its heat to the turbine working
fluid via a heat exchanger.

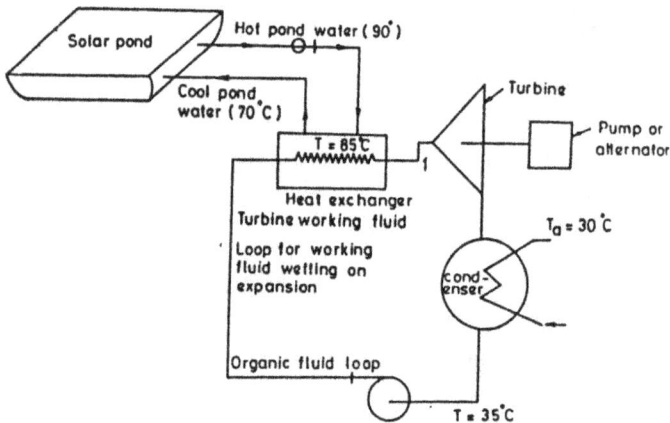

FIG.32:   A TYPICAL SOLAR POND POWER PLANT LOOP EMPLOYING
          AN ORGANIC FLUID.

The annual power production from a solar pond can be
calculated easily with some approximations. The efficiency of
1.2m deep nonconvective layer solar pond may be about 25 percent
and it will give a temperature of about 90°C. Assuming a carnot
engine working in the temperature range of 90°C and 27°C will
give an overall conversion efficiency of about 4.2 percent. If
the actual heat engine efficiency is 70 percent then the overall
conversion efficiency will be 2.94. Assuming an average insolation
of 6KWhr/m^2 day, and 2.94 efficiency, we obtained an annual
output of 64 KWhr/m^2/yr. Thus the output from 1 Km^2 area solar
pond is equivalent to a conventional fossil fuel power plant
which burns approximately 18000 tons/yr. of oil or coal or 19
million m^3/yr of methane. At 15 dollars per barrel of oil, this
amounts to about 1080 thousand dollars saving in fuel oil.

10.3   Industrial Process Heat

Industrial process heat is the thermal energy used directly
in the preparation and/or treatment of materials and goods
manufactured by industry. Several scientists have determined
the economics of solar pond for the supply of process heatin
industries. According to them the solar pond can play a significant
role in supplying the process heat to industries thereby saving
oil, natural gas, electricity, and coal. Styries et al[57] have
computed the cost of ponds for process heating in two industries
i.e. for crop drying and paper processing. According to them
a heat of 3.5x10^3 KWh/day (1.25x10^{10} J/day) with air at a
temperature of 38°C flowing at about 20m^3/min is required to
dry 10 tonnes of crop per day. Styries et al[56] have calculated
the unit cost and annual cost for a 10 tons/day crop drying
facility using oil, natural gas, and solar pond. This comparison
is shown in Table 17.

Here the pond having an area of 580m^2 with the total
cost of $8680 has been assumed. This pond produces 3.5x10^{3}KWhr/
day for 40 days which is the drying time (66 percent of a 60
day drying season). The annual cost of solar pond comes to

Table 17: Cost comparison of 10 tons/day crop drying facility

Thermal Energy Source	Unit Cost ($/KWh$_t$)	Annual Cost (100 $)
Oil	0.007	1.0
Natural gas	0.005	0.7
Solar pond	0.0077	1.1

about $1080 per year by assuming a 20 years life, 10 percent
interest rate, $ 200 per season maintenance cost, and a salt
salvage value equal to 70 percent of salt cost. From this table
it is seen that the thermal energy supplied by solar pond for
crop drying is competitive to that of oil operated crop drying
unit.

Similar calculations have been made for supplying process
heat to a typical paper industry which require about 20 thermal
MW. If 50 percent of the need is met by solar pond then the
total area required would be $4.5 \times 10^8 m^2$. The total cost of the
pond including the salt will be about 4.63 billion dollars. The
annual cost of this solar pond comes to about 542 million dollars
per year, by assuming a 20 ιyear life, 10 percent interest rate,
80 million dollars per year maintenance cost, and a salt salvage
value equal to 70 percent of salt cost. The cost comparisons
of using heat from solar pond, oil, and natural gas, assuming
that half the requirement is met by them is shown in Table 18.

Table 18: Cost comparisons of process heat for a typical paper
Industry.

Thermal Energy Source	Unit Cost ($/KWh$_t$)	Annual Cost* (million $)
Oil	0.007	875
Natural gas	0.005	625
Solar pond	0.004	542

* based on 25 percent of the 1.8×10^{18} J/year used by paper
  and food industries.

From this table it is seen that using heat from solar pond for a paper industry is highly competitive with oil and natural gas.

10.4   Desalination

Multiflash desalination units alongwith a solar pond is an attractive proposition for getting distilled water because the multi-flash desalination plant below 100°C which can well be achieved by a solar pond. This system will be suitable at places where potable water is in short supply and brackish water is available. It has been estimated that about $4700m^3$/day distilled water can be obtained from a pond of 0.31 Km^2 area with a multi-effect distillation unit. Taking the pond cost at an upper figure of $ 7 x 10^6 per Km^2 and using costs for a locally produced desalination plant, the system competes with fuel at $42 a ton. These figures are based on rather high capital charges of 15 percent on the pond and 20 percent on the plant. The cost of distilled water comes to about $0.67/$m^3$ which appears to be high for industrialised countries but can be used in developing countries where there is a shortage of potable water. Moreover this desalination plant produces five times more distilled water than the conventional basin type solar still.

11.0   FURTHER RESEARCH POTENTIAL

Although research work on the Physics, Chemistry and Engineering of solar ponds has been going on, the hydrodynamics of solar pond is not yet clearly understood. The interaction of the NCZ and LCZ is still to be explained.

Transmission of solar radiation through operational solar pond is still to be correctly modelled. Reliable and economic methods of controlling salt diffusion upwards areto be developed. Suitable methods should be developed to inject the salt at appropriate position and flushing the pond surface and finally

disposing the flushed brine without any environmental hazard.
Economics and effectiveness of suitable heat removal system
from the LCZ is still to be worked out. Considerable research
is required in developing and identifying suitable pond materials
like salt, liner , cooker material, heat exchanger material,
pump, measuring and monitoring instruments and equipments etc.

Moreover, the implementation of solar ponds should be
attempted only after-technically and economically viable designs
have been developed; the implementation should be then tried
in places where it seems to other maximum advantage. Research,
development and demonstration should be conducted at different
places in the country which have different climates and have
the infrastructure to take up the responsibility. A nationally
recognized organization should be identified and entrusted with
the lead responsibility of evaluating the techno-economic
aspects of different designs and validating the models through
extensive field testing. The large scale adoption of the
technology will then be a natural consequence with financial
incentives.

12.0   REFERENCES

1.   A.K. Kalecsinsky (1902), "Ungarische warme und Heisse
        Kochsalzeen, Ann. D. Physik, 7(4), 408-416.

2.   C.G. Anderson (1958), 'Limnology of shallow saline
        meromistic lake, Limnology and Ocenography, 3, 259-269.

3.   H. Tabor (1959), 'Solar collector developments, Solar
        Energy, 3(3), 8-9.

4.   H. Tabot (1961), 'Large-scale solar collectors (solar
        ponds) for power production; Proc. UN Conf. New Sources
        of Energy. Paper No. S/47, Rome,1961.

5.   H. Weinberger (1964), 'The physics of the solar pond,
        Solar Energy, 8(2), 45-46.

6.   C. Elate and O. Levin (1965), 'Hydraulics of the solar
        pond, Cong. Int. Assoc. Hydraulic RCS. 11th, Leningrad,
        USSR.

7.   H. Tabor and R. Matz (1965), 'Solar pond; status report',
        Solar Energy, 9(4), 177-182.

8.   J. Hirschman (1961), 'Suppression of natural convection
        in open ponds by a concentration gradient' UN Conf. New
        Sources of Energy, Rome, 1961, p. 487.

9.   J. Hirschman and W.E. Gaete (1962), 'Collector solar de
        Poza abiert con soluciones de cloruro de magnesio, Rev.
        Sci. No. 120, 43-95.

10.   K.D. Stolzenbach, J.M.K. Dake and D.R.F. Harleman (1968),
        'Prediction of temperatures in solar ponds, Annual
        Meeting, Solar Energy Society, Palo Alto. California,
        Oct. 21-23, 1968.

11. Yu U. Usmarov, U. Eliseev and Zakhidov (1969), 'Salt water ponds on solar energy accumulators, Applied Solar Energy, 5(2), 49-55.

12. S.B. Savage (1977), 'Solar Pond' Chapter in Solar Engineering. Edited A.A.M. Sayeigh, Academic Press, Inc., 217-232.

13. C.E. Nelson (1980), 'Non-convective salt-gradient solar ponds, Chapter 11 in Solar Energy Technology Handbook, Part A. Edited W.C. Dickinson and P.N. Cheremisinoff, Marcel Dekker. Inc., New York.

14. H. Tabor (1981), 'Review article; Solar ponds, Solar Energy, 27(3), 181-194.

15. H. Tabor and Z. Weinberger (1981), 'Non-convective solar ponds' Chapter 10 in Solar Energy Handbook, Edited J.F. Kreider and F. Kreith, McGraw Hill Book Co., New York.

16. H.P. Garg (1985), 'Solar Ponds, Treatise on Solar Energy, Vol. II: Industrial Applications of Solar Energy (Part-I), John Wiley and Sons, Inc., England.

17. A. Rabl and C.E. Neilson (1975), 'Solar ponds for space heating, Solar Energy, 17(1), 1-12.

18. E. Wilkins and K.L. Pinder (1979), 'Experiments with a model solar pond, Sunworld, 3(4), 110-117.

19. H.C. Bryant and M.K. Rothmeyer (1980), 'Solar pond studies: phase III', Fifth semi-annual progress report to US Deptt. of Energy, DOE Contract No. EG-77-5-D4-3977.

20. L.J. Wittenberg and M.J. Harris (1981), 'Construction and startup performance of the Miamisburg salt gradient solar pond', J. Solar Energy Engg., Trans. ASME,103, 11-16.

21. C.F. Kooi (1979), 'The steady state salt-gradient solar pond', Solar Energy, 23, 37-45.

22. M.N.A. Hawlader and B.J. Brinkworth (1981), 'An analysis of the non-convecting solar pond', Solar Energy, 27(3), 195-204.

23. N.D. Kaushik and P.K. Bansal (1981), 'Transient behaviour of salt-gradient stabilized shallow solar ponds' Applied Energy, 10(1). 63.

24. N. Chepurniy and S.B. Savage (1975), 'The effect of diffusion on concentration profiles in a solar pond, 'Solar Energy, 17, 203-205.

25. N. Chepurniy and S.B. Savage (1974); 'An analytical and experimental investigation of a Laboratory solar pond model, ASME. 74-WA/Sol-3.

26. J.S. Turner (1968), J. Fluid Mechanics, 33, 183.

27. D.G. Daniels and M.F. Merriom (1975), 'Fluid dynamics of selective withdrawal in solar ponds', ISES Congress Session, 35, Los Angeles, California.

28. C. Elata and O. Lavin (1962), 'Selective flow in a pond with density gradient', Hydraulic Laboratory Rep., Technion, Hafia, Israel.

29. J.R. Hull (1980), 'Computer simulation of solar pond thermal behaviour', Solar Energy, 25, 33-40.

30. A. Akbarzadeh and G. Ahmadi (1980), 'Computer simulation of performance of a solar pond in the southern part of Iran', Solar Energy, 24(2), 143-151.

31. P.K. Bansal and N.D. Kaushik (1981), 'Salt gradient stabilized solar pond collector', Energy Conv. and Magmt., 21, 81-85.

32. M.S. Hipsher and R.F. Boehrn (1976), 'Heat transfer considerations of a non-convecting solar pond exchanger', ASME, 76-WA/Sol. 4.

33. M.J. Harris and L.J. Witterberg (1979), 'Heat extraction from a large salt-gradient solar pond', 2nd Ann. Solar Heating and Cooling Conf., Colorado Springs, Nov.1978.

34. M. Edesess (1980), 'Solar pond economics' Proc. Non convecting solar pond workshop, Desert Research Institute, University of Nevada System, July, 30-31, 1980.

35. G.D. Mehta et al (1976), 'Engineering and economics of a solar pond system', 16th Annl. ASME Symp. on Energy Alternatives, Albuquerque, New Mexico, Feb. 1976.

36. G.C. Jain (1973), 'Heating of solar pond', Int. Congress-Sun in the service of mankind, UNESCO, Paris.

37. S.M. Patel and C.L. Gupta (1981), 'Experimental solar pond in a hot humid climate', Sun World, 5(4), 115-118.

38. S.L. Sergent and D.L. Neeper (1980), 'Overview of the DOE National and International Program for salt gradient solar ponds', Proc. AS ISES Annl. Meet. Phoenix, Arizona, Vol. 3. 1, 295-399, June 1980.

39. J.R. Hull (1978), 'The effects of radiation absorption on convective stability in salt gradient solar ponds', Proc. An. Soc. of ISES Meet., Denver, Colorado, USA. 37-40.

40. H.C. Boyant and I. Colbeck (1977), A solar pond for London', Solar Energy, 19, 321-322.

41. T.R.A. Davey (1968), 'The Aspendale Solar Pond', Rep. R. 15, CSIRO, Melbourne, Australia.

42. H. Tabor (1966), 'Solar Ponds', Sci. J., 66-71.

43. S. Shahar (1968), US Pat 337291.

44. M. Edesess, J. Henderson, and T.S. Jayadev (1979); A simple design tool for sizing solar ponds', SERI/RR-351-347, Golden, Calorado.

45.   H.P. Garg, B. Bandyopadhyay, U. Rani and H. Brinhakennn,
        (1982), 'Shallow solar pond: State of the art' Energy
        Conversion and Management.

46.   A.F. Clark and W.C. Dichinson (1980), 'Shallow solar ponds'
        Chapter 12 of Solar Energy Technology Handbook, Part A,
        Edited W.A. Dickinson and P.N. Cheremisinof, Marcel Dekker,
        Inc., New York.

47.   N.D. Kaushika, P.K. Bansal and M.S. Sodha (1980),'Partitioned
        solar pond collector/storage system', Applied Energy,
        7, 169-190.

48.   L.H. Shaffer (1978), 'Viscosity stabilized solar ponds',
        SUN, Mankinds future Source of Energy, Proc. Int. Solar
        Energy Soc. Congress, New Delhi, 1970, 1171-1175 p.

49.   J.R. Hull (1980), 'Membrane stratified solar ponds',
        Solar Energy, 25, 317-325.

50.   T.L. Ochs and J.O. Bradley (1979), 'Stability criteria
        for saturate solar ponds' Proc. 14th Int. Soc. Energy
        Conversion Engineer Conf., Aug. 5-10, 1979, Boston USA.

51.   J.M.K. Dake (1973), 'The solar pond: Analytical and
        Laboratory Studies', UNESCO Congress, Sun in the Service
        of Mankind, Paris, Paper No. E-17.

52.   R.A. Tybout (1967), 'A recursive alternative to Wienberger's
        model of solar pond', Solar Energy, 11(2), 109-111.

53.   K.D. Stolzenbach, J.M.K. Dake and D.R.E. Harleman (1968),
        'Prediction of temperatures in Solar Ponds', Annual
        Meeting of Solar Energy Society, Palo Alto, California,
        Oct. 21-23.

54.   M.S. Sodha, N.D. Kaushik, and S.K. Rao (1981), 'Thermal
       analysis of three zone solar pond', Energy Research,
       5,321-340.

55.   K.A. Mayer (1981), 'A one dimensional model of the
       dynamic layer behaviour in a salt gradient solar pond'
       Solar Rising Conference and Exposition, Philadelphia
       Civic Centre, May 26-30, 1981.

56.   D.L. Styris and O.K. Harling, (1975), 'The Non-convecting
       solar pond, an overview of technological status and
       possible pond application', Report No. BNWL-1891/UC-13,
       Prepared for US Atomic Energy Commission, 1975.

57.   D.L. Styris, O.K. Harling, R.J. Zawerski and J. Leahuk
       (1976), 'The Nonconvecting Solar pond applied to building
       and process heating', Solar Energy, 18, 245-251.

58.   R.R. Isaac and C.L. Gupta (1982), 'A Parametric Design
       Study of Solar Ponds', Applied Energy', 35-49.

59.   C.L. Gupta, R.R. Isaac and S.M. Patel (1980), 'Design
       procedure for a solar Pond', Reg. J. of Energy Heat Mass
       Transfer, Vol. 2, No.3, 193-203.

STATUS OF SOLAR THERMAL POWER SYSTEMS

Prof. H.P. Garg
Centre of Energy Studies
Indian Institute of Technology
Hauz Khas, New Delhi 110 016
India

1.0 INTRODUCTION

Availability of cheap power is an index of technological advancement and standard of living of a country. Conversion of solar energy into mechanical power or electrical power has been a subject of research for nearly last three centuries. Most of the early research conducted on solar mechanical power generation was for small power generation and was abandoned not due to technological reasons but more due to economically more viable and cost-effective power options. It is hoped that solar generated power will play a significant role by the end of this century. There are several options of converting solar energy into electrical energy. The first option is the direct conversion of solar energy into electricity which can be done either by photovoltaic method or photogalvanic method or photoemissive method or photomagnetic method.

The direct conversion is not practically adopted for commercial purposes for large power demands due to high cost of cells. Thermal energy obtained from solar energy can also be directly converted into electricity by any of the methods: thermoelectric, thermionic, ferro-electricity, magnetohydrodynamics, electrogasdynamic. The conversion efficiency of all the above method is generally very low and, therefore, the system is not cost-effective.

There is yet another very important way of converting solar energy into electricity which is known as thermodynamic way in which solar energy is converted into thermal energy; thermal energy into shaft work through heat engines based on the principle of either Rankine cycle, Stirling cycle, or Brayton cycle; and shaft work (mechanical energy) into electricity using alternator. In this paper only the thermodynamic way of converting solary energy into mechanical energy or electrical energy will be discussed.

A solar thermal power system mainly consists of a solar energy collector field, some kind of fossil fuel combustor (auxiliary system), a fluid flow distribution system, some kind of suitable thermal energy storage device, a heat engine, electric generator, and a control system, as shown in Fig.1.

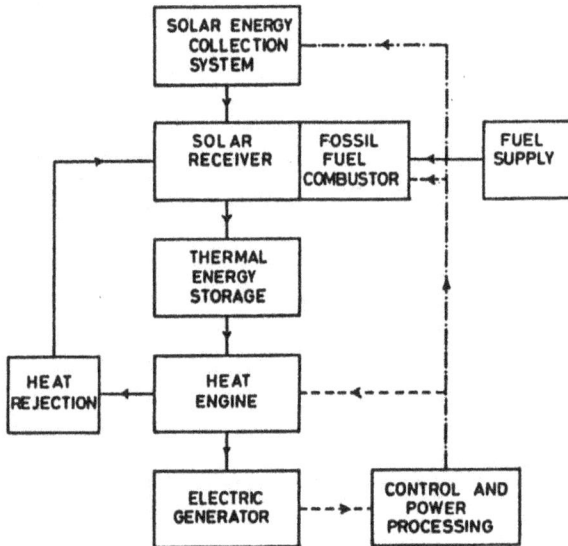

FIG.1 - SIMPLIFIED BLOCK DIAGRAM OF SOLAR THERMAL POWER SYSTEM

Amongst the many available systems, the two generic type of systems, the central receiver thermal electric power system and the distributed solar thermal electric power system are considered suitable because of their comparatively high efficiencies and cost effectiveness. In the central receiver concept large arrays of sun-tracking mirrors known as heliostats reflect the solar flux on to the central receiver boiler at the top of the tower. Here concentration ratios of the order or 1000 are used and turbine (steam or organic fluid type) operates at about 600°C. In the dispersed or distributed power system, a large collector array consisting of line focus collectors such as parabolic troughs or linear fresnel reflectors or linear fresnel lenses or parabolic dish, spread over to a large area is used. Energy is collected through pipes interconnecting the dispersed collector units and then supplied to the heat engines directly or through a heat exchanger where two different fluids are used.

The type of heat engine and the associated thermodynamic cycle depends on many parameters including, temperature of operation, cycle efficiency, working fluid, cost, reliability, conversion efficiency, transport fluid, size of load, collector type, storage size, etc. Depending on the power produced, the solar thermal power systems are divided into the following three classes:

(i) Small units in the watts range. These units operate at very low thermodynamic efficiencies and are costly due to low efficiency, high heat and mechanical losses. If the power requirements are only in the watts range then photovoltaic systems will be more feasible.

(ii) Medium range units or KW range units. Here generally dispersed or distributed solar system is used and can be feasible. Photovoltaic systems may compete under certain situations.

(iii) Large range units of MW capacity where the central receiver solar system is preferred.

In the recent past, successful demonstrations of solar thermal power systems in the above three ranges, i.e. watts, KW and MW, are made and many technical and operational problems solved except of the suitable thermal storage device which still remains the unsolved technical problem particularly for large units.

In this paper we briefly describe the history of solar power units, power cycles and different technical limitations in power production, types of heat engines and turbines, examples of dispersed and central power systems, etc. The systems like solar collectors, storage systems, etc., used will not be discussed here as these are discussed by other speakers.

2.0 HISTORY OF SOLAR THERMAL POWER

Several surveys have recently been made on the use of solar energy for generating power[1,2] and for lift irrigation[3]. This historical review is due to Pytlinski[3]. Although solar energy has been used for variety of applications since time immemorial but the first successful application of solar energy for raising water was made by a French engineer[4]

Solomon de Caux (1576-1625) who, through the expansion of air using solar
energy, was able to pump water. Over 200 years later (1854-1873),
G. Guntner of Austria[5] used 10.0 m^2 of narrow mirror strips to generate
steam for producing 746 watts. August Nouchot[6], a Frenchman, during
the years (1980-1878) built several solar engines; the first solar engine
was made in 1866, the second in 1875 and the third in 1878.

An American, John Ericsson[7] (1803-1889) worked for 20 years on
solar engines and developed several solar steam power engines and solar
hot air engines. In his experiments, he used parabolic troughs of various
dimensions, and in one of the experiments at New York, he used a parabolic
trough of 3.3 m long and 4.9 m wide which produced steam at a pressure of
240 KPa driving a reciprocating engine and generated 1.2 KW. In 1876 W.Adams
in Bombay, India, operated the first solar steam engine in India where
the solar energy was collected using a spherical mirror of 12.2 m diameter
and the system was able to produce steam at a pressure of 207 KPa producing
a power of about 1.9 KW. Abel Pifre of France in 1880 used a parabolic
reflector of 9.3 m^2 area to power his rotary pump raising about 100 litres
of water through 3 m head in 14 minutes time. C.L.A. Tellier[10] of France
in 1885 used 20 m^2 flat plate collectors (metal roof top collectors) and
produced less than one horse power using a small vertical engine with
ammonia as a working fluid. According to a report published in Scientific
American[11] an American, A.G. Eneas (1901-1905) built several solar power
plants using truncated cone shaped reflectors and water as a working fluid
and a compound condensing engine connected to a centrifugal pump. In one
of the demonstrations in California in 1901, he used a truncated cone of
60 m^2 of collecting surface and produced steam at a pressure of 1000 KPa
and pumped water at a rate of 5.3 m^3/min. through a 3.6 m head developing
about 7.6 KW.

During the years 1902-1908, H.R. Willsie[12] and John Boyle, Jr.,
built several solar engines. Probably they were the first to use two-fluid
system in solar engines. They used flat plate collectors with double
glazing and low temperature working fluid such as ammonia, ether or sulphur
dioxide. In 1904, they developed a 4.5 KW sun powered plant at St. Louis
in which they have used a double glazed flat plate collector of 55.7 m^2 area
using an ammonia engine. In 1905, they built a solar engine of 15 KW capacity
using sulphur dioxide as working fluid near Needles, California. Another
solar engine of 11.5 KW capacity was built by them in 1908 at Needles which

used 9.3 m^2 flat plate collectors. An American, Frank Shuman[13], in 1907 developed a solar pump in which 1200 m^2 flat plate collectors were used to evaporate ether from water and the vapours were used to drive a vertical vapour engine of 2.6 KW at Tacony, Pennsylvania. In 1911 Frank Shuman developed[14] a 24 KW solar engine in which double glazed flat plate collectors with flat booster reflectors with a total collecting area of 956.5 m^2 were used. This system pumped 11.3 m^3 of water per minute to a height of 10 m. One of the most successful and largest solar pump was operated at Meadi, Egypt, by F. Shuman and C. Boys[15] in 1913. In this system parabolic troughs with a total area of 1263 m^2 were used, which generated steam and in turn developed 37.3 KW power. This pump was used for pumping irrigation water from the Nile river but was later dismantled. During the years 1919-1934, R.H. Goddard[16] published several papers related to solar power production. In 1919 Goddard described[16] one of the largest solar dish electric power plant. J.J. Harrington in New Mexico[17] in 1920 developed a solar steam engine and pumped 8.9 m^3 of water through a height of 6 m.

C.G. Abbot[18] in 1936 contributed greatly towards solar power generation and used equatorially mounted parabolic trough reflectors along with an ingenious single tube solar flash boiler which produced steam within five minutes of sun's exposure. The system had produced saturated steam at 647°K and ran a steam engine of 370 watts. In 1940, F. Molero[19] of the Heliotechnical Laboratory at Tashkent, USSR, used a 10 m diameter parabolic dish for generating steam at a pressure of 203 KPa which was used for pumping water for irrigation and livestock watering. M.L. Ghai and M.L. Khanna[20] of New Delhi, India, during 1951-1955 used a small reciprocating piston, hot air engine and operated with small paraboloidal solar reflector at about 800°K and produced power between 100 to 125 watts. This was used for demonstration purposes and pumped water through a depth of 5.0 m.

In 1961, Tabor and Bronicki[21] in Israel described a 4 KW solar turbine using a binary Rankine cycle with monochlorobenzine as a working fluid. These turbines were later modified and known as ORMAT Rankine power units[22] and are now available in the power range of 100 watts to 15 KW. These turbines are now in use in many parts of the world.

In Dakar, Senegal, Masson and Girardier[23] during 1962-1966 developed two solar motors. The solar motor operating since 1962 lifted 8 to 10 litres water per minute from a depth of 13 metres which is equivalent to 21 watts.

In another experiment in 1966, flat plate collectors of 300 m^2 were used and the solar engine pumped water at the rate of 40 m^3/hr from a depth of 8 to 10 m.

The first photovoltaic power plant of 0.25 KW using concentrated solar radiation on photovoltaic cells was installed in September 1964 near Gelendzhik, USSR[24]. The total area of the silicon solar cells was 3.6 m^2 and the panels could rotate equatorially. Plane mirror boosters were used to concentrate solar radiation on solar cells.

During the years 1974-1980, a French company, Societe Francaise d'Etudes Thermiques et d'Energie Solaire, SOFRETES[25,26] installed several solar irrigation pumps in many countries of the world. The first solar pump (1975) of 25 KW capacity using 1500 m^2 flat plate collectors was installed in San Luis de la paz, Mexico, which had given an output of 2600 m^3/day of water from a head of 10 m. Later a 50 KW solar pump was also developed by this firm. Here water is heated through flat plate collectors which in turn through heat exchanger boils the organic liquid such as butane or Freon operating a Rankine cycle reciprocating engine or turbine.

A solar irrigation pump of 37.3 KW capacity was installed[27] in April 1977 at Gila Bend Ranch southwest of Phoenix, Arizona, which was designed and built by Columbus Laboratories of Battelle Memorial Institute, USA, and Northwest Mutual Life Insurance Company of Milwaukee. The system consists of a turbine, boiler, condenser, regnerator, and preheater; and a low-lift, high volume flow propeller pump. The working fluid used is Freon-113 and the pump was able to pump about 37 m^3 of water per minute at peak operation. The concentrating collectors were manufactured by Hexcel Corporation, Casa Grande, Arizona, and the Freon turbine by Barber-Nichols Engineering Company, Arvada, Colorado, USA.

In April 1977, an 18.7 KW solar pump was designed and tested by Barber-Nichols Engineering Company of Colorado[28,29] with the design specifications supplied by Sandia Laboratories and the solar pump delivered 3.3 m^3/min of water from a well of 23 m deep at Willard, New Mexico. The system consists of field of tracking parabolic trough collectors (625 m^2) supplied by Acurex Solar Corporation, a thermocline thermal storage system (22.7 m^3) and a Rankine cycle Freon turbine. The Freon turbine runs at 36300 rpm and developed 1760 rpm at the output shaft of the gearbox.

A 7.5 KW solar photovoltaic solar pump was put into operation in August 1977 near Mead, Nebraska[30] which was the joint effort of MIT/ Lincoln Laboratory and the University of Nebraska and sponsored by ERDA. The photovoltaic panel consists of 120000 individual silicon solar cells with a peak power rating of 25 KW and driving a 7.5 KW pump. Power from the photovoltaic panels is stored in large lead acid batteries capable of storing 85 KW. This pump can pump 3.8 m^3/min water from a reservoir for 12 hours a day.

Under a joint programme of Government of India and Government of West Germany, a 10 KW solar thermal power station was built[31] and operated at Madras, India, in 1978. This experimental power station uses flat plate collectors (756 m^2) with mirror boosters (756 m^2), storage unit of 35 m^3, vapour generator, a Freon-114 operated screw expander, and an alternator generating 440 volts 3 phase 50 Hz AC power. The system generated 10 KW net at peak load and the designed output per day is 35 KWH on a standard day with 1000 watts/m^2 peak insolation at noon.

A solar power generation[32-34] unit of 150 KWe with a technical support from Sandia Laboratory, Arizona Solar Energy Commission and University of Arizona in October 1979 on the Dalton Cole farm southwest of Coolidge in Central Arizona was set up. The plant was completely designed by Acurex Corporation, California and used 2140.5 m^2 line-focus collector system, a storage of 114 m^3 tank of hot oil, and an organic Rankine cycle turbine engine built by Sundstrand Corporation. This plant was operated for three years by the University of Arizona to characterize energy performance, identify needed equipment improvements and quantity operating and maintenance requirements. This plant was later deeded to the owner of the farm on which it is sited, Dalton Cole, Jr.

Under the National Sunshine Project of Japan, a 100 KW solar thermal electric power plant[35] was installed in March 1981 at Nio-cho in Kagawa prefecture of Shikoku. In this system a combination of plane mirrors and cylindrical parabolic mirrors (hybrid mirror system) with a total mirror area of 11160 m^2, a steam accumulator, molten salt heat storage (KCL-LiCl), and a Rankine cycle steam turbine is used.

A 100 KW solar thermal electric power station is in operation in Sulaibyh, Kuwait[36] since June 1981 which is a joint venture of Kuwait Institute for Scientific Research (KSIR), Kuwait, and Messrs Chmitt-Bolkow-Blohm (MBB) of

Germany. The solar power station consists of 56 point focusing parabolic dishes, each of 5 m diameter with total collector area of 1025 m^2, a thermal storage tank and an organic Rankine cycle engine using a toluene as working fluid and provides both electric and thermal energy for heating and cooling of greenhouses, desalination, irrigation and pumping of brackish water.

On July 31, 1981, a 500 KWe solar power plant under the auspices of the International Energy Agency (IEA) by eight countries (Germany, USA, Spain, Greece, Switzerland, Sweden, Belgium, Austria) was made operational in the remote village of Tabernas, Almeria, Spain[37,38]. The operating agent for this project was DFVLR (Space Agency) of Germany and the design, construction and commissioning was done by a consortium consisting of Tecnicas Reunidas, S.A. of Madrid, Spain; Acurex Solar Corporation of California, USA; and MAN-Neue Technology of Munich, Germany. The solar-thermal electric generation plant consists of two sets of linear parabolic collectors, one set single axis tracking type supplied by Acurex Solar Corporation of 2674 m^2 area and second set two-axis tracking type supplied by MAN of 2688 m^2, a thermal storage of 0.8 MWhe capacity with Santotherm 55 - a synthetic heat transfer oil; an oil-to-water Baeltz steam generator, and a Stal-Laval condensing steam turbine generator.

In Shenaddoah[39], Georgia, 40 Km of south of Atlanta, a 400 KWe parabolic dish power plant started in April 1982 to supply electricity to the Georgia Power Company grid and at the same time cogenerate electricity and process steam to a nearby garment factory operated by Bleyle of America, Inc. It employs 114 parabolic dishes each of 7 m diameter with a total collecting area of 4330 m^2, a heat exchanger producing superheated steam, a thermocline heat storage tank filled with silicon heat transfer fluid, a steam turbine, and an alternator. The overall efficiency, solar to electrical, is about 10 per cent.

Australian National University, Canberra in March 1982 designed and built a 25 KWe and 140 KW low quality heat solar power station at White Cliffs, New South Wales, Australia[40]. The power station comprises 14 modular semi-autonomous parabolodial tracking collectors, each of 5 m diameter with a total collecting area of 277 m^2, generating steam up to 550°C of 7 MPa; and a reciprocating steam engine which powers an alternator to produce 25 KW of 240 volt electricity and the exhaust steam used for desalination, with minor amounts for hot water and for space heating and cooling.

A 350 KW photovoltaic power system[41] was installed near the villages of Al Jubay-Lah, Al Uyaynak, and Al Hijra, which are about 50 Km northwest of Riyadh, Saudi Arabia, in 1982, sponsored by the Saudi Arabian-United States Program for Cooperation. The design and fabrication of the system was largely carried out by Martin Marietta Corporation, Denver, Colorado. The photovoltaic system consists of 160 single pedestal concentrator arrays, each array containing 260 circular solar cells of 5.71 cm diameter and providing 2.2 KW of power under optical concentration of 33 suns. This has automatic 2-axis tracking arrangement. The battery subsystem consists of four batteries and each battery has 116 commercial lead acid cells of 1600 Ah each in series. The inverter converts 215/300 VDc power to 277/480 VAC, 3-phase 60 Hz AC power.

A hybrid solar/diesel power station of 100 KWe using parabolic troughs with a total aperture area of 920 m^2 was built at Meekatharra, Australia[42] in April 1982. Thermal oil is used to collect and store the heat and through a heat exchanger generate steam which is fed into the two stage screw expansion engine which drives an electric generator.

Photovoltaic power plant[43] of 1-MWE capacity was installed in December 1982, in California's high desert 160 Km northwest of Los Angels which contains 900000 single crystal solar cells mounted on computer-controlled two-axis trackers. One-megawatt inverter system converts the $\pm$300 VDC from the field to 12 KVAC, for delivery to the Edison grid near the site. The photovoltaic power station is commercial, unmanned and fully automated. ARCO Solar Industries of USA has also announced a 6-MW photovoltaic plant to be installed in phases on the Carrisa Plain east of Bakersfield, California. Simple reflector units on the trackers and high efficiency solar cells are supposed to be used. The Sacramento Municipal Utility District (SMUC), USA, has announced a 100-MW photovoltaic plant to be installed near Ranch Seco Nuclear Power Plant to be completed in 10 stages.

Non-convective solar ponds can be used as an effective means for collect-ing and storage of solar heat and can be used for power production. Israel[44] is the first to develop a solar pond power station. A 6-KWe solar pond power facility was created in 1978 at Yavne, Israel[1] in which 1500 m^2 density gradient solar pond along with a low temperature turbogenerator was used. Another solar pond power plant was made in December 1979 in Ein-Bokek near

Dead Sea, Israel, in which 7500 m^2 solar pond with a depth of 2.5 m was
used where the bottom temperature reached to $93^{o}C$. With the success of
Ein-Bokek power plant, Israel has made an ambitious plan as follows:

- A 5-MW power plant to be operational by 1982 using 0.25 Km^2
  solar pond.
- By 1983 the above pond will be expanded to 1 Km^2 providing
  20 MW power and additional construction of 100000 m^2 solar pond.
- An additional 20 MW power plant and 1 Km^2 solar lake
  (peaking/intermediate) to be completed by 1985.
- Construction of 50 MW unit and 4 Km^2 solar pond to be completed
  by 1986.

One of the most promising methods of collecting solar energy and
converting into electricity is the central receiver system, in which a
field of large individually controlled mirrors (heliostats) reflects
solar radiation on a receiver at the top of a tall tower, heating a
fluid at very high temperature, producing steam and driving the conventional
steam engine or turbine. The concept of central receiver system was first
given (1957) by V.A. Baum, R.R. Aparsi and B.A. Garf in USSR[45] in which
heliostats are moving on circular railroad cars and reflecting solar
radiation on an elevated cavity receiver boiler. The cavity was to be
rotated to face the heliostats throughout the day to achieve better
performance. Later, Francia[46] in 1967 built a pilot model of central
receiver system at the University of Genoa, Italy, in which a clock driven
field of 271 heliostats were used reflecting the solar radiation on a
receiver at the top of the tower producing useful steam at temperatures
up to $650^{o}C$ at a rate equivalent to 150 KWt.

An advanced components test facility (ACTF) of 400 KW capacity is
operated by the Georgia Institute of Technology[47], Engineering Experiment
Station, Atlanta, Georgia, for the US Department of Energy since September
1977. The ACTF consists of a 550 tracking heliostat field each of 111 cm
in diameter with total area of 532 m^2, a tower 21.3 m tall located in the
centre of the mirror field, instrument and control building, a computerized
data collection system, and a heat rejection system. The heliostats are
electrically driven by their mechanical supports without feedback which is
a unique facility.

A 5-MW solar thermal test facility[48] (STTF) for testing receivers of
solar tower system was constructed in October 1978 by US Department of
Energy about 32.2 Km southeast of downtown Alburquerque, New Mexico, and

operated by Sandia Laboratories. The STTF uses 222 heliostats with a total reflector area of 8250 m^2. The tower provides test platforms at 36.6, 42.7, 48.8 m on north face and 61 m at the top. The operating conditions for the receiver are 516°C, 10.4 MPa, and 1.5 Kg/s with feed water being supplied to the receiver at 288°C.

A 2.5 MWe central receiver system of the THEMIS project is in operation at Targasonne in Southern France since March 1981. Solar radiation is focused on a cavity type receiver on a 100 m tall concrete tower by 200 heliostats each with a mirrored surface area of 54 m^2. Energy absorbed in the receiver heats moltensalts (sodium and potasium nitrates and nitrites) to 450°C from where it is piped to ground level where it generates steam which drives the conventional steam turbine. This work was undertaken by the French National Centre of Solar Tests, which is funded by Commissariat a l'Energie Solaire (a government organisation) and Electricite de France (the French national utility).

EURELIOS is a central receiver plant located in Sicily, Italy[49], sponsored by the Commission of European Communities and producing 1 MWe since May 1981. An European consortium of Italian, French and German companies designed and built the plant. Eurelios uses two types of heliostats with a total area of 6216 m^2 with total 182 heliostats. 70 heliostats each of 52 m^2 are made by Cethel and 112 heliostats each of 23 m^2 are made by MBB (Measserschmitt Boelhow-Blohm). These heliostats are arranged in subfields beneath the cavity type receiver, which is mounted on a 55 m high tower. Steam exiting the receiver at 512°C enters the steam turbine without going through an intermediate heat exchanger. Hitec salt is used as a heat storage material which can provide 30 minutes of energy to smooth out cloud transients.

A central receiver electrical generating plant of 1.0 MWe capacity is located on the island of Shikolku[50], (Nio) Japan and is generating electricity since August 1981. The plant uses 807 heliostats each with a reflecting area of 16 m^2, surrounding a conical-cavity receiver and steam drum on top of a 69 m high tower. The receiver produces steam at roughly 250°C and 4 MPa and enters the turbine generator at a temperature of 187°C and at flow rate of 7940 Kg/hr. The energy is stored in tanks containing pressurized water which is equivalent of 1 MWe for 3 hrs.

A 500 KWe central receiver system is in operation at Almeria[51],

Spain since September 1981 (SSPS) which is sponsored by nine member
countries of the International Energy Agency (Germany, Spain, Italy,
Austria, Sweden, Belgium, Switzerland, Greece and USA). The system
employs a field of 93 heliostats having a total reflecting area of
3660 m^2, a cavity receiver using sodium as the heat transfer fluid at
an operating temperature of 530°C, a steam driven piston engine coupled
to a three-phase-current generator and a hot tank cold tank sodium
storage system.

The Solar-one is the world's largest solar powered central receiver
plant[52] which is of 10 MWe capacity and is in operation since April
1982 near Barstow, California, USA. This pilot solar facility is
sponsored by US Department of Energy and Cooperating agencies are:
Southern California Edison Company, the Los Angeles Department of Water
and Power, and the California Energy Commission. The collector field
consists of 1818 Martin Marietta sun tracking heliostats with a total
reflecting area of 72538 m^2. An external type receiver is used at the
top of 90.8 m tall steel tower and water-steam is used as the heat
transport fluid. Energy is stored in one tank of 13.7 m high and 19.8 m
diameter containing heat transfer oil and 6798 tons of rocks. Stored
hot oil is used to generate steam at 274°C and 2.7 MPa which drives the
conventional steam turbine.

A 1.2 MWe central receiver plant (CESA-1) funded by United States
and Spanish Joint Committee for Scientific Technological Corporation and
built by Centro de Estudios de la Energia (CEE) of Spain is located in
southern Spain near the town of Almeria, Spain and working since June 1983.
It used 300 heliostats each of 38 m^2 and a cavity receiver on a 60 m tall
concrete tower. The energy is stored in 300 tons of molten salt. The
receiver uses water as the working fluid and steam at 520°C drives a
Rankine cycle turbine engine.

A central receiver plant, known as 20 MWe Gasgekuhltes Sonnenturn-
Kraftwerk GAST, is built[53] in 1983 in Germany by a German industrial
consortium and sponsored by the German Government (BMFT). The GAST
pilot plant employs 3000 heliostats each of reflecting area 40 m^2
concentrating solar radiation on to two receiver modules mounted on top
of tower at a height of 200 m. The heat transfer medium is air which is
heated to 800°C. A combined gas/steam thermal energy conversion process

using two open-cycle gas turbine generator sets and one steam turbine generator set is employed. The plant does not employ any storage system but equipped with an auxiliary fossil fuel-fired system.

A central receiver electric plant of 30 MWe is planned[54] by a team drawn from Rockwell International's Energy Systems Industries the Pacific Gas and Electric Company, and ARCO Solar Industries and the same will be operational by the end of 1986 at the Carrizo plain in Central California, USA. The collector field will consist of 1904 advanced third generation ARCO heliostats each of 95 m^2 reflective area. These heliostats reflect the solar radiation on to the receiver at the top of 122 m tall tower. Sodium heated to more than 600°C would be piped to ground level, where it would generate steam to drive a turbine generator. The warm storage tank will contain about 1419420 litres of sodium.

It is reported that USSR is also planning to build a 5 MWe central receiver plant in the Crimea but its details are not available.

3.0 PRINCIPLES OF SOLAR ENGINES

The heat energy can be converted into mechanical power through any of the energy conversion cycles: the Rankine cycle, the Stirling cycle, and the Brayton cycle. The heat engine based on any of the above cycles is a thermochemical device operating between a high temperature heat source and a low temperature heat sink, extracting some of the thermodynamic heat energy from the working fluid and converting it into shaft power. Only a small portion of heat energy added to the cycle can be converted into mechanical power and Carnot cycle efficiency defines the maximum amount of energy which can be converted into mechanical power. If Q_H heat is added at temperature T_H and after performing the work the fluid rejects Q_C heat at a low temperature T_C then the Carnot efficiency, η_{carnot}, is obtained by dividing the difference between heat input and rejected heat by the heat input, i.e.

$$\eta_{carnot} = \frac{Q_H - Q_C}{Q_H} \tag{1}$$

This concept was developed by a Frenchman, Sadi Carnot and is named as Carnot efficiency. The Carnot efficiency η_{carnot} is defined as the ratio of temperature (absolute) difference between the source T_H and sink temperature T_C divided by the source temperature T_H, i.e.

$$\gamma_{carnot} = \frac{T_H - T_C}{T_H} \qquad (2)$$

From this expression it is seen that: (1) the engine efficiency increases as the source temperature increases and the rate of increase is greater at low temperature than at high temperature; (2) the engine efficiency increases with the decrease in sink temperature at a faster rate than if the source temperature increases by the same amount; (3) if the sink temperature reaches to $0^{\circ}K$ the Carnot cycle efficiency reaches to 100 per cent, which is an impossible situation according to the third law of thermodynamics. For a given temperature range the Carnot efficiency is the maximum possible cycle efficiency and is limited by the second law of thermodynamics.

The schematics of heat engine cycles and their corresponding p-v diagrams are shown in Fig.2. In the Carnot cycle, as shown in Fig.2(a), there are two reversible isothermal processes at temperatures T_H and T_C respectively, connected by two reversible adiabatic processes. When the working fluid is condensable vapour, the two isothermal processes are easily obtained by heating and cooling at constant pressure while the fluid is a wet vapour. Saturated water in State 1 is evaporated in the boiler at constant pressure to form steam in State 2. The steam is expanded adiabatically to State 3 while doing work in a turbine or reciprocating engine. After expansion the steam is partially condensed at constant pressure and the heat is rejected. The condensation is stopped at Stage 4. Finally the steam is compressed adiabatically in a rotary or reciprocating compressor to State 1. The efficiency of a practical heat engine is always lower than the Carnot efficiency due to many reasons such as heat losses through insulation, friction between moving parts, compressors not working at 100 per cent efficiency, the expansion device not working at 100 per cent efficiency, problems with the actual working fluids, etc. If the engines are carefully designed then efficiency can be about 50 to 80 per cent of the Carnot efficiency depending on the temperature difference. The actual cycle efficiency obtained with flat plate collectors working at temperature difference of about $90^{\circ}C$ is about 8 to 10 per cent while with concentrating collectors operating at a temperature difference of $400^{\circ}C$ the cycle efficiency is about 20 to 25 per cent.

The ideal Rankine cycle is shown in Fig.2(b). It differs from the

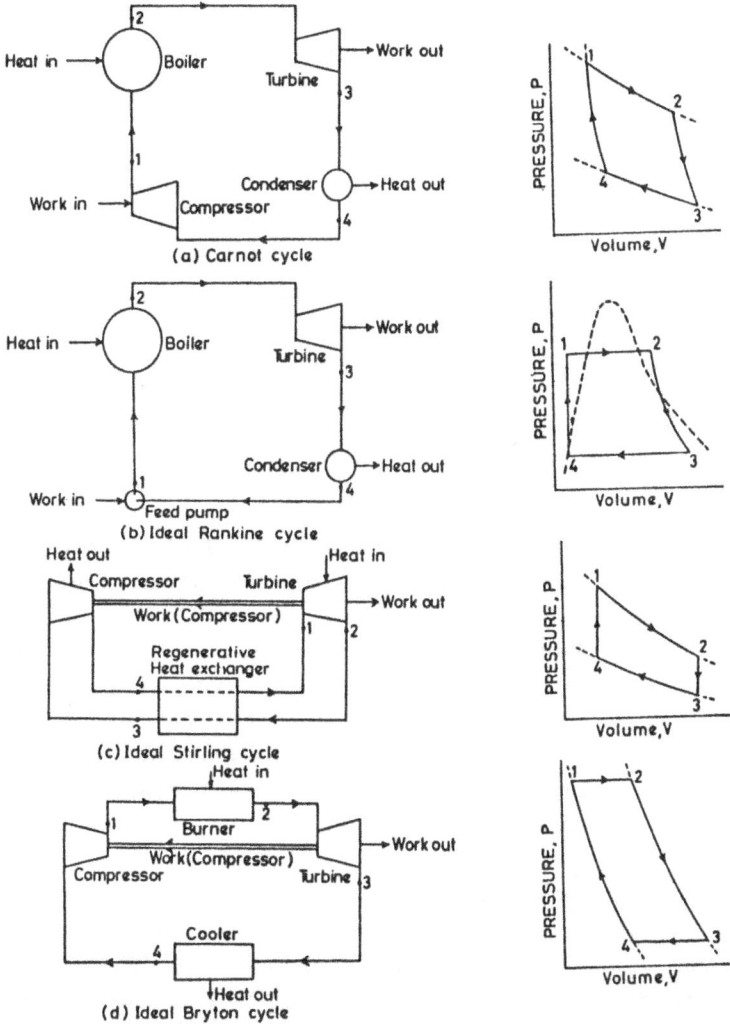

FIG.2 - SCHEMATICS AND P-V DIAGRAMS OF HEAT ENGINE CYCLES

Carnot cycle in that the heat-addition process does not occur at constant temperature. In an ideal Rankine cycle the working fluid is heated in the boiler, and the vapour so produced is expanded in the turbine to do mechanical work. The exhaust from the turbine consists of mixture of vapour and liquid droplets at a much lower temperature and pressure than at the inlet of the turbine. Exhaust vapour is liquified in the condenser rejecting the heat. This liquid residue is pumped up by a feed pump to high pressure and fed to the boiler to complete the cycle. It can easily be shown that the efficiency of Rankine cycle is less than the Carnot cycle operating between the same temperatures, because all the heat supplied is not transferred at the upper temperature. In spite of the low efficiency of a Rankine cycle, it has high work ratio and the steam consumption is less compared to Carnot cycle. Thus the size of the Rankine cycle can be increased by superheating the vapour or by partially expanding it and then reheating it several times. The efficiency can also be increased by using a part of rejected heat in heating the liquid by using a regenerator before the liquid enters the boiler. Some improvements can also be made by using more than one working fluid. In this case the heat rejected by a high temperature cycle is used as input to a cycle using a low boiling point fluid.

The Stirling cycle is similar to a Carnot cycle except that two adiabatic steps are replaced by two constant volume steps (Fig.2(c)). Here some suitable gas or air is used as working fluid and the turbine, compressor and heat exchanger are very closely coupled in a single housing and not as shown in Fig.2(c). Here the heat addition and rejection takes place at constant temperatures. The heat supplied during the process 4-1 is equal in quantity to the heat rejected during process 2-3. Moreover, the temperature of the working fluid varies between the same limits during these two processes. It is, therefore, theoretically possible that the heat rejected is returned to the working fluid. This heat transfer is accomplished reversibly in a regenerator consisting of a matrix of wire gauze or small-small tubes. During the passage of the fluid through the regenerator the volume of the working fluid does not change in either of the directions. The Stirling cycle has higher efficiency than the Rankine cycle between the same two temperatures because in the Stirling cycle heat is delivered at high temperature and rejected at the low temperature as in the Carnot cycle. The main difficulty with the Stirling cycle lies in making efficient regenerator of reasonable size which can operate at temperatures comparable to temperature used in internal combustion engines.

The Brayton cycle as shown in Fig.2(d), uses a gas or air as the working fluid and works at temperature well in excess of 500°C. In the Brayton cycle the expansion and compression processes are reversible and adiabatic

and heat addition and rejection take place at constant pressure. In the
Brayton cycle the hot compressed gas is allowed to expand through a
turbine producing work. The exhaust gas from the turbine is fed to the
heat exchanger where the heat is rejected and then compressed by the
compressor to complete the cycle. The Brayton cycle is less efficient
compared to the Stirling cycle but larger power plants can be made using
Brayton cycle than those using a Stirling cycle. The performance of the
Brayton cycle can be improved by inserting a regenerator between the
turbine exhaust and the cooler for preheating the compressed gas prior
to the heater. These power cycles are described in detail in text books
on thermodynamics[55] and in a paper by Howe[56].

The efficiency of different thermodynamic cycles can be compared on
a temperature-entropy (T-S) diagram. The temperature-entropy diagram of
these cycles is shown in Fig.3.

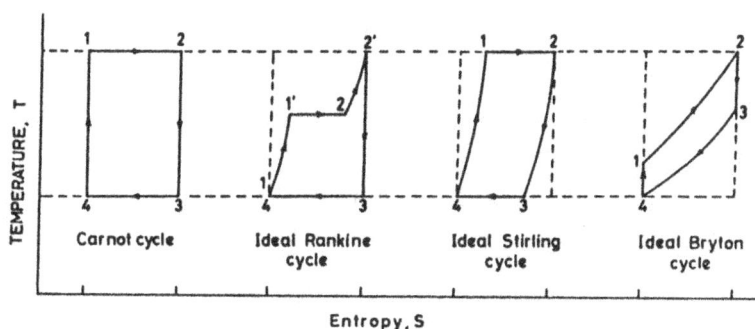

FIG.3 - COMPARISON OF HEAT ENGINE CYCLES

It is seen from this figure that the Carnot cycle can be represented by a
rectangle and its area will give the cycle efficiency. The areas within
solid lines give the efficiency of corresponding cycles. It is seen from
this figure that the Stirling cycle is better than the ideal Rankine cycle
and also the ideal Brayton cycle. The main limitation as mentioned earlier
with the Stirling cycle is the design of a suitable efficient regenerator.
Moreover, in the large power plants, Stirling cycle is not preferred due to
many practical problems. In solar energy applications the Rankine cycle is
generally preferred and widely used due to its superior overall cycle
efficiency and component sizes. Net engine efficiency of various peak cycle

temperatures for all the four power cycles is compared[57] in Fig.4 which also indicates the superiority of the Stirling cycle.

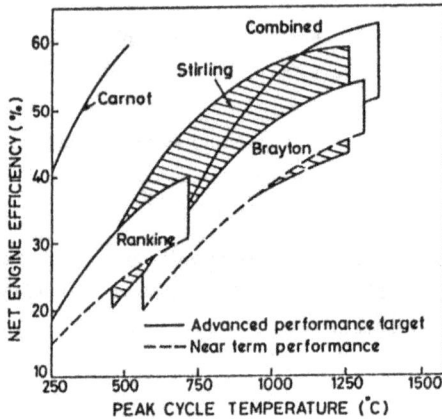

FIG.4 - ADVANCED ENGINE PERFORMANCE POTENTIAL FOR RANKINE, STIRLING, BRAYTON AND COMBINED BRAYTON/RANKINE CYCLES

A diagram between the enthalpy (H) and entropy (S) generally known as Mollier diagram for a working substance is used for analysing the performance of a cycle. The Mollier diagrams are available in heat and thermodynamics text books.

4.0 IDEAL WORKING FLUID

The maximum output of any energy conversion device is limited by the second law of thermodynamics and depends on heat sink and heat source temperatures. The actual efficiency of any practical system is governed by the properties of the working fluid. The choice of working fluid depends on the operating temperatures in the boiler and condenser and the type of engine. Steam is the most widely used working fluid in heat engines due to its low cost, high chemical stability, universl availability and better cycle efficiency. But due to some technical and other operational reasons, generally organic fluids are used as working fluids and are selected based on their physical and thermodynamic properties.

The working fluid should be cheap, thermally and themically stable, non-flammable, non-corrosive, non-toxic, small number of stoms in the molecule, and of high molecular weight. A list of working fluids is given by Curran[58].

5.0 LIMITATIONS OF SOLAR MECHANICAL POWER CONVERSION

There are several limitations in the effective conversion of solar energy into mechanical power. Some of them are:

1. The main problem is that the efficiency of the collection system decreases as the collection (operating) temperature increases while the efficiency of the engine increases as the working fluid temperature increases.

2. The theoretical efficiency that can be attained by any heat engine operating between two temperatures is well understood and provide fixed fundamental barriers.

3. The heat engine used is not reversible and, therefore, efficiency obtained would be less than the maximum limiting theoretical efficiency.

4. A part of the heat is lost from the working fluid during its passage from collector or boiler.

5. Due to the intermittant nature of the solar radiation some kind of thermal storage device is required to operate the heat engine continuously. Generally the heat storage materials degrade with time and are not completely reversible.

6. Like in many other fields the materials of construction of heat engine and suitable working fluid and their interaction pose a problem. The construction materials should withstand high temperature and pressure.

7. In solar mechanical power generation both solar collectors and engines pause problems. Solar collectors are generally more expensive than the engines. Moreover, they require large areas for installation.

6.0 RANKINE CYCLE CHARACTERISTICS

As mentioned earlier the Rankine cycle solar power pumps are preferred because their efficiencies are little smaller to Carnot cycle while the steam consumption is less and the work ratio is high. Moreover, in the temperature range of solar systems, 50-300°C, the Rankine cycle is superior to other cycles in terms of component sizes. Following guidelines should be followed to convert efficiently the heat into the shaft power:

(i) Collect as much of heat as possible from the solar collector.

(ii) Supply heat to the engine at a temperature as high as possible.

(iii) Reject heat at a temperature as low as possible.

(iv) All the components like boiler, preheater, turbine, regenerator, condenser, feed pump, etc. should be designed for maximum efficiency.

(iv) All the components like boiler, preheater, turbine regenerator, condenser, feed pump, etc. should be designed for maximum efficiency.

(v) Minimize the pressure drop in all components and pipes.

A typical Rankine cycle power plant is schematically shown in Fig.5.

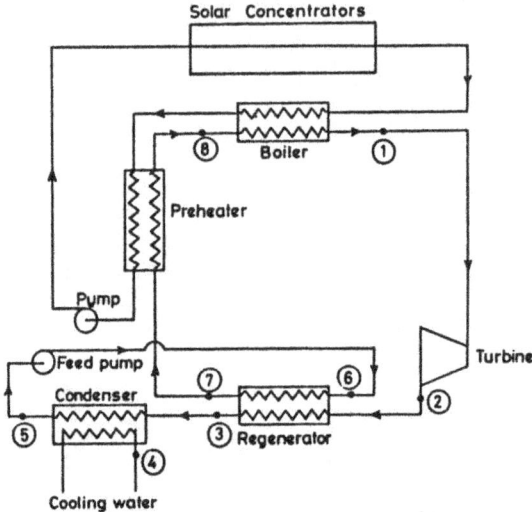

FIG.5 - SCHEMATIC OF SOLAR POWERED RANKINE CYCLE

Here two small electric pumps are shown. One pump circulates hot water through the solar concentrators, boiler, preheater and back to solar concentrators. The second pump circulates the working fluid (Refrigerant-113) through various heat exchangers. As is seen from Fig.5, the working fluid in the liquid condition is pumped at high pressure by the feed pump through regenerator and preheater where it is partially heated and then to the boiler. In the boiler, the working fluid gets vaporized by hot water coming from solar concentrators. The high pressure vapour gets expanded through turbine producing shaft power which can be used for generating electricity using alternator or for pumping water using a pump. The low pressure and low temperature vapour passes through regenerator transferring some of its heat to the high pressure working fluid and then liquified through condenser.

The performance of a Rankine cycle can be determined if the thermodynamic fluid properties at various points are accurately known; efficiencies of pump, turbine and other mechanical components and effectiveness of regenerator and heat exchanger are known; and component pressure drops arc known. In the literature several definitions of cycle efficiency are given. Here we define the cycle efficiency as the ratio of useful output to the heat added to the cycle from the source. The thermodynamic cycle efficiency is defined[59] as:

$$\eta_{TC} = \frac{\text{work done by expander} - \text{work done by pump}}{\text{heat added}}$$

$$= \frac{(H_1-H_2) - (H_6-H_5)}{(H_1-H_7)} \tag{3}$$

where H is the enthalpy at a particular cycle point.

The cycle efficiency η_C is defined as:

$$\eta_C = \frac{\text{useful shaft power}}{\text{heat added}}$$

$$= \frac{\text{work done by expander} - \text{work done by pump} - \text{mechanical losses}}{\text{heat added}}$$

$$\tag{4}$$

The solar collector area, $Ac(m^2)$ for the maximum power output, P, can be calculated from the following equation:

$$Ac = \frac{P}{I\,\eta\,\eta_{rc}} \tag{5}$$

where

   p = power of the engine

   I = incident solar radiation on collector aperture at solar noon on the design day (KW/m^2)

   η = collector efficiency (instantaneous)

   η_{rc} = thermal efficiency of Rankine cycle system

$$= (1 - \frac{T_c}{T_H})\,\eta_e\,\eta_r \tag{6}$$

   η_e = engine efficiency, and

   η_r = ratio of Rankine to Carnot cycle efficiency.

The cycle efficiencies for various flow rates, fluid properties, heat transfer requirements, etc. can be accurately obtained using computer programs as described by Abbin et al.[60] and Badr et al.[61].

7.0 SOLAR HEAT ENGINES

For converting solar energy into shaft power or mechanical energy any heat engine like single or multistage reciprocating steam engines, steam engines of rotary type (not successful) turbomachines (using steam or organic vapours as working fluid), Stirling hot air engines, Brayton engines can be employed[62,63]. The selection of a particular solar (heat) engine depends on many parameters including power requirement (size), type of working fluid, temperature of operation and solar energy collection device. For low power requirements (< 50 KW), reciprocating engines, rotary displacement engines and Stirling hot air engines are preferred. For larger power capacities, turbines (rotodynamic) are favoured due to several advantages such as:

1. Wear and tear and maintenance cost in case of turbine is low and, therefore, their reliability is high compared to reciprocating engines.

2. High thermal efficiency particularly in large sizes can be obtained. Efficiencies of around 80 per cent has already been obtained while in case of steam engines the thermal efficiency is of the order of 20 to 25 per cent.

3. Turbines are most appropriate for use with generators.

4. High speed turbines up to 4000 rpm can be obtained thereby increasing the power output per unit volume of the working fluid, while speed more than 250 rpm in steam engines is not available.

5. Since no internal lubrication is required in case of turbine except on main bearings, there is no chance of mixing lubricant oil with the working fluid.

6. Very large capacity turbines up to 5 MW capacity have already been built while steam engines of this capacity is impossible to build and operate.

7. In turbines the motion in the form of rotating shaft is directly obtained and hence perfect balancing of the whole system is theoretically possible.

8. In case of turbines the condenser pressure can be at low vacuum, converting the energy of working fluid into useful work almost up to maximum.

9. The operation and regulation of turbine is easy.

10. There is a possibility of large expansion to condenser pressure.

There are some problems also with the turbines, like small turbines are less efficient, costly, and the moisture content in the expanded vapour can result in the erosion of turbine blades.

As mentioned earlier, both steam and organic vapours can be used to drive the turbine. The minimum vapour pressure of a working fluid required to drive a turbine is about 700 KPa, and the preferred pressure is around 2000 KPa for useful work. These pressures are possible with water (steam) when it is heated to a temperature of 150 to 200°C. Sometimes such high temperatures are not possible with solar energy such as by flat plate collector or simple low cost concentrating collectors where temperatures around 100°C can be obtained. In such cases, turbines, along with a suitable organic working fluid can be employed and, therefore, water becomes completely inappropriate. However, there are disadvantages of using organic fluids such as these are expensive, poisonous, flammable, and leakage can be a great problem. Moreover, large parasitic energy due to high amount of organic fluid is required for pumping to produce the same power output compared to a steam turbine. The energy requirement (parasitic energy) in case of organic fluid turbine can be 5 times compared to steam operated turbines.

7.1 Steam Engines

In the past, several attempts[7,16,18] have been made to use reciprocating type steam engines for producing power using solar energy. Steam engines were earlier considered to be of great significance but their use in modern industries is declining with the advent of turbines due to their large capacity, higher efficiency and superior economy.

The work done by the steam engine can thus be determined by evaluating the area of theoretical indicator diagram and then multiplying it with the diagram factor. The actual work done (W) by a steam engine can be calculated from the following expression:

$$\text{Work done per minute} = f P_m \, LAN \quad (Nm) \tag{7}$$

where f = diagram factor

       L = stroke of piston, m

       A = area of piston, m^2

       N = number of strokes per minute (n for single acting and 2n for double acting engine)

       n = rpm

      P_m = mean effective pressure on piston (N/cm^2)

$$= \frac{P_1}{r} (1 + \log_e r) - P_3 \tag{8}$$

P_1 = absolute pressure of steam at entry to the steam engine (N/cm^2)

P_3 = absolute exhaust pressure of steam (N/cm^2)

r = expansion ratio, V_2/V_1.

Small solar steam engines are not promising because they are expensive and inefficient while small internal combustion engines running on gasoline are widely used for pumping of water, lawn movers, motor boats, automobiles and diesel power sets. The internal combustion engines are inexpensive, efficient and operated throughout the world without much difficulties and technial expertise. Attempts have been made to convert the internal combustion engines into steam engines with high efficiency[17]. Farber[64] converted a lawn-mover gasoline operated engine into a solar steam engine. Recently, Inall[65] converted a diesel engine into a 25 KW solar steam engine which is used at the White Cliffs Solar Power Station in Australia. This engine is made from parts of two diesel engines[65]. The engine starts by a standard electric motor and solar generated steam is supplied to a chamber in the head of each cylinder. As a piston reaches 15° before top dead centre, the pins in its crown lift the three hall valves from their seats and steam enters the cylinder until the valves seat again at 15° past dead centre. The steam expands while pushing the piston until it reaches to the exhaust parts in the cylinder. The parts like crankcase, crankshaft, flywheels, connecting rods, starter, and pump are taken from a Lister Diesel Model HR3. The cylinder liners and piston are taken from GM diesel Model 53 and other parts like cylinders, cylinder heads, valve seats, and steam chambers are locally made. The specifications of the engine are as follows:

Bore	: 98.4 mm
Stroke	: 114.3 mm
Number of cylinders	: 3
Maximum steam pressure	: 70 Kg/cm^2 (abs.)
Condenser pressure	: 0.25 Kg/cm^2 (abs.)
Maximum steam temperature	: 450^oC
Expansion ratio (adjustable)	: 1.25 (used)
Lubrication	: As in Lister engine
Lubricant	: Mobil oil XRN 1301C
Measured efficiency (Steam pressure 4.2 Kg/cm^2, temperature 425^oC)	: 21.9 per cent

7.2 Turbines

Turbine is a machine which converts enthalpy to kinetic energy and then to mechanical work. Generally steam[66] or organic vapour[67] turbine both based on Rankine cycle can be used to produce mechanical work using solar energy. The basic principles of construction for these two types of turbine are generally similar, but owing to the difference in characteristics of steam and organic vapours, several design deviations exist. A turbine in its simple form consists of the following four parts:

1. A set of nozzles, where the working fluid (steam or organic vapours) expands to increase the kinetic energy.

2. A rotor assembly consisting of some form of blades (or buckets) on a rotating shaft.

3. The blades or vanes or buckets which can either be moving or fixed type. The fixed type vanes are fitted with the casing and meant to direct the working fluid towards the moving vanes. The moving vanes are mounted on the rim of the rotor.

4. The casing which encases the rotor, vanes, shaft, bearings, glands, etc. and prevents the leakage of working fluid.

Turbines are basically of two types: impulse turbines and reaction turbines. But turbines are classified in several ways as follows:

1. By shaft or casing arrangement: It can be of a single casing, cross compound (two or more shafts not in line), tandem compound (two or more casings with the shaft coupled together in line).

2. By exhaust stages: It can be double flow or triple flow.

3. By class design: It can be on impulse or reaction principle.

4. By steam supply and exhaust conditions: It can be condensing or non-condensing type, automatic extraction type, or reheat type.

5. By direction of steam flow: It can be axial flow, radial flow, or tangential flow type.

6. Multi-stage or single-stage.

In the impulse type turbines, the entire pressure drop takes place across the nozzles or stationary elements, the flow through rotor blades then being substantially at constant static pressure. In the reaction turbines the entire pressure drop is equally divided between the stationary blades and the rotor blades. Multistage turbines often incorporate both impulse and reaction stages. The efficiency of an impulse turbine can be improved by lowering the condenser pressure which can be achieved by using more than one set of nozzles, blades, rotors in series. This process is known as compounding of impulse turbines which can be achieved by three ways:

1. Compounding for velocity,
2. Compounding for pressure,
3. Compounding for velocity and pressure.

More than three-fourth of the electric power produced in the world is generated using some kind of steam turbines. Steam turbines as large as 1000 MW capacity are presently feasible. For solar power generation also, steam turbines have been used. Hsu and Leo[68] developed a simple and inexpensive reaction turbine in which the steam is rejected from a nozzle to rotate a wheel at high velocity. The steam pressure used is 450 KPa producing 200 watts at an efficiency of 1.2 per cent only. Higher efficiencies were obtained when exhaust steam from nozzle was condensed to water in a condensor rather being discharged into the air.

Practically in all the central receiver solar power plants, steam turbines which are commercially available are used. In the 1-MWe experimental solar thermal electric power plant[66] installed at Sicily (Italy), 7-stage impulse turbine working at a steam inlet pressure of 6000 KPs at temperature 510°C with a flow rate of 5200 Kg/hr and exhaust pressure 6 KPa at 36°C is employed. The capacity of the turbogenerator unit is about 1 MWe.

Another steam turbine-generator unit with a capacity of 1.5 MWe was selected for the pilot plant[69] of central receiver solar thermal power plant in California. The turbine selected for the pilot plant is a tandem-compound, single flow, single automatic admission condensing industrial turbine with a rating of 1.5 MWe (gross) when exhausting at 17 KPa absolute. The steam conditions are 10000 KPa at 477°C, steam flow 14.2 Kg/sec.

Organic Rankine cycle turbines are preferred for solar power generation in the low and medium power levels, i.e. up to 1000 KW due to their high cycle efficiency. Due to high molecular weight of the organic working fluid, the vapours can give some kinetic energy ($1/2 \ mv^2$) at lower velocities compared to that when steam is used as working fluid. Because the organic fluids becomes superheated as they expand through the turbine, all re-generative heating can be accomplished with a re-generator located between the turbine exhaust and the condensor. Moreover, the vapour generator is to produce saturated vapour only and thereby avoiding the compexity of the vapour generator. But due to low enthalpy change during expansion, large flow rate of vapour through the turbine will be required for the same power output. This may be advantageous in small turbines only, since the blades of the turbine will be made larger bringing the nozzles nearer to full admission. Due to the large dimension and mass flow rate, the efficiency of the small turbine will be high. Moreover, due to positive slope of saturation curve for organic vapours, there will not be erosion of blades while expansion. As discussed earlier, many organic working fluids have been used in solar thermal power plants.

A survey of organic Rankine engines was conducted by Curran[58]. He collected data on 2150 Rankine engines which use 16 different organic fluids providing power in the range of 0.1 to 1120 KW and were operational for different durations up to September 1979. This data is correlated by Badr et al.[61] in the form of the power output with rotational speed (rpm) and observed that the design data collected or available is not sufficient to draw any specific conclusion except that low speed engines ($<$ 5000 rpm) are predominantly positive displacement engines providing power output up to 10 KW and turbines with high rotational speed ($>$ 5000 rpm) are used for meeting greater demands ($>$ 10 KW). The data as compiled by Curran[58] and in the format as presented by Badr et al.[61] is shown in Table 1. It is observed from the table that power output for positive displacement expanders rises with increasing speed whereas in case of turbine the behaviour is reversed. This data was collected from more than 20 leading manufacturers and observed that the major manufacture of organic Rankine turbine is Ormat Turbines of Israel, who have produced few thousands of such engines and the recommended working fluid is trichlorobenzene.

The concept of similarity can be employed to compare the performance of different types of expander. It is shown by Balje[70] that four parameters like specific speed, specific diameter, Mach number and Reynolds number can be employed to represent the maximum obtainable efficiencies and the optimum design geometry of turbines. The specific speed N_s is a measure of the rotational speed of the expander, for a given volume flow rate and a given enthalpy change through the expander. The specific diameter D_s can be a measure of the size of the machine. Reynolds number R_e represents the

TABLE 1 - SURVEY OF RANKINE CYCLE ENGINES (FROM BADR et al.[61])

Type	Expander Power output (KW)	Speed (rpm)	Energy source	Equipment driven	Working fluid	Maximum working fluid temperature (°C)	Number of engines operating	Total operational period
1	2	3	4	5	6	7	8	9
Single-stage radiation inflow turbine	1.7	35000	Solar insolation	Vapour compressor	R113	80	2	-
Single-stage radial inflow turbine	1.7	60000	Solar insolation	Vapour compressor	R11	95	4	-
Two-stage multi-vane	2.3	1625	Solar insolation; electricity[a]	Vapour compressor;[a] dynamometer	Fc-88	175	1	1000h[a]
Multi-vane	2.3	1200-1800	Electricity[a]	Vapour compressor;[a] dynamometer	R11	105	None at present	1200h[a]
Reciprocating	3.0	1800	Gas	Electricity generator	Cp-34	290	1	> 100h
Reciprocating	3.0	1800	-	Electricity generator; sweeper vehicle	F-85	290	3	>100h
Reciprocating	3.0	3600	-	None	R-22	230	1	> 100h
Single-stage turbine	3.0	70000	-	Electricity generator; vehicle	F-85	290	1	>100h
Single-stage reaction turbines	4.0	1200	Solar insolation; geothermal heat	Water pump; electricity generator	Tetra-chloro-ethylene	80	1	1000h

TABLE 1 (cont'd)

1	2	3	4	5	6	7	8	9
Screw	5-100	1500-1800	Solar insolation	Water pump; electricity generator	R11	95	70	4000h
Multi-vane	7.5	1800	Gasoline	Electricity generator; dynamometer	Aocohol/Water	340	-	1800h
Multi-vane	7.9	1625	Solar insolation electricity[a]	Vapour compressor; dynamometer[a]	FC-88	175	2	1000h[a]
Single-stage radial inflow turbine	12	30000	Solar insolation	Vapour compressor	R11	95	2	-
Single-stage radial inflow turbine	15	24000	Solar insolation	Vapour compressor	R113	80	7	750h
Turbine	15	42000	Solar insolation	Electricity generator	R113	110	3	500h
Single-stage radial inflow turbine	16	40000	Solar insolation	Vapour compressor	R11	150	1	25h
Single-stage radial inflow turbine	19	36300	Solar insolation	Irrigation pump; electricity generator	R113	165	1	600h
Helical screw	20	7500	Solar insolation; oil	Electricity generator	R114	200	5	1000h
Single-stage radial inflow turbine	20	20100	Solar insolation	Vapour compressor	T113	150	1	-
Single-stage impulse turbine	32	19400	Solar insolation; gas	Electricity generator	CP-25	300	1	2400h

TABLE 1 (cont'd)

1	2	3	4	5	6	7	8	9
Single-stage radial inflow turbine	34	11950	Solar insolation	Vapour compressor	R11	86	1	200h
Single-stage radial inflow turbine	35	5500	Solar insolation	Irrigation pump	R11	86	1	>100h
Single-stage radial inflow turbine	37	18000	Solar insolation	Vapour compressor	R11	95	2	-
Single-stage radial inflow turbine	37	30700	Solar insolation	Irrigation pump	R113	135	1	500h
Single-stage turbine	38	35000	Diesel engine exhaust gases	Automobile	E-50	315	3	>100h
Three-stage turbine	38	60000	Diesel engine exhaust gases	Automobile	F-50	315	1	>100h
Turbine	40	6700	Exhaust gases	Electricity generator	Tetra-chloro-ethylene	115	1	300h
Turbine	45	6700	Solar insolation	Electricity generator	Flutec pp3	280	1	-
Turbine	50	6600	Geothermal	Electricity generator	Tri-chloro-ethylene	70	1	-
Turbine	60	-	Steam produced by diesel engine exhaust	Electricity generator	R113	70	1	-
Single-stage radial inflow turbine	63.4	22400	Solar insolation	Vapour compressor	R113	135	2	100h

509

TABLE 1 (cont'd)

1	2	3	4	5	6	7	8	9
Reciprocating	112	1800	-	Automobile	F-85	315	1	>100h
Turbine	150	20000	Solar insolation; oil	Electricity generator	Cp-25	450	1	-
Radial inflow	335	9500	Steam produced by diesel engine exhaust	Electricity generator	R11	88	1	-
Single-stage impulse turbine	375	1800	R114 vapour from process	Compressor	R114	120^c	1	6 years
Single-stage impulse turbine	375	1800	R114 vapour from process	Compressor	R114	120^c	1	10 years
Six-stage turbine	450	12500	-	Electricity generator	F-85	290	1	>100h
Single-stage turbine	450	18000	-	Electricity generator	F-85	290	1	> 100h
Single-stage impulse turbine	600	9300	Waste gas	Electricity generator	Cp-25	240	5	-
Turbine	670	11100	Furnace exhaust	Electricity generator	F-85	290	1	-
Turbine	1000	-	Geothermal heat	Electricity generator	-	-	-	-
Single-stage reaction turbine	1050	4800	R114 vapour from process	Compressor	R114	120^c	1	3 years
Six-stage impulse turbine	1120	1200	R114 vapour	Compressor	R114	120^c	1	2 years

a = Laboratory test;   b = Experimental unit with simulated solar input;   c = Value estimated by Curran[58].

physical properties of the working fluid. The Mach number M is the ratio
of the velocity of the fluid to the acoustic velocity in the fluid. It
was observed by Barber and Prigmore[59], that only the two parameters,
i.e. the specific diameter D_S and the specific speed N_S can be used to
determine the performance of the expander and the other two parameters,
i.e. Reynolds number and Mach number, have only secondary effects on the
performance of expander. These two similarity parameters, i.e. specific
speed N_S and specific diameter D_S are given as:

$$N_S = \frac{NV^{1/2}}{(\Delta h)_{is}^{3/4}} \qquad (9)$$

$$D_S = \frac{D(\Delta h)_{is}^{1/4}}{V^{1/2}} \qquad (10)$$

where

$$
\begin{aligned}
N &= \text{expander rotational speed (rpm)} \\
V &= \text{expander exit flow rate (m}^3\text{/sec)} \\
(\Delta h)_{is} &= \text{adiabatic enthalpy drop across the expander (J/Kg)} \\
D &= \text{diameter of the expander (m)}
\end{aligned}
$$

The available performance data on expanders was used by Balje[70] who
used the similarity concepts and computed the optimal geometries and maximum
obtainable efficiencies for different types of expander. This information
was plotted by Balje[70] in the form of N_S-D_S diagrams. The data presented
by Balje was used by Barber and Prigmore[59] who plotted N_S-D_S diagram for
all expander types, and the same is shown in Fig.6. From this figure it is
seen that in different specific speed ranges there can be a particular type
of expander which can give better performance. It is observed that in the
low specific speed range of 30-100, the performance of radial turbines is
similar to full admission axial turbines. The optimal performance chart for
axial flow turbines showing efficiency as a function of N_S and D_S is shown[59,70]
in Fig.7.

Tabor and Bronicki[21] examined 16 different organic fluids for Rankine
cycle organic vapour turbine and developed a 2 KW monochlorobenzene turbine
operating at 150°C and at 18000 rpm using a 6 to 1 reduction gear. The overall
efficiency of converting heat to shaft power was 10 to 15 per cent. Tabor also
built 3.8 KW solar operated organic vapour turbine and demonstrated the same
for pumping of water in Rome in 1961. Efficiency calculations are also
made for a 10 KW turbine with 10 nozzles and 100 per cent admission using

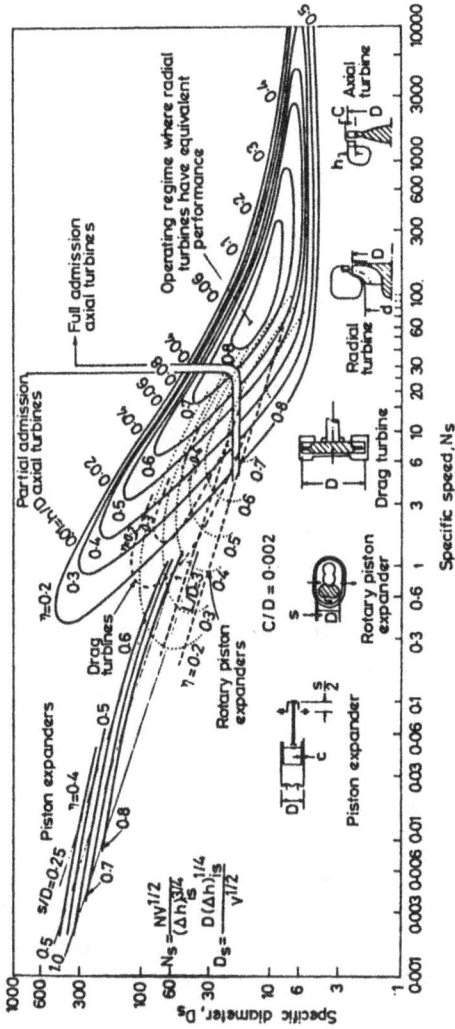

FIG. 6 – PERFORMANCE CURVES OF VARIOUS TYPES OF EXPANDERS

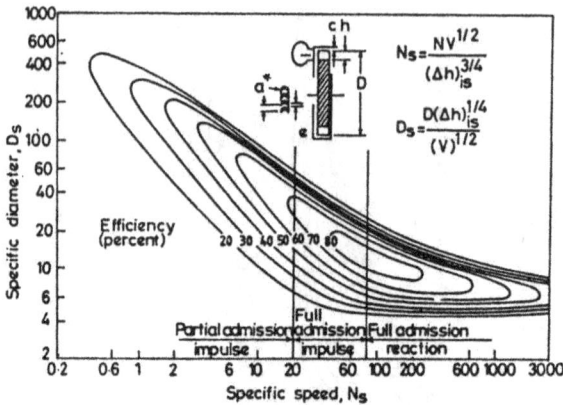

$$N_s = \frac{N V^{1/2}}{(\Delta h)_{is}^{3/4}}$$

$$D_s = \frac{D(\Delta h)_{is}^{1/4}}{(V)^{1/2}}$$

FIG.7 - PERFORMANCE OF AXIAL FLOW TURBINES

monochlorobenzene as the working fluid. In the temperature range of 160-180°C the efficiencies were found to be 15-20 per cent.

The Barber-Nichols Engineering Company of Colorado, USA, has developed[71] an organic Rankine cycle power conversion subsystem in 1980 under a subcontract from Ford Aerospace and Communications Corporation(FACC). The working fluid of organic vapour turbine is toluene which is heated to about 371°C in the receiver of a parabolic disc through a secondary fluid Caloria HT-43 and is expanded through a single stage axial flow turbine. The exhaust vapour from the turbine is passed through an integral recuperator or regenerator and into a forced air cooled condenser which forms the outer annuls of the converter assembly. This generator is a heat exchanger which transfers a part of exhaust waste heat to the low temperature incoming fluid. An alternator of permanent magnet type which is of high speed is directly coupled to the turbine. Several simulation tests are conducted on this power converter and found that when the engine input is 56 KWt, the power output is about 16.5 KWe. The engine is capable of operating up to 83 KWt and the net efficiency (heat to electric engine) was observed to be about 25 per cent. Barber and Prigmore[59] described the working of a 19 KW Rankine cycle engine developed by Barber-Nichols Engineering Company which was installed at Willard, New Mexico, in April 1977. The Rankine power system is sketched in Fig.8. The working fluid is Refrigerant-113 which is vapourized at a temperature of 160°C and supplied to the turbine. The collector working fluid

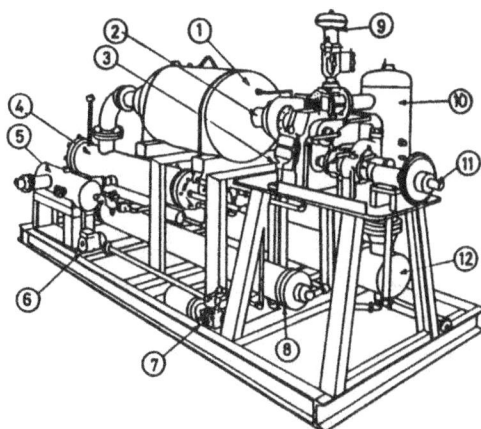

(1) Regenerator
(2) Turbine
(3) Gear box
(4) Condensor
(5) Float tank
(6) Boost pump

(7) Startup pump
(8) Preheater
(9) Speed control valve
(10) Demister
(11) Output shaft
(12) Boiler (heat exchanger)

FIG.8 - ORGANIC RANKINE CYCLE ENGINE (19 KW)

is Caloria HT-43 which transfers its heat through a heat exchanger in the boiler to R-113. The turbine shaft speed is 36300 rpm and the output shaft speed is 1730 rpm. The predicted turbine efficiency is 75 per cent and the pump efficiency is 40 per cent. The overall cycle efficiency is calculated to be 15.3 per cent.

Badr et al.[61] in their recent review pointed out the limitations of the positive displacement type engines and turbines particularly in the low power range. It is concluded that for low power requirements the organic Rankine cycle engines like the reciprocating piston, the rotary screw (i.e. the helical system) and the rotary multivane expander (MVE) will show high efficiency compared to turbines.

7.3 Stirling Engine

Stirling engine as discussed earlier has the highest theoretical efficiency reaching to the Carnot engine due to isothermal heat addition and removal during expansion and compression, and through isothermal regenerative heat addition and removal at constant volume[78]. But, unfortunately, practical Stirling engines suffer from many defects and are, therefore, not a good approximation to the theoretical engines.

The Stirling engine was first invented by Robert Stirling of Scotland in 1816 and used by Erricson[72] in 1870. These engines were earlier called air engines up to 1950's. From 1860 to 1920 several thousand air engines with a low efficiency (up to 3 per cent) were made in sizes up to 4 KW. The N.V. Philips Company of Netherlands in the year 1937 revived the interest in air engines and used modern engineering and concepts and obtained an efficiency of 30 per cent. Dr. R.J. Meijer used helium or hydrogen as a working fluid and reached to an efficiency of 38 per cent and called these air engines as Stirling engines. Today several firms like General Motors and Ford Motors of USA, United Stirling Engines of Sweden, Sun Power Inc. of USA and a firm in Germany are making Stirling engines of various sizes[73].

Farber and Prescott[74] studied a Stirling hot air engine and obtained an efficiency of 9 per cent at 100 rpm with a brake horse power of about 150 watt. The principle of a hot air engine is shown in Fig.9.

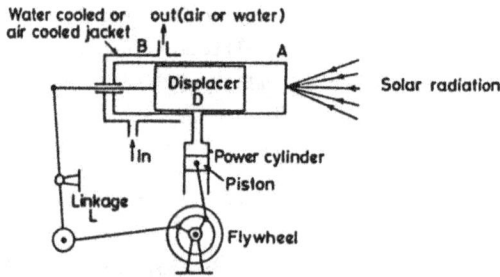

FIG.9 - PRINCIPLE OF STIRLING ENGINE

The focused sunlight heats the air contained in the right side of the cylinder, which then expands and forces down the piston P turning the flywheel clockwise. When the flywheel turns it moves the displacer D towards left in the cylinder through linkage L. During the upstrock of the piston, the displacer moves to the right leaving the hot air in the left section of the cylinder B from where the heat is taken away by water-cooled jacket or air-cooled jacket. The simplified version of the four stages of hot air engine is shown in Fig.10.

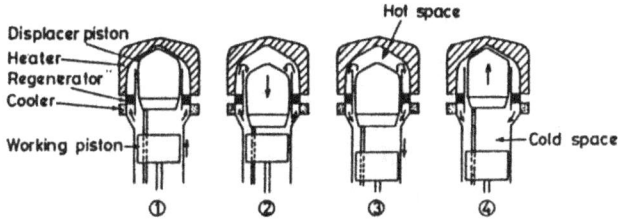

FIG.10 - STIRLING ENGINE CYCLE

In process one, the air is first compressed in the cold space of the engine
and then by means of a displacer piston forced to flow through a regenerator
into a hot space surrounded by the heat source. In the second process, the
air gets heated and at a high pressure passes through the regenerator to
the cold space where the high pressure air expands and activates the piston.
The actual p-v diagram of this engine is shown in Fig.11 which is quite
different from the ideal Stirling cycle shown in Fig.2(c).

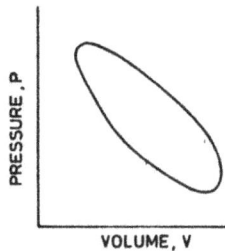

FIG.11 - ACTUAL STIRLING CYCLE

The main limitation with the Stirling hot air engine is the poor heat
transfer across the head of the cylinder, due to its small area and of
metallic construction. An improvement was made by Finkelstein[75] and
also by Trayser and Eibling[76] who used transparent quartz windows for
the cylinder head to focus the solar radiation directly inside the engine.
Thus the solar radiation is absorbed directly by the inside air without any
loss. An efficiency of about 32 per cent was reported in this improved
Stirling engine.

A good review of solar Stirling engine is made by Martin[77] and who
in his recent paper[73] classified the Stirling engine into three main
engine types: the alpha type, the beta type and the gamma type.

Basically solar operated Stirling engines are of two types[78,79], e.g. Free Piston Stirling Engine (FPSE) and the Kinetic Stirling Engine (KSE). The free piston Stirling engine was first invented by William Beale[80] and is basically a thermally driven mechanical oscillator and with linear alternator hermetically sealed. In this arrangement the piston remains stationary and the displacer moves under the influence of pressure differential between the working space and the bounce space in the cylinder. The movement of the cylinder due to pressure differential performs work. Since the entire pressure enclosure is to move as a unit, the free piston Stirling engine eliminates the need for high pressure shaft seals or lubricants. The free piston Stirling engines have high mechanical efficiency but suffers from many problems[79] such as reproducible and tuning performance, starting and power take off. Because of the required masses and resulting inertial effects, the free piston Stirling engines are not supposed to be practical for large power demands ($>$ 100 KWe). The kinematic Stirling engines require a mechanical drive external to the gas cycle, with a working seal interface[78]. These engines are bulky and, therefore, not suitable for mounting at the focus of solar concentrating collector. Moreover, due to their poor mechanical efficiency, the overall performance gets reduced.

The Stirling engine as discussed above differs from gasoline engine in two ways: firstly, the working fluid (gas) is recycled and secondly, the heating and cooling takes place by heat transfer successively. The Stirling cycle differs from Rankine cycle in the sense that the working fluid in the Stirling cycle remains only in the gaseous phase which is generally air, hydrogen or helium. The only disadvantage is that it operates at high temperatures ($> 600^{\circ}$ C).

Mechanically[81] there are two classes of Stirling engines: single acting (displacer) and double acting. In the single acting unit, two pistons are used for each power unit. In the double acting type, one piston per cylinder is used which acts as compressor, expander and displacer by interaction between neighbouring cylinders. There is no limit for the number of cylinders but generally for a single power unit, 3 to 7 cylinders are used. Most commonly, four cylinders are used which are arranged either in line or axially in a square. The Stirling engine has two parts, one external and another internal. The external system supplies heat either from sun or by combustion of fuel to the engine heater. The internal system is filled with

a working gas like air or helium or at an elevated pressure and consisted
of two variable volumes and three heat exchanger called the heater, regenerator and cooler.

The Sun Power Inc., Ohio (USA), which was founded by William Beale
who is the inventor of Free Piston Stirling Engine, started manufacturing
it since 1971. As mentioned earlier, these free piston Stirling engines
have several advantages over the kinetic Stirling engines. The firm has
developed[82,83] three models: Model-10 (10 watt capacity), Model-SD100
(100 watt capacity) and Model-RE1000 (1150 watt capacity). The working
of the free piston engine, as described by Taylor[83], is shown in Fig.12.

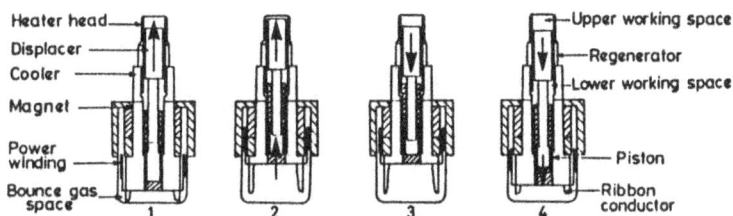

Heater head — Displacer — Cooler — Magnet — Power winding — Bounce gas space — Upper working space — Regenerator — Lower working space — Piston — Ribbon conductor

FIG.12 - WORKING PRINCIPLE OF FREE PISTON STIRLING ENGINE

It is shown in part 1 of the figure that when the gas pressure in the lower
space pushes the displacer up, the gas in the upper hot end space is forced
through the heater head and regenerator into the cold lower space. The
displacer continues to move up till the pressure in the bounce space equals
the pressure of the cool working gas thereby reducing the total working gas
volume and compressing the gas as shown in part 2. Now the head of the
displacer gets heated, the pressure force difference forces the displacer
down as shown in part 3 of the figure. It pushes the working gas through
the regenerator and the heater head thereby heating the gas. This hot working
gas, as shown in part 4, pushes down the displacer via the remaining cold
end gas the power piston. Now the gas pressure in the upper portion drops
and the gas in the lower portion pushes the displacer up again and thereby
completing the cycle. Extensive data have been collected on Stirling engine
Model-RE1000 and its output power is plotted verses piston stroke for different
heater head temperatures as shown in Fig.13. Details of a free piston Stirling
engine is shown in Fig.14.

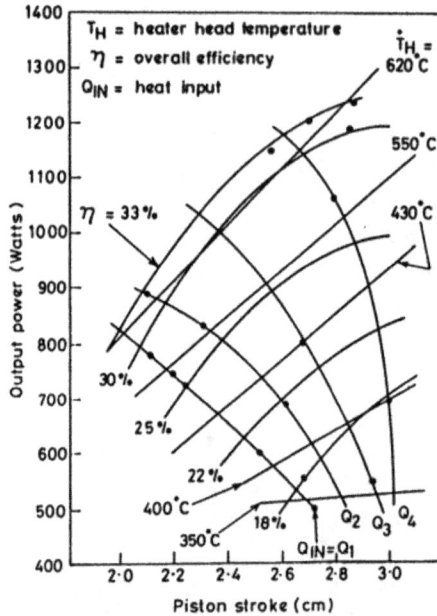

FIG.13 - PERFORMANCE OF FREE PISTON STIRLING ENGINE
RE-1000

Mechanical Technology, Inc., of USA analysed several free piston Stirling
engines. Their design calculations for a 50 KWe Stirling engine with linear
alternator have shown an efficiency of about 38 per cent.

The firm United Stirling, located in Malma, Sweden (USS), is one of the
largest and most experienced[84,85] one in the field of Stirling engines.
The first commercial Stirling engine of 150 KWe capacity, with 4 cylinder
in-line engine with a rhombic drive mechanism, was made in 1971 which was a
single acting displacer type engine having two piston per cylinder. Later
in 1972 the United Stirling decided to work only on double acting systems.
Since then this firm has made hundreds of Stirling engines of different
applications. With the interest on solar parabolic dish for power production,
the interest on Stirling engines was revived and United Stirling started
concentrating on compact and efficient engines for parabolic disc.

On the request of Jet Propulsion Laboratory, USA, the USS tested the
performance of their Stirling engine p-40 in the inverted position so that
it can be kept at the focus of parabolic dish concentrator. This engine ran

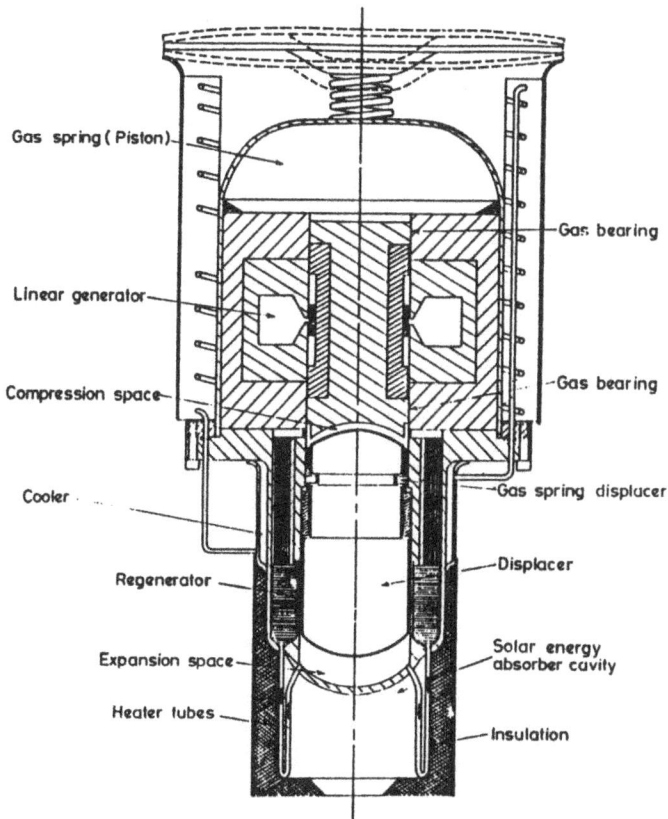

FIG.14 - SPUNPOWER 1 KILOWATT ENGINE SPIKE
(COURTESY SUNPOWER INC., OHIO)

successfully[71] for 3 hours at 720°C, 11 MPa at 1500 rpm producing about
16 KWe. This engine had later given an efficiency of 38.1 per cent. This
p-40 Stirling engine, known[85,86] as 4-95 engine, is a four-cylinder, 95 cc
displacement, double acting, twin crank drive machine. Its details are
described by Christer in an early publication[81]. In this engine, the
four cylinders are arranged in a square. Eight regenerators and coolers
are placed in a ring outside the cylinders which minimise the dead volumes
in the hot end manifolds as well as dead volumes in the ducts connecting the
cooler with the compression space. The heater is suitable for direct solar
energy absorption and is a two pass cross flow turbular type with rotational
symmetry.

Another 85 KW capacity Stirling engine 4-275, called p-75, has a
maximum speed of 1800 rpm. The displacement per cylinder in this case
is 275 cc. Studies are also in progress on designing of 400 KW and 500 KW
stirling engines suitable for solar power generation based on the concepts
of 4-95 and 4-275 engines.

The general equation to know approximately the power output (W in watts)
as given by Martini[73] is as follows:

$$W = 0.035 \ fpm \left(\frac{Te-Tc}{Te+Tc} \right) \left(\frac{\pi}{2} \ \frac{Vc \ Ve}{Vm} \right) \sin F \qquad (11)$$

where

W = power output (watts)

f = operating frequency (Hz)

P_m = mean cycle pressure (bar)

Te = expansion space (heater) temperature (K)

Tc = compression space (cooler) temperature (K)

Vc = volume swept by power piston (cm^3)

Ve = volume swept by displacer (cm^3)

Vm = gas volume of mid-stroke of power piston or
of both power pistons (cm^3)

F = phase angle between displacer and power piston motion,

7.4 Brayton Engines

As discussed earlier, the engines based on Brayton cycle can be of large
capacity, more efficient (> 30 per cent) compared to Rankine engines, but
require high temperature (> 600°C) for operation. The cycle consists of
adiabatic compression, constant pressure heating and adiabatic expansion.
Work is done by the hot gas while expansion, which is more than the work of
compression. By using a regenerator in the exhaust of the gas turbine,
intercooler in the compressor and reheating the working fluid during expansion,
the performance of the turbine can be improved.

Although the thermal efficiency of a Brayton cycle mainly depends on
compressor ratio ($R = P_2/P_1$), the turbine inlet temperature, and the parasitic losses (like efficiency of turbine and compressor), but for an ideal
cycle, the thermal efficiency depends only on pressure ratio.

$$\text{Thermal efficiency} = 1 - \left(\frac{1}{R} \right) \left(\frac{\gamma-1}{\gamma} \right) \qquad (12)$$

where γ is the ratio of specific heat of air at constant pressure to constant volume. Figure 15 shows the effect of compressor pressure ratio on the thermal efficiency of Brayton cycle at different turbine inlet temperatures.

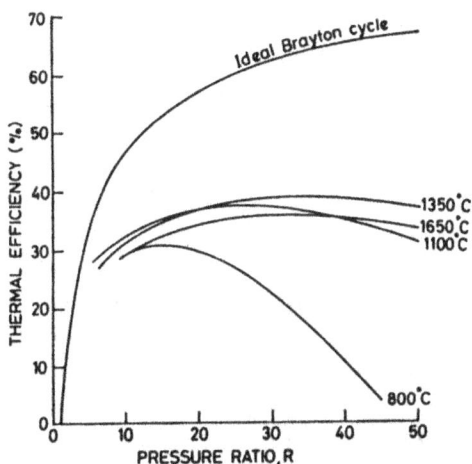

FIG.15 - BRAYTON CYCLE EFFICIENCY FOR DIFFERENT PRESSURE RATIO

The actual Brayton cycle differs from the above ideal cycle because of:

(i) the air properties (K, C_p) are not constant over the range of operating temperatures;

(ii) compression and expansion processes are not frictionless and take place with increase in entropy;

(iii) the internal losses are difficult to control;

(iv) the mass of gas flowing through the turbine is more than the mass of air flowing through the compressor;

(v) the specific heat of combustion gas is higher than air.

If the Brayton engine operates at a compression ratio less than the optimum value as shown in Fig.15, then the temperature of the turbine exhaust gas becomes more than the compressor exhaust. In such a case the performance can be improved by using a regenerator transferring some exhaust heat to the compressed air stream. There are two general types of gas turbine power plants as shown in Fig.16. In the open cycle plant, Fig.16(a), the fuel is injected into the combustion air subsequent to compression and the resulting gas expands in the turbine and finally exhausted into the atmosphere. The open cycle plant is generally preferred due to its simplicity, no need of cooling unit,

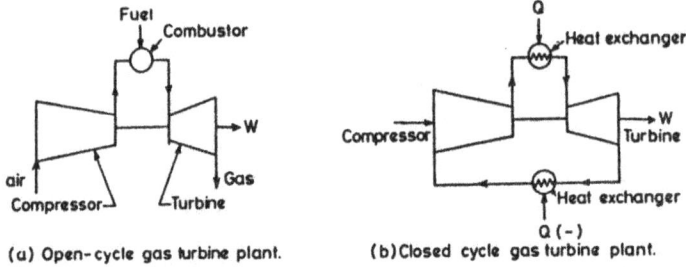

(a) Open-cycle gas turbine plant. (b) Closed cycle gas turbine plant.

FIG.16 - SIMPLE GAS TURBINE CYCLES

simple controls, and possible design for high power-weight ratio. In the closed cycle plant, Fig.16(b), the working gas is continuously recycled. Fuel is burned externally to the system, and the energy liberated in the combustion reaction is transferred as heat to the circulating gas. The advantages of a closed cycle plant are: low grades of fuels may be burned, clean working gas, ability to control the density of working gas. The density control helps in changing the power output without changing the compression ratio and turbine inlet temperature. The closed cycle power plant is suitable with nuclear reactors since fluids other than air like helium can be used at high temperatures. The disadvantages of closed cycle being the large size and additional cost of heat exchangers.

The simple gas turbine cycle with intercooling, reheat and exhaust heat exchange described by Gupta and Prakash[87] is shown in Fig.17 along with the T-S diagram showing various positions of the cycle.

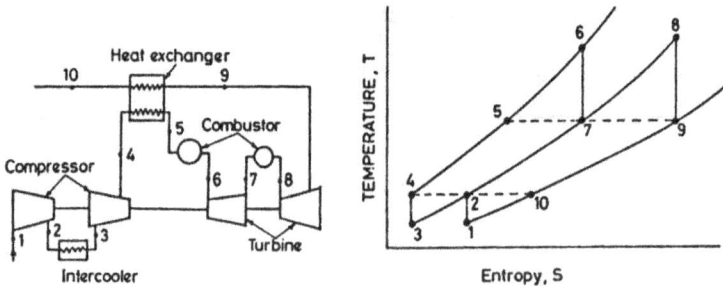

FIG.17 - BRAYTON CYCLE WITH DIFFERENT MODIFICATIONS

The thermal efficiency of this cycle is given as:

$$\text{Thermal efficiency} = \frac{(T_6-T_7)+(T_8-T_9)-(T_2-T_1)-(T_2-T_1)-(T_4-T_3)}{(T_6-T_5)+(T_8-T_7)} \qquad (13)$$

The effect of various modifications in the simple Brayton cycle on the work output and thermal efficiency are   also computed by them for a turbine with pressure ratio of 4, T_{max} = 684 K, T_{min} = 288 K and P_1 = 1 atmosphere. The comparison is given[87] in Table 2.

TABLE 2 - EFFECT OF VARIOUS MODIFICATIONS IN THE SIMPLE BRAYTON CYCLE TURBINE

Modification	Percentage effect on efficiency	Percentage effect on output
Heat exchanger	+ 50.0	nil
Intercooling	- 6.5	+ 10.2
Reheat	+ 10.4	+ 24.5
Reheat + heat exchanger	+ 66.7	+ 24.5
Intercooling + heat exchanger	+ 68.0	+ 10.2
Reheat + intercooling	- 18.2	+ 34.7
Reheat + intercooling + heat exchanger	+ 80.0	+ 34.7

The main components of the Brayton power plant are as follows:

1. Turbine - All gas turbines are axial flow type except in small sized installations where radial inward flow type turbines are used. Although pressure in the turbine is quite low but due to very high gas temperature special kind of cooling arrangements are to be made and special materials for turbine blades and other hot parts are to be employed. High temperature alloys like nickel and cobalt are used. Research is in progress for special kind of high temperature coatings and use of ceramic materials for turbine blades etc.

2. Regenerator - The purpose of the regenerator is to transfer a part of heat from the turbine exhaust to the air stream from the compressor resulting in increase in efficiency, increase in volume and weight of the plant, and increase in cost of the system. Generally, Rotary regenerators which show high performance and of low weight are preferred. The regenerator should be

able to handle large amount of working fluid with low pressure drop and with rapid large temperature change. The intercoolers have also similar problems.

3. Combustor - The main purpose of the combustor is to bring the gas at constant uniform temperature with little loss of pressure. Combustors of very high rates are available (2×10^{11} J/hr.m^3 atm.).

4. Compressor - Generally axial-flow compressors are employed because of their high efficiency and capacity. Compressors which can handle air capacity of 380 m^3 are available.

The open cycle Brayton engine with parabolic dish shows higher efficiency and is cost effective compared to the organic Rankine cycle engines. Under a contract from Lewis Research Centre, USA, the Garrett Airesearch of Torrance, California, USA, is building several advanced gas turbines suitable for parabolic dish. The turbine is made of ceramic materials[88] to operate in the temperature range of 1100-1400°C and all metallic components to operate in the range of 700-900°C. The first generation engine developed by Garrett is shown in Fig.18, which consists of two parts,

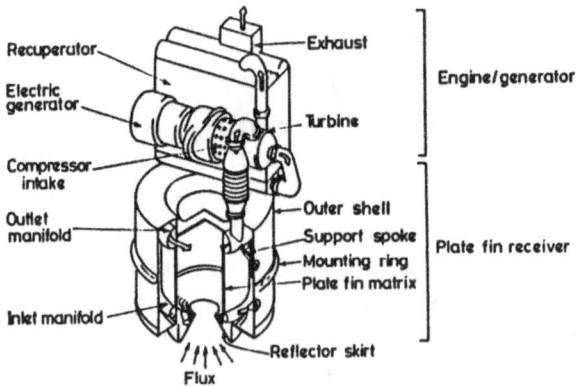

FIG.18 - AIR-BRAYTON ENGINE, GENERATOR AND RECEIVER

a receiver and an engine/generator. The engine is a recuperated[71,88], open cycle of 20 KW capacity with overall efficiency (heat to electricity) of 30 per cent. In the second generation engines, where ceramic materials are used and which will operate at higher temperatures, higher efficiencies are

expected. Under a contract from JPL (Jet Propulsion Laboratory), the
Garrett also built an all-metal receiver suitable for the above machine.
The receiver is designed for an outlet temperature of 815°C and inlet
temperature of 565°C. Design and performance data of Brayton engines
with maximum power less than 97 KW is compiled by Fujita et al.[89]
from the publications made by Rockely[90] and Helins[91]. The data is
reproduced in Table 3.

TABLE 3 - COMPONENT EFFICIENCY OF BRAYTONE ENGINES

Component	Current Technology (all metal)		Advanced Technology (ceramic)	
	CCPS-40	Solar	Garrett	Allison
Turbine inlet temp. (°C)	816	816	1371	1178
Pressure ratio	1.9	2.7	5.0	4.5
Recuperator effectiveness	0.90	0.94	0.93	-
Compressor efficiency	0.76	0.76	0.80	0.80
Turbine efficiency	0.86	0.86	0.87	0.83
Loss factor	0.94	0.90	0.91	-

8.0 SOLAR POWER PLANTS

Several solar power plants ranging from a few KWe to 10 MWe are built in
many countries for field studies. A few of them are very successful and are
competitive to conventional power plants. A contract was given to Solar
Energy Research Institute, Colorado, USA, by US Department of Energy to
conduct a comparative analysis and ranking of eight generic solar thermal
systems in the 0.1 to 10 MWe capacity range[92]. A simple optimization
procedure for solar thermal electric power plant design is discussed by
Bohon and Lavy[93]. This optimization technique is based on suboptimization
in which the performance (energy output) to each power plant selected for
the study is maximized and then the best design via differential cost estimates
(minimum cost of electricity) is selected. It is not possible to discuss
all the solar thermal power plants and the optimization procedure, but a few
typical power plants which differ in concepts or size are discussed briefly
in this section.

8.1 Coolidge 150 KWe Power Plant (Ref.33)

The Coolidge 150 KWe solar thermal power plant[32-34],which is operational since November 1979, is a joint effort of University of Arizona, Sandia Laboratory, US Department of Energy, Acurex Corporation, Sandstrand Coporation and Sullivan and Masson Consulting Engineers. This plant was constructed to study the feasibility of using solar energy for driving irrigation pumps. The plant was installed at the Dalton Cole farm, south of Coolidge, Arizona.

The Coolidge solar thermal electricity plant, as shown schematically[33] in Fig.19, consists of an array of solar collectors, thermal energy storage

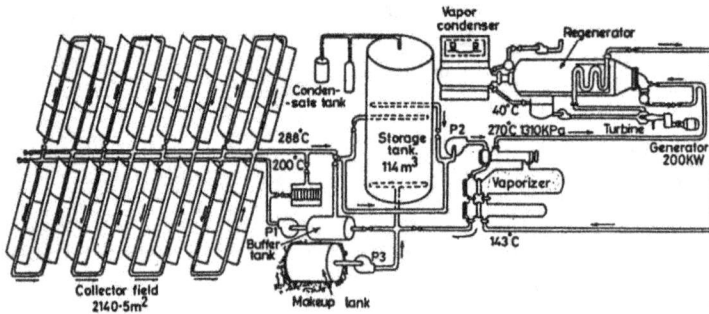

FIG.19 - FLOW DIAGRAM OF COOLIDGE 150 KWe SOLAR POWER PLANT

unit, and a power conversion subsystem. The details of the plant are listed in Table 4. The collector field with a total area of 2140.5 m^2 of line-focusing parabolic trough collectors, manufactured by Accurex Corporation, California, is arranged in 8 north-south oriented loops, each containing 48 collectors. Each collector trough is of about 1.8 m wide and 3 m long and originally had aluminium reflective surface which was later (spring 1981) laminated with aluminium acrylic film (FEK-244) to improve their performance. The collector receiver tubes, located at the solar collector focus, are coated with black chrome selective coating and surrounded by a pyrex glass tube. The concentration ratio of the collector receiver system is about 36. A heat transfer oil, Caloria HT-43, is pumped through the receiver tube by a pump at a controlled flow rate such that the outlet oil temperature reaches to 288°C.

TABLE 4 - SUMMARY OF THE 150 KW COOLIDGE SOLAR THERMAL POWER PLANT
(FROM LARSON[33])

1. Collector

Type	:	Parabolic trough collectors (1.8 m to 3 m)
Reflector material	:	Polished aluminium, aluminized mylar (FEK-244)
Number and area of collectors	:	384 collectors in 48 groups, 2140.5 m^2.
Orientation of collectors	:	N-S axis
Receiver coating	:	Black chrome
Concentration ratio	:	36
Collector fluid	:	Caloria HT-43
Temperatures	:	Inlet temperature = 200°C Outlet temperature = 288°C
Design conditions	:	q_i = 600 W/m^2 $\dot{m}^i$ = 7167 Kg/hr system efficiency = 38.6 per cent

2. Storage

Type	:	Stratified liquid (thermocline) (sufficient for 5 hrs operation)
Size	:	4.16 m diameter and 14.93 m high tank (114 m^3 usable storate)
Storage medium	:	Caloria HT-43
Storage temperature	:	200°C to 288°C
Insulation thickness	:	30 cm (Fibre glass)

3. Cooling System

Type	:	Vapour condenser
Water (makeup)	:	2270 litres/hr
Condensing temperature	:	40°C

4. Power Generation

Type	:	Organic Rankine cycle
Working fluid	:	Toluene
Gross efficiency	:	20 per cent

Heated caloria at 288°C is returned to the top of the 114 m^3 insulated
storage tank which is 4.16 m in diameter and 14.93 m high and provides
sufficient energy to operate the power conversion subsystem for more than
5 hours. A thermocline separates the heated caloria at the top of the tank
from the cooler caloria at the bottom of the tank.

The power conversion unit consists of a heat exchanger transferring
heat from heated caloria to the working fluid toluene, single stage impulse
turbine made by Stanstrand Corporation, synchronous generator for generating
electric power, and evaporative cooling tower for condensing the toluene.

Thus basically there are three closed heat transfer loops. In the first
loop warm caloria from the bottom of the storage tank is extracted, circulated through the receiver tubes of the collector field and the heated caloria
is returned at the top of the storage tank. In the second loop hot caloria
from the top of the tank is extracted, circulated through the vaporizer heat
exchanger and returned to the bottom of the storage tank. In the third heat
transfer loop, the high pressure vaporized toluene from the heat exchanger
is sent to the turbine to produce shaft work, the vapour expands which is
condensed in an evaporative cooling tower and then sent to the inlet of the
vaporizer heat exchanger for further vaporization.

An automatic control subsystem monitors and controls the tracking and
collection of solar energy, flow rates of caloria and storage of heat, flow
rates of caloria and heat exchanger and flow rates of toluene in the power
conversion unit, and the generation and supply of electric power. The
control unit also protects the solar thermal electric power plant arising
due to system related anomalieis or due to natural means. A natural gas
fired auxiliary heater is also provided for testing purposes in case of low
insolation values.

This plant was operated daily except during periods of collector testing,
equipment modification activities and other breakdowns and repairs. During
the years 1980, 1981 and 1982, the collector system operated 89, 93 and 98
per cent of the possible operating hours respectively. The power conversion
subsystem operated about 90, 97 and 97 per cent of the possible operating hours
during the year 1980, 1981 and 1982 respectively. Daily data on available
solar energy (compiled only that direct radiation which is more than 300 W/m^2),
collected thermal energy,and electrical energy generated have been compiled
for 1980, 1981 and 1982.

It was found that average monthly collection efficiency ranges from
7 per cent in winter to 27 per cent in spring and fall to 35 per cent in
summer. Some collector efficiency tests are conducted on clear days near
winter and summer solastrice; and spring and autumnal equinex. The total
energy produced during the years 1980, 1981 and 1982 is 114930, 163410 and
178030 KWh respectively. This increase in electric power production is due
to some equipment and collector improvements and operating experience.
The maximum electrical power was produced in June 1982, which was 27350 KWh
and 17000 KWh in September and only 3000 KWh in December. On a single
clear June day in 1982, the sytem produced about 1300 KWh electricity.
The average thermal to electrical energy conversion efficiency ranges from
9 to 12 per cent in winter months and 12 to 18 per cent in summer months.
During January 1980 tests, the thermal to electrical conversion efficiency
at 200 KW design point was 19.7 per cent. The parasitic power requirement
(for running various pumps, auxiliary equipment and cooling tower operation
etc.) was about 24 KW thereby showing a net cycle efficiency of 17.3 per cent
for power conversion subsystem.

8.2 Solar Thermal Electric Generation Plant of 500 KWe at Almeria, Spain
(Ref.94)

A 500 KWe capacity solar thermal electric plant was commissioned
[37,38,94,95] on July 31, 1981 which is built by an international consortium
and sponsored by nine countries and operated by Spanish Utility, Sevillana
Electricidad y gas, under the direction of DFVLR (Space Agency) of Germany.
The plant was installed on the Spanish platform solar in Almeria, Southern
Spain with the specific objective of generating electric power using solar
energy for supply to established electric grid system or in communities
where electricity is not available and difficult to supply by conventional
means.

The simplified flow diagram of the solar thermal plant is shown[94] in
Fig.20 and its details are listed in Table 5. The plant utilizes two sets
of line focus collectors; 2674 m^2 Accurex (USA) model 3001 type E-W oriented
single axis tracking collector and 2688 m^2 MAN (Germany) model 3/32 type two
axis tracking type collector, a stratified liquid storage tank with a working
volume of 114 m^3 and Santotherm 55 heat transfer oil as the storage media;
oil to water Baeltz steam generator; and a stal-laval condensing steam turbine
generator. Apart from the above main subsystems there are many other[37] systems

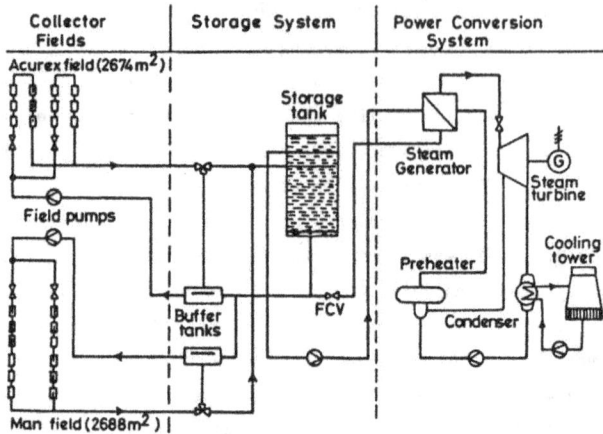

FIG.20 - FLOW DIAGRAM OF 500 KWe SOLAR THERMAL POWER STATION AT ALMERIA, SPAIN

like: an automatic control system, nitrogen ullage system to protect hot oil from fire, water treatment plant, solar collector cleaning system, a computer for master control functions and data collection, a constant power supply for protection, etc. The whole system is designed around three heat transfer loops. In one loop, cold heat transfer oil is extracted from the bottom of the storage tank circulated through the collector field and returned to the top of the storage tank. In the second loop, hot oil is extracted from the top of the tank, circulated through the oil to water heat exchanger unit and returned to the bottom of the storage tank. In the third loop high pressure steam is supplied to the steam turbine where it is expanded and cooled in a cooling tower and supplied to the inlet of the steam generator unit.

As described earlier, the collector field consists of two subfields and each field can be operated separately. One field consists of line focus parabolic trough type collectors oriented east-west with one axis tracking (Model 3001) made by Acurex Solar Corporation, California (USA) arranged in 10 parallel loops with each loop consisting of four groups of Acurex 12-module collector, model 3001. These collectors use extra thin glass reflectors (0.6 mm) on a metal support with a reflectivity of 0.94 and intercept factor of 0.95. The second field consists of two axis tracking type of collectors (model 3/32 Helioman) made by a German firm MAN and arranged in 14 parallel loops with each loop having 6 collectors. These

TABLE 5 - SUMMARY OF 500 KWe DISTRIBUTED COLLECTOR SOLAR THERMAL ELECTRIC PLANT
AT ALMERIA, SPAIN

1. Collector

     Accurax Model 3001

Type	:	Parabolic trough collectors (1.83 m x 3.05 m)
Reflector material	:	Glaverbel thin glass reflector (0.6 mm)
Number and area of collectors	:	480 collectors in 10 parallel loops with each loop having 4 groups of 12 module collectos (2674 m^2)
Orientation of collectors	:	East-west oriented with single axis tracking
Land use factor	:	0.27
Receiver coating	:	Black chrome
Concentration ratio	:	35.5
Collector fluid	:	synthetic oil - santotherm 55
Temperatures	:	Inlet temperature = 225°C Outlet temperature = 295°C

     MAN Collector Helioman Model 3/32

Type	:	Parabolic trough collectors (5.16m x 7.96m)
Reflector material	:	Glass reflector (3 mm)
Number and area of collectors	:	Azimuth orientation with two axis tracking
Land use factor	:	0.32
Receiver coating	:	Solartex selective coating

2. Storage

Type	:	Stratified liquid (thermocline) 4.2 m diameter and 15 m high tank (114 m^3 working fluid)
Storage media	:	Santotherm 55 heat transfer oil
Capacity	:	0.8 MWhe
Storage temperature	:	225°C-295°C
Working fluid	:	Water (steam)

3. Steam Generator

Maximum steam output	:	4230 Kg/hr
Steam pressure	:	2870 KPa
Steam outlet temperature at the oil inlet temperature of 295°C	:	285°C

4. Power Generator

Type	:	Multistage condensing axial flow steam turbine generator (Stal-Laval steam turbine)
Working fluid	:	Water
Inlet steam pressure	:	2500 KPa
Inlet steam temperature	:	283°C
Inlet steam flow	:	3767 Kg/hr
Exhaust steam pressure	:	7 KPa
Exhaust steam enthalpy	:	2277 KJ/Kg
Exhaust steam flow	:	3215 Kg/hr
Generator rating	:	713 KVA

5. Power at Design Point

Solar insolation	:	4933 KW
Thermal collection	:	2580 KW
Gross electric	:	577 KW
Net electric	:	500 KW
Thermal/gross electric efficiency	:	22.4 per cent
Thermal/net electric efficiency	:	19.4 per cent
Insolation/net electric efficiency	:	10.1 per cent

6. Cooling Conditions : Wet cooling
81 m^3 per day cooling water

7. Data acquisition System and Weather Station : Pyrheliometer for insolation temperature
Wind speed and direction
15 seconds intervals
Averaging to 5 minutes intervals

---

collectors use self-supporting glass troughs of 3 mm thickness with a reflectivity of 0.86 and intercept factor of 0.97.

Energy is stored in an insulated tank, 4.2 m inside diameter and about 15 m high, filled with Santogherm-55 heat transfer oil with a layer of N_2 at the top to protect the oil from oxidation. Two buffer tanks, one for each collector field, are provided for the circulation of oil at start-up. Oil make-up tank along with a pump is also provided to compensate any oil loss. For each collector field, a main piping system is provided. A three-way valve is provided to bypass the storage for circulation of the oil. For each field, a pump is provided on the cold oil side. In the steam generator oil circuit a pump is provided on the hot oil side and steam flow control valve is provided on the cold side.

A steam generator having maximum steam output of 4230 Kg/hr at 2870 KPa and at $285^{\circ}C$ consists of a superheater with hot oil inlet, a drum and an evaporator with cold oil outlet, a blow down tank, sample cooler and design unit. The turbine-generator set consists of a multistage condensing turbine, a generator, ejector, deaerator and feed water storage tank.

The solar thermal electric power plant is designed using a 'design point' which means that for that insolation and other conditions expected at that moment the plant will perform the year round. Originally this design point was the winter solastice noon which means that the plant will work with better operational conditions for the whole year. This design point was later changed to equinox noon with solar insolation of 920 W/m^2. Later it was realised that the method of using a design point is not optimum. Using the design point, the performance predictions may be too optimistic and the actual average performance may be far from these predictions.

The plant was tested for one and a half years from September 1981 to May 1983. During the year 1982, about 28 per cent of the days were clear, 45 per cent of the days were hazy with intermittent clouds and for the rest 27 per cent of the days the plant was not operated due to very low insolation. In Table 6[95], the expected performance of the plant is compared with the actually measured plant performance adjusted analytically to design conditions. It is seen from this table that actual efficiency of the Acurex Collector field is 59.1 per cent which is 3.7 per cent higher than the predicted efficiency. The MAN collector field has given a lower efficiency than the predicted (48.3 per cent instead of 50.1 per cent). The actual overall plant efficiency was only 9.1 per cent compared to the expected 10.1 per cent. This lower efficiency is due to the low power conversion efficiency (19.1 per cent instead if 22.2 per cent).

During the commissioning and operation of the plant, several management, site specific, technical and material problems were encountered. Some of the problems which required special attention included: water removal from thermal oil, soiling of collector field, overheating and damage of some collectors, water and power failure, low insolation, turbine condenser water level control problems, breaking of reflector mirrors and receiver covers, leakage through

TABLE 6 - ACTUAL AND EXPECTED PERFORMANCE OF 500 KWe SOLAR THERMAL ELECTRIC
PLANT (From Kalt[95])

	Actual performance	Expected performance
Power Output		
Net electric output (KWe)	456	507
Range of net electric output (KWe)	0-600	50-557
Collector's field thermal output to the storage device (KWt)	2685	2602
Acurex field thermal output (KWt)	1486	1362
MAN field thermal output (KWt)	1199	1240
Thermal power to steam generator for 500 KWe electric net output (KWt)	2953	2599
Efficiency		
Overall plant efficiency for 500 KWe net electric output (per cent)	9.1	10.1
Collector field's efficiency (per cent)	53.1	52.7
Acurex collector field efficiency (per cent)	59.1	55.4
MAN collector field efficiency (per cent)	48.3	50.1
Storage efficiency for 24 hrs (per cent)	-	92.2
Stream generator/power conversion system efficiency for 500 KWe (per cent)	19.1	22.2
Storage Capacity		
Useful thermal energy content after changing (KWH_t)	5316	-
Equivalent electric capacity (KWhe)	900	-
Useful thermal energy content after 24 hrs (KWh_t)	4900	4160
Equivalent electric capacity (KWhe)	830	923
Parasitic Consumptions		
Total plant consumption (KWe)	60	70
Other Performance		
Minimum effective insolation for a collector field operation (W/m^2)	400	300
Cooling water consumption (m^3/day)	81	81
Land use factor	0.30	0.30

different valves, changes and often failure of control system, steam generator
leakage, defects in tracking control system, etc. All these problems were
overcome during one and a half years of operation and the system worked well
to satisfaction later.

8.3 The White Cliffs (Australia Solar Power Station (Ref.40,96)

The White Cliffs solar power station was built and put to use in March 1982
with the objective to study the feasibility of paraboloidal dish system to
provide electric power (25 KWe) and thermal energy (140 KWe) to an isolated
small community at White Cliffs (1100 Km west of Sydney) of 40-50 people
who have no existing power supply and have an extreme and hostile climate.
The main electrical load at White Cliffs consists of some eight houses,
community hall, street lights, post office, school and hospital, all within
1 km from the power station. Electrical power is supplied continuously on
a stand alone basis with diesel back-up.

The power station mainly comprises 14 modular semi-autonomous paraboloidal tracking collectors, a reciprocating uniflow steam engine with an
alternator and a central controller giving instructions to modular units
for starting, offsteering, stopping and parking. The details of the plant
are shown[96] in Table 7.

The collector field consists of two north-south parallel rows of total
14 modular semi-autonomous paraboloidal tracking collectors, each of 50.2 m
diameter, 70°C rim angle - rim supported on fibreglass substrate of 6 mm
thick - pasted about 2300 mirrors of 2.5 mm thick back silvered glass with
dimensions 100 mm x 100 mm. Each dish is mounted on a frame pivoting on a
horizontal axis (absorber is mounted on this axis) which in turn can rotate
about a vertical axis carried on the pedestal pipe. Each collector carries
its own battery supply, charged from the central plant, and each collector is
tracking the sun by printed circuit motors through actuators and controlled
by a dish-mounted sun sensor. When there is a 'start' signal from the central
control room clock, the dish starts tracking the sun all day until it receives
a 'park' signal in the late afternoon, when it parks facing horizontally south.
In case of cloudy weather, the dish control system generates time pulses
keeping the dish approximately facing the sun and when the sun reappears the
sensor signals override these timed pulses.

TABLE 7 – DETAILS OF 25 KWe AND 140 KWt SOLAR PLANT OF WHITE CLIFFS, AUSTRALIA

1. Collector

Type	:	Parabolic dish collector of 5.02 m diameter
Reflector material	:	Fibreglass substrate paraboloidal shell 6 mm thick on which about 2300 back silvered glass reflectors of 2.5 mm thick and 100 mm x 100 mm are pasted. Reflectivity 0.86.
Number and area of collector	:	14 dishes each of 19.8 m^2 aperture area
Focal length	:	1.808 m
Rim angle	:	70°
Intercept factor	:	0.95
Geometric concentration ratio	:	0.95
Tracking mode	:	Semi-automatic, Altitude/Azimuth
Absorber type	:	Coil type, semi-cavity, 160 mm diameter and 160 mm long
Working fluid	:	Steam
System pressure	:	7 MPa
Absorber pressure drop	:	7 KPa
Collector fluid inlet temperature	:	50^oC
Collector fluid outlet temperature	:	550^oC
Rated mass flow	:	50 ml/s total
Design output power	:	14.0 KWt at 1000 W/m^2
Collector efficiency	:	72.5 per cent at 840 W/m^2 insolation and steam at 415^oC
Typical heat transfer coefficient from absorber	:	0.2 W/cm^2K

2. Storage

Thermal energy	:	Nil
Electrical energy	:	Lead acid battery

3. Poer Conversion

Type	:	Rankine reciprocating uniflow steam engine
Maximum steam temperature	:	7 MPa
Maximum steam temperature	:	450^oC
Condenser pressure	:	25.4 KPa

Measured heat-to-mechanical work conversion efficiency (at 4.1 MPa and 415°C)	:	21.9 per cent
Thermal to net electric efficiency	:	9-10 per cent
Mode of operation	:	Stand alone-continuous + Battery, diesel

In case of windspeed exceeding 80 Km/hr, the dish automatically gets parked facing vertically. Each dish carries a semi-cavity tube type absorber with 160 mm diameter 160 mm long and connected to inlet water and outlet steam tubes. The water and steam pipes are connected through horizontal and vertical axis rotary joints to the main ducts and then to the engine feed water pump and steam is conveyed to the engine room through insulated ducts. Maximum steam pressure allowed in the absorbers is 550°C at a pressure of 7 MPa and a total water flow of 50 ml/s at an insolation of 1000 W/m^2. The quality of the steam is controlled by the speed of the positive displacement 3-cylinder feed water pump which is driven by a thyristor controlled servomotor acting in response to control signals from an otpimizing unit which takes account of insolation and other variables.

The steam from the solar collectors array is used to drive a high performance uniflow reciprocating steam engine. This steam engine accepts all the energy from the array till the steam quality is adequate to run the engine and till the power produced by the engine is greater than to run the auxiliaries. Since steam engine[65] of proper specifications and high efficiency was not available, a 3-cylinder lister diesel engine model HR3 was converted for steam operation - retaining engine block, crankshaft, oil pump and filter, starter, flywheels and connecting rods, but adding General Motors (diesel model 53) pistons, rings and cylinder liners. Three new cylinder shells were made out of 420 stainless steel, each with valve seats, guides and steam chamber and 3 impulse pins were fitted to the top of each piston.

After 10-25 minutes of sunrise, the cycle starts automatically, the collectors take the direction of the sun, feeding the water to the collector array, generating the steam. The bypass valve closes as soon as the steam temperature reaches to 180°C. As the steam pressure reaches to 2.5 MPa, the throttle valve opens and the engine starts and delivers useful energy. Engine starts delivering useful power within 45 minutes from the start signal. About 30-40 minutes before the sunset the clock signal causes the array to park horizontally facing south. During cloudy or intermittent weather conditions, depending on the quality of the steam, the engine stops and starts automatically. During days of high insolation or if the power demand is less , the excess power, via dc machines is stored in lead acid heavy traction type battery. During periods of low insolation, the battery/dc machine drive the alternator and meeting the requirement. In case the battery gets discharged, a back up diesel unit starts automatically to supply the load. It was observed that unless the solar insolation reaches 400 W/m^2, there will be no net useful electric power.

The power plant was tested with dummy load for one complete year, i.e. from June 1982 to June 1983 to assess the reliability operation and maintenance procedures. The power plant was connected to the down load in November 1983 on a continuous, stand-alone basis with diesel back up. The collectors are tested at different insolation levels and steam temperatures. At a steam temperature of 400°C and insolation levels of 1000, 800, 600 and 400 W/m^2, the collector efficiencies are 73, 71, 68 and 66 per cent respectively. The energy delivered by the collector array to the engine room is also measured at different steam temperatures and insolation levels. At a steam temperature of 100°C and insolation levels at 1000, 800, 600 and 400 W/m^2, the energy delivered to the engine room is 165, 107.5, 57.5 and 7.5 KW respectively. It is also observed that significant amount of heat output from collector array is not used by the steam engine particularly during the morning and evening hours since the insolation is not sufficient to produce the useful net electric power. This heat and other losses are being employed for water desalination. The electric power consumed by auxiliaries is of the order of 3 KWe. The gross electric power produced at different insolation levels is also measured. The gross electric power is 29, 18, 10 and 3 KW at the insolation levels of 1000, 800, 600 and 400 W/m^2 respectively thereby showing solar to net electric power conversion efficiency of 9-10 per cent. The overall efficiency can be made double-fold by using better quality and precisely made glass reflectors, reducing the steam duct losses, improving the steam engine efficiency, improving the generator efficiency and otpimizing the dish dimensions.

It has been reported[96,97] that except of cleaning of dishes and various other components due to dusty conditions of White Cliffs which otherwise may reduce the thermal performance up to 20 per cent, there has been no maintenance and operational problems for two years. The system has proved to be of

a level of sophistication well able to be handled by local personnel with
largely automotive and agricultural machinery type skills.

4.8.4 Central receiver electric plant of 1 MWe capacity (EURELIOS)
(Ref.49,98)

A project for constructing 1.0 MWe helioelectric power plant was
sponsored by the Commission of the European Communities (CEC) in the year
1975. The site selected for the erection of the plant is some 40 Km northwest of Catania, Sicily at the village of Adrano (latitude $37.64^{\circ}N$, longitude
$14.80^{\circ}E$). The plant has been built by an industrial consortium consisting of:

- Messerchmitt-Bolkow-Blohm (MBB), F.R. Germany,
- ANSAIDO SPA and Ente Nazionale per l'Energia Electrica (ENEL), Italy,
- CETHEL (Combining Renault, Five-Cail-Babcock, Saint-Gobain Point-a-
  Mousson and Heurtry S.A.), France.

These firms along with CEC completed the design specifications of all the
subsystems of the plant including the testing of the prototype models by
November 1978 and the construction completed in December 1980. In April 1981,
the plant was fully tested and started delivering power to the Italian
Electricity Generating Board grid. The plant is designed to supply 1.0 MWe
power at an insolation of 1 KW/m^2 at equinox noon at the site of Adrano,
Sicily.

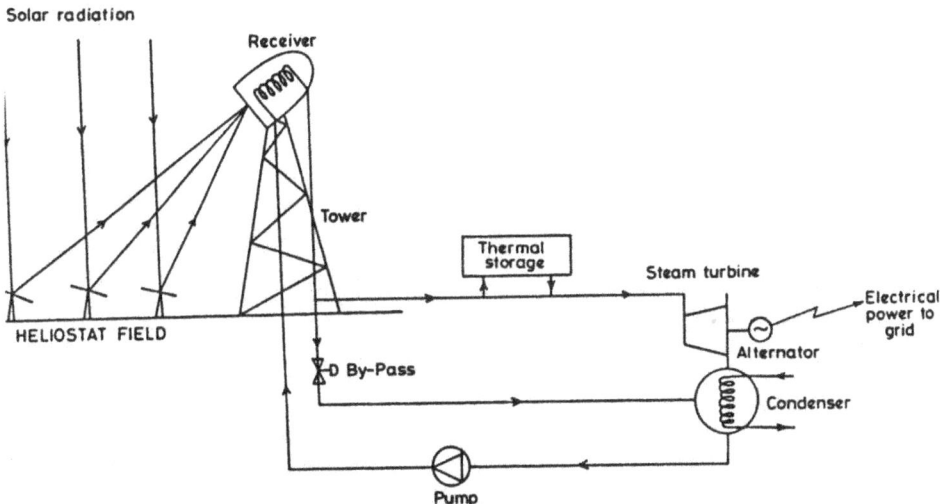

FIG.21 - SCHEMATIC OF EURELIOS CENTRAL RECEIVER SYSTEM AT ADRANO, ITALY

The power plant as is shown schematically in Fig.21 and is based on the principle of central receiver and consists of large field of heliostats concentrating direct solar radiation on to a receiver mounted on the top of tower converting water into high pressure steam which is used to run a turbine coupled to an alternator. The electrical energy produced is fed to the existing electric grid. A thermal buffer storage is used so that the plant can continue to operate for a maximum period of 30 minutes in case of cloud cover and a bypass is used for starting and shutdown operations. Table 8 shows the details of the EURELIOS.

TABLE 8 - MAIN CHARACTERISTICS OF 1 MWe SOLAR POWER PLANT (EURELIOS) OF ANDRANO, SICILY, ITALY

1. Heliostats

Type	:	Two types (CETHEL and MBB) with a total area of 6202 m^2, two axis controlled, focusing	
Configuration	:	CETHEL	MBB
		8 modules of 6 mirrors of 1.8 x 0.6 m with an area of 51.8 m^2	16 mirrors of 1.2 x 1.2 m with an area of 23 m^2
Number of heliostats	:	70	112
Total mirror area	:	3626 m^2	2576 m^2
Mirror type	:	6 mm backsilvered float glass	3 mm backsilvered gloat glass sandwiched
Reflectivity	:	0.80	0.85
Overall inaccuracy	:	$\pm$ 4 m rad	4 m rad
Panel dimensions	:	8.84 m x 7.34 m	5.63 m x 5.01 m
Panel rotational speeds (azimuth)	:	12° per min. fast, 0.6^o per min. slow	4.79^o per min. fast, 0.48^o per min. slow
Panel rotation speeds	:	12° per min. fast, 0.6^o per min. slow	4.79^o per min. fast, 0.63^o per min. slow
Range of rotation	:	$\pm 12^o$ Az + 75° - 90° El	$\pm 120^o$ Az, $\pm 90^o$ El
Position of heliostats	:	West field	East field
Spacing of helostats	:	11.7 m	7.8 m

2. Receiver/Tower

Type	:	Once through cavity type receiver, 4.5 m aperture at 55 m height, 110° inclination
Coil length	:	Two identical parallel and independent branches each of 482 m long and 55 mm diameter

Volume : 3.4 m^3

Outlet steam conditions : 512°C, 6.48 MPa, 4860 Kg/hr

3. Steam Cycle

Feed water temperature
at receiver inlet : 36°C

Cooling water
temperature : 25°C (maximum)

Nominal power : 1.2 MW (mechanical) with steam at 510°C,
6.8 MPa

Connection : Direct connection of turbine with the
receiver (no heat exchanger)

4. Thermal Storage

Capacity : Steam 300 KWh, Hitec 60 KWh capable of
operating plant for 30 minutes at
reduced power

Water storage : Pressurized water storage for 4300 Kg
vapour from 1.9 MPa to 0.7 MPa

Molten salt storage : Two tanks, each of 1 m^3 volume containing
1250 Kg HITEC salt (an eutectic mixture:
53% KNO; 40% NaNO_2; 7% NaNO_3) and two
steam (salt) heat exchanger for 1.9 MPa,
480°C and 410°C steam temperature

5. Turbine

Type : Rankine cycle 7-stage steam turbine

Steam conditions at
inlet : 6.8 MPa, 510°C, 5200 Kg/hr

Steam conditions at
exhaust : 6 KPa, 36°C

Turbine speed : 8195 rpm

Turbine output at
turbine generator joint : 1.2 MW

6. Alternator

Type : GSN 500 Y4 Alternator (3 phase) made by
ANSALDO with 1500 KVA at 0.8 phase

Speed : 1500 rpm

Efficiency : 94.9 per cent at full load

Two different types of heliostats each having roughly equal 'subfields' supplied by CETHEL, France, and MBB, Germany, which are considerably different in size, are placed within two subfields divided by a line going North-South: 70 CETHEL heliostats each of area 51.8 m^2 located in the western part, and the 112 MBB heliostats each area 23 m^2 in the eastern part. The total reflecting area of heliostats is 6202 m^2 in the eastern part. Each heliostat has its own microprocessor-based local electronics unit which is controlled by a central unit. The central unit can control a single heliostat, a group of heliostats or a subfield of heliostats. The heliostat field is designed in such a way as to avoid the shadowing and blocking effects at equinox noon. The MBB heliostat uses 16 mirrors each of 1.2 m x 1.2 m x 3 mm back silvered float glass (average reflectivity 85 per cent) mounted on a frame with a swiveling pedestal. The CETHEL heliostat consists of 8 modules of 6 mirror strips. Each mirror strip is 1.8 m x 0.6 m with a thickness of 6 mm and is silvered from back side showing an average reflectivity of 80 per cent. The heliostats can be driven at two speeds - fast speed to reach a desired direction and slow speed to track the sun.

The solar receiver is placed on the top of 55 m high steel tower, the centre line being inclined downwards 22° from horizontal towards the heliostat field and is constructed by ANSALDO which is based on the results and experiences gained by Prof. Francia with the 100 KW plant. The receiver is a once-through circular cavity type steam generator, conical in shape. It consists of two parallel tubes, through which pressurized water flows, rolled up in a coil to form the conical internal wall of the cavity into which the solar radiation is focused by the heliostats. Each tube is of about 482 m long and the pipe work is divided into four sections:

1. An economiser, an open bottomless basket like structure in the centre of the cavity which is designed to heat the incoming feed water to about $200-280^oC$.

2. A truncated conical evaporator section, which makes the wall of the receiver and is designed to boil the feed water without superheating.

3. A truncated conical primary superheater, where the steam is superheated between 350 to 400^oC and is mounted above the evaporator.

4. A cylindrical secondary superheater, consisting of circular and straight pipes forming a honeycomb like structure bringing the steam to its final state generally at 512° and a pressure of 6.48 MPa.

The tubes of the receiver are blackened and finned so as to absorb maximum solar radiation and behaves like a black body. The heat loss from the receiver is reduced by using an anti-radiating structure made of pyrex sheets (transparent only to solar radiation) located from the inside over the tubes.

The outlet steam temperature is maintained constant in spite of the solar flux variation by using four automatically controlled direct contact attemperators spray valves, two in each branch.

The steam cycle is single superheating Rankine cycle consisting of a receiver boiler, a steam turbine, a condensor and other appropriate pumps, valves safety system, controls, etc. A thermal storage is used as an auxiliary system designed to operate the plant up to 30 minutes without solar insolation in order to protect the turbine against thermal shocks. Under clear sky conditions, this system takes about 45 minutes to provide steam at a temperature of 350°C which is the temperature required to operate the turbine from the feed water temperature of 25°C. The steam produced during this period is taken through a bypass valve and attemperator/desuperheater into a Flash tank before sending it to the condensor and pumped through the system again. The unsteam pressure is maintained by adjusting the bypass valve which is done automatically. The water level and pressure in the Flash tank is also automatically adjusted. The condensor is an evacuated four-pass water cooled unit. As soon as the steam temperature reaches to the design level it is automatically admitted to the turbine. The exhaust of the turbine is passed directly to the condenser.

A 7-stage impulse steam turbine runs at a speed of 8195 rpm and through the reduction gear is connected to the alternator with a flexible joint providing the alternator a speed of 1500 rpm. The turbine receives steam at a temperature of 510°C and at a pressure of 6.8 MPa.

The thermal storage system consists of two parts and provides steam at a temperature of 510°C. The storage system consists of a tank providing saturated steam at a pressure decreasing from 1.9 MPa to 0.7 MPa and a molten salt storage subsystem containing HITEC with two steam/salt heat exchangers and two salt storage tanks superheating the saturated steam from the hot water tank up to 410°C.

The control is an essential part of the whole system. There is a master control centre whose main function is to control and supervise the whole plant including receiver, steam cycle, heliostat field and storage. Basically the plant control system consists of two parts: the heliostat control system and the turbine and the steam cycle control system. Each heliostat is automatically controlled and can take any of the three positions: storage, standby and tracking mode. The turbine and steam cycle control system control: steam temperature at the outlet of the receiver, steam pressure at the outlet of the receiver and the steam-water transition point. The bypass circuit in case of start-up and shut-down of the plant is controlled manually.

The power plant supplies electricity to the local grid. In the beginning, to start the plant, electricity from the grid is used to start the auxiliaries of the plant through transformers and used to operate the heliostat field, the feed water system, receiver, turbine and alternator. After the plant comes to its full operation, the auxiliaries consumes a small fraction of the generated power and the balance is fed to the 20 KW grid through breakers.

8.5 Carrisa Plain Solar Photovoltaic Power Plant (Ref.99)

Perhaps the world's first largest,commercial and impressive example of the photovoltaic plant of 1.0 MWe capacity was built in December 1982, by ARCO Solar, Inc., on Southern California Edison (SCE) property adjacent to the Utility's Lugo substation in Hesperia, California. On November 14, 1983, a 6 MWe plant located on the Carrisa Plain in San Luis Obispo County, California, was interconnected to the Pacific Gas and Electric Company (PG&E) grid. The main objective of this commercial plant is to show the feasibility of the photovoltaically generated electricity to PG&E.

The 6-MWe photovoltaic power plant is located at 35°N latitude on an elevation of 615 m on PG&E property adjacent to a 115 KV transmission line. The main components of the power to a plant are: photovoltaic modules to produce electricity, two-axis trackers to track the sun, inverters to convert DC into AC; switchgear to direct flow of current and distributed data acquisition components. The facility utilizes 756 computer-controlled two-axis trackers, each tracker with 128 photovoltaic modules augmented by reflective glass panels. The cells used are single crystal silicon solar cells. The 65 hectare array field is divided into nine independent segments,

each with 84 two-axis trackers. It has been estimated that by tracking
the photovoltaic module, up to 40 per cent more kilowatt hours on the
average over a year's time can be produced against a fixed mounted
module. By using laminated glass reflectors, an increase in power output
up to 50 per cent on an average yearly basis is expected.

The facility uses the third generation dual axis tracking with
central pedestal unit of 95 m^2 with independent azimuth and elevation
drive unit. The power requirement is about 0.4 KWh per tracker. Each
tracker contains eight 1.22 m x 4.88 m panels of unframed 41 W square cell
laminates, and 4.88 x 4.88 panels of laminated reflective glass mounted at
a 60° angle. Solar tracking is done using microcomputer based systems
using a special clock/calendar program without the requirement of sun
sensors. This tracking accuracy is found to be adequate using flat plate
photovoltaic panels. During periods of night or high wind velocity,
trackers are driven to a vertical stow position or horizontal stow
position respepectively.

REFERENCES

1. R.C. Jordan and W.E. Ibele (1955), 'Mechanical energy from solar energy Proc. World Symp. Appl. Solar Energy, Phoenix, Arizona, Nov. 1955, pp 81-101.

2. R.C. Jordan (1975), 'Solar energy powered systems - history and current status', Standardization News, 3(8), 13-20.

3. J.T. Pytlinski (1978), 'Solar energy installations for pumping irrigation water', Solar Energy, 21, 255-262.

4. S. de Caux (1615), 'The cause of motive power', E.I. Norton, Frankfurt, Germany.

5. C. Guntner (1906), Scientific American Supplement, 61, 25409-25412.

6. A.B. Mouchot (1978), 'Comptes Rendus de l'Academica des Sciences', Paris, France, 86, 132.

7. J. Ericsson (1884), Nature, 29, 217.

8. W. Adams (1878), Scientific American, 38-39, 376.

9. A. Pifre (1882), Nature, 26, 503-504.

10. C.L.A. Tellier (1885), 'Equipment for the activity of a pump apparatus for solar heat', German Patent 34749.

11. Charles F. Hodlen (1901),'Solar motors', Scientific American, 84, 169.

12. H.R. Willsie (1909), 'Experiments in the development of power from the sun's heat', Engineering News, 61(19), 511.

13. A.S.E. Ackermann (1915), 'The utilization of solar energy', J. Royal Soc. of Arts, London, 538-565, April 1915.

14. A.D. Blake (1911), 'Producing power from the sun's rays', Power, 34(2), 506.

15. F. Shuman (1914), 'Sun power plant', Proc. Manchester Association of Engineers, Discussion Session, 1913-1914, Manchester, England, 405-443.

16. R.H. Goodard (1929), Popular Science Monthly, 115, 22-23.

17. F. Danials (1929), Direct Use of the Sun's Energy, Yale University Press, New Haven, USA.

18. C.G. Abbot (1943), 'Solar radiation as a power source', The Military Engineer, 35(208), Feb. 1943.

19. Anonymous (1946), USSR Taskent plant uses solar energy, Power, 90, 85, Jan. 1946.

20. K.N. Mathur and M.L. Khanna (1954), Proc. Symp. on Solar Energy and Wind Power in the Arid Zones, New Delhi, India, Oct. 1954.

21. H. Tabor and L.Y. Bronicki (1961), 'Small turbine for solar energy power package', Proc. U.N. Conf. on New Sources of Energy, Rome, Vol. 5, Paper No.5/54, 68-79.

22. L.Y. Bronicki (1972), Proc. 7th Inter. Soc. Energy Conversion Conf., San Diego, California, Paper 729057, p. 327.

23. H. Masson and J.P. Girardier (1966), 'Solar motors with flat-plate collectors', Solar Energy, 10(4), 165-169.

24. V.A. Baum (1969), 'Semiconductors solar energy converters', Consultants Bureau, p. 37, New York.

25. J.D. Walton, Jr., A.H. Roy and S.H. Bomar, Jr. (1978), 'A state of the art survey of solar power irrigation pumps, solar cookers and wood burning stoves for use in Sub-Sahara Africa', Final Technical Report, June 1 to November 30, 1977, Georgia Institute of Technology, Atlanta, Georgia (USA).

26. Anonymous (1977), 'Report on the use of solar energy for water pumping in Arid areas', Report from SOFRETES, Zone Industrialla d'Amilly, Montargis, France.

27. George M. McClure (1977), 'Solar powered 50-horsepower irrigation pump', Proc. Solar Irrigation Workshop, Albuquerque, New Mexico, July 7-8, 1977, p. 28, SAND-77-0992.

28. R.E. Barber (1977), 'Design and development of the 25 HP Rankine engine for use on the Estancia Valley irrigation pump', Proc. Solar Irrigation Workshop, Albuquerque, New Mexico, July 7-8, 1977, p. 38, SAND-77-0992.

29. G.H. Abernathy and T.R. Mancini (1977), 'Design and installation of a solar powered irrigation pump', Agr. Engng., 58(10), 39.

30. R.W. Matlin (1977), '25 Kilowatt photovoltaic powered irrigation experiment', Solar Irrigation Workshop, Albuquerque, New Mexico, July 7-8,1977, SAND 77-0992.

31. Anonymous (1978), 10 KW Solar Power Plant, Leaflet from Bharat Heavy Electricals Ltd., New Delhi, India.

32. D.L. Larson (1981), 'Operational evaluation of a 150 KW solar thermal power plant', Proc. 1981 Annual Meeting AS/ISES (edited B.H. Glenn and G.E. Franta), Philadelphia, May 1981, pp 309-313.

33. D.L. Larson (1983), 'Final report of the Coolidge Solar Irrigation Project', Report from University of Arizona, Tucson, Arizona.

34. D.L. Larson (1983), 'Coolidge solar power plant', Proc. Solar World Congress (edited S.V. Szokolay), Vol. 3, pp 1828-1832.

35. T. Horigome, T. Saishoji and T. Uto (1981), 'Solar thermal electric power generation system in Japan', Solar World Forum (edited D.O. Hall and J. Morton), Proc. Int. Solar Energy Society Congress, Vol. 4, Pergamon Press, Inc.

36. H. Newen, G. Schmidt and S. Moustafa (1983), 'The Kuwait solar thermal power station: Operational experiences with the station and the agricultural application', Proc. Solar World Congress (edited S.V. Szokolay), Vol. 3, pp 1527-1532, Pergamon Press, Inc.

37. F.A. Schraub and H. Dehne (1983), 'Electrical generation system design: Management startup and operation of IEA distributed collector solar system in Almeria, Spain', Solar Energy 31(4), 351-354.

38. W. Grasse and C.J. Winter (1980), 'Two 500 KWe solar power plants in Almeria/Spain', Colloques Internationaux du CNRS No.306-Systems Solaires Thermodynamiques, STS 80-3.

39. Solar Thermal Report, 3(3), 3-5, Jet Propulsion Lab., Pasadena, USA.

40. S. Kaneff (1983), 'The White Cliffs solar power station', Proc. Solar World Congress (edited S.V. Szokolay), Vol. 3, Pergamon Press, Inc.

41. B.H. Khoshaim (1982), 'Performance characteristics of 350 KW photovoltaic power system for Saudi Arabian villages', Int. J. Solar Energy, 1, 91-103.

42. G. Hanselmann, E. Hellweg, T.S. Crawford and V.A. Power (1983), 'Combined solar and waste heat utilization in a power plant at Meekatharra, Western Australia', Proc. Solar World Congress (edited S.V. Szokolay), Vol. 3, pp 1598-1601, Pergamon Press, Inc.

43. R.J. Arnault, E. Berman, C.F. Gay, R. Tolbert and J.W. Yerkes (1983), 'The ASI one-megawatt photovoltaic power plant', Proc. Solar World Congress (edited S.V. Szokolay), Vol. 3, pp 1624-1628, Pergamon Press, Inc.

44. Y.L. Bronicki (1981), 'A solar pond power plant', IEEE Spectrum, 56-59, February 1981.

45. V.A. Baum, R.R. Aparase and B.A. Garf (1957), 'High power solar installations', Solar Energy, $\underline{1(1)}$, 6-12.

46. G. Francia (1968), 'Pilot plants of solar steam generating stations', Solar Energy, 12(1), 51-64.

47. H.L. Teague, C.T. Brown, C.J. Swafford and G.C. Lewis (1979), 'Identification of alignment and tracking errors in the open loop, time based heliostat system of the 400-KW advanced components test facility', Sun II (edited K.W. Boer and B.H. Glenn), Proc. International Solar Energy Society Congress, Vol. II, pp 1315-1318, Pergamon Press, Inc.

48. C. Eugene and Earl Rush (1978), 'High pressure steam generation at 5 MW solar thermal test facility', Proc. 1978 Annual Meeting of Am. Soc. of ISES, Denver, Colorado, Vol. 2-1, pp 811-814.

49. G. Cefratti and J. Gretz (1981), 'Eurelios' Sun World, $\underline{5(4)}$, 106-110.

50. T. Tanka (1980), 'Solar thermal electric power systems in Japan', Solar Energy, $\underline{25}$, 97-104.

51. W. Frasse (1981), 'Small solar power systems (SSPS)', Sun World, $\underline{5(3)}$, 68-72.

52. The Solar Thermal Report, 2(9), 3-9, 1981, from Jet Propulsion Laboratory, Pasadena, USA.

53. W. Meinecke and P. Wehowshy (1980), 'The 20 MWe gas cooled solar power plant GAST, An overview', Colloques Internationaux du CNRS No.306 - Systems Solaires Thermodynamicques - STS-80-7.

54. Anonymous (1982), 'Thirty MWe plant for California', Sun World', 6(4), 103-105.

55. J.P. Hollman (1969), Thermodynamics, McGraw-Hill Book Company, Inc., New York.

56. E.D. Howe (1981),,'The cycles', Sun World, $\underline{5(3)}$, 65-67.

57. J.W. Stearns, Jr., Y.S. Won, E.W. Chow, P.T. Poon and R. Das (1979), 'Solar stirling system development', AIAA Terrestrial Energy Systems Conference, June 4-6, 1979, Orlando, Florida, USA.

58. H.M. Curran (1981), 'Use of organic working fluids in Rankine engines', J. Energy, 5(4), 218-223.

59. R. Barber and D. Prigmore (1981), 'Solar powered heat engines', Chapter 22, Solar Energy Handbook (edited J.F. Krieder and F. Kreith), McGraw-Hill Book Co., Inc., New York.

60. J.P. Albin and W.R. Lenenberger (1974), 'Program cycle: A rankine cycle analysis routine', Report No. SAND 74-0009, Oct. 1974, Sandia Laboratories, New Mexico.

61. O. Badr, P.W. O'Callaghan and S.D. Probert (1984), 'Performance of Rankine cycle engines as functions of their expanders efficiencies', Applied Energy, 18, 15-27.

62. H.R. Cox (1975), 'Gas turbines principles and practice', George Newnes Ltd., London.

63. A. Stodola (1927), Steam and Gas Turbines, Vol. 1 & 2, McGraw-Hill Book Co., New York.

64. E.A. Farber (1974), 'Solar energy conversion research and development at the University of Florida', Building Systems Design, February/March 1974, 195-206.

65. E.K. Inall (1984), 'The White Cliffs solar power station steam engine and condensate treatment system', Solar World Congress (edited S.V. Szokolay), pp 1478-1482, Pergamon Press, Inc.

66. J. Gretz, A. Strub and W. Palz (1984), 'Thermo-mechanical solar power plants', D. Reidel Publishing Co., Dordrecht, Holland.

67. C. Casci, G. Angilino, P. Ferrari, M. Gaia, G. Giglioli and E. Macchi (1980), 'Experimental results and economics of a small (40 KW) organic Rankine cycle engine', Proc. 1980 IECEC, Aug. 18-22, pp 1008-1014.

68. S.T. Hsu and B.S. Leo (1958), 'A simple reaction turbine as a solar engine', Solar Energy, 2(3&4), 7-11.

69. R.W. Hallet, Jr., and R.L. Gervais (1976), 'Central receiver solar thermal power system phase I (final report)', Vol. 1, Book 2, System Analysis and Design Report No. MDC G6040, McDonnell Douglas Aeronautics Company, California, January 1976.

70. O.E. Balje (1962), 'A study on design criteria and matching of turbomachines: Part A - Similarity relations and design criteria of turbines', Trans. ASME, J. of Engg. for Power, 84(1), 83-102.

71. V.C. Truscellow and A.N. Williams (1981), 'Power converters for parabolic disches', 1981 Annual Meeting of American Section of ISES, May 1981, Philadelphia, USA.

72. J. Ericson (1870), 'Sun power: the solar engine', Contributions to the Centennial, pp 521-577, Philadelphia.

73. W.R. Martini (1982), 'Stirling engines are coming', Alternative Sources of Energy, 57, 26-29.

74. E.A. Farber and F.L. Prescott (1965), 'Closed cycle solar hot air engines - Part 1, a 1/4 H.P. engine, Solar Energy, 9(4), 170-176.

75. T. Finkelstein (1961), 'Internally focusing solar power systems', Am. Soc. of Mech. Engineers, Paper 61-WA-297, Annual Meeting.

76. D.A. Trayser and J.A. Eibling (1966), 'A 50-watt portable generator for remote area use employing a solar powered Stirling engine', Solar Energy, 3(4), 153-159.

77. W.R. Martini (1978), Stirling Engine Design Manual, University of Washington, DOE/NASA/3152-78/1, NASA CR-135382.

78. J.W. Stearns, Jr., Y.S. Won, E.Y. Chow, P.T. Poon and R. Das (1979), 'Solar Stirling system development', AIAA Terrestrial Energy Systems Conference, June 4-6, 1979, Orlando, Florida.

79. J.P. Albin, Jr., (1980), 'Binary Rankine cycle engines for solar thermal power systems', Paper 80-WA/Sol-19, Annual Meeting.

80. W.T. Beale (1973), 'Free piston stirling engines - A progress report', SAE Paper 730647, June 18-22, 1973.

81. R.L. Pons and W. Perçival (1980), 'A Stirling cycle engine for use with solar thermal electric systems', Paper 80-Pet-32, ASME.

82. J.R. Senft (1979), 'Advances in Stirling engine technology', Proc. Cong. IECES, 1979.

83. G. Taylor (1979), 'Free piston Stirling engines increase solar/thermal efficiency', International Power Generation, July/August, 1979.

84. B. Christer et al. (1981), 'The Stirling engine - ready candidate for solar thermal power', SAE Technical Paper Series 810456.

85. D. Wells, W. Percival, C. Bratt, K. Rosenquist and J. Berntell (1982), 'Stirling engines for solar power generation in the 50 to 500 KW range', Proc. 17th IECEC, pp 1749-1754.

86. B. Christer (1980), 'Design characteristics and test results of the United Stirling P-40 engine', Proc. 15th IECEC, Seattle, August 18-22, 1980, Vol. 3, pp 1964-1966.

87. C.P. Gupta and Rajendra Prakash (1981), Engineering thermodynamics, Nem Chand & Bros., Roorkee, India.

88. V.C. Truscellow and A.N. Williams (1980), 'Power from parabolic dishes: Progress and prospects', Paper prepared for U.S. Department of Energy, November 1980.

89. T. Fujita, J.M. Bowler and B.C. Gajanana (1982), 'Composition of advanced engines for parabolic dish solar thermal power plants', J. of Energy, 6(5), 293-297.

90. R.A. Rockley (1979), 'Advanced gas turbine power train development project', Paper presented at Automotive Technology Development Contractors' Meeting, Garrett Airesearch, Michigan, October 1979.

91. H.E. Helins (1979), 'Advanced gas turbine power train system development project', Paper presented at Automobile Technology Development Contractors' Meeting, Detroit Diesel Allison, Division of General Motors, Michigan, October 1979.

92. J.P. Thornton, K.C. Brown, J.G. Finegold, J.E. Gresham, F.A. Herievich, J.S. Kowalik and T.A. Kriz (1980), 'A comparative ranking of 0.1 to 10 MWe solar thermal electric power systems, Vol. 1: Summary of results', Report SERI/TR-351-461, August 1980, Solar Energy Research Institute, Colorado.

93. W.M. Bohon and S.L. Levy (1979), 'An optimization procedure for solar thermal electric power plant design', Sun II (edited K.W. Boer and B.H. Glenn), Proc. Int. Solar Energy Society Congress, Vol. II, pp 1229-1232, Pergamon Press, Inc.

94. A. Kalt, M. Loosme and H. Dehne (1982), 'DCS construction report', Report No. SSPS SR1, IEA-SSPS Operating Agent DFVLR, Cologne, F.R. Germany.

95. A. Kalt (1983), 'SSPS-DCS First year of operation', Report No. SSPS SR3, IEA-SSPS Operating Agency DFVLR, Cologne, F.R. Germany.

96. S. Kaneff (1984), '14 dish community electric system, White Cliffs collector system performance', Paper for presentation in the IEA Workshop on the Design and Performance of Large Solar Thermal Collector Arrays.

97. S. Kaneff (1983), 'On the practical and potential viability of paraboloidal dish solar thermal electric power systems', Proc. Solar World Congress (edited S.V. Szokolay), Vol. 4, pp 2767-2771, Pergamon Press, Inc.

98. J. Gretz, A. Strub and W. Palz (1984), Thermomechanical Solar Power Plants, D. Reidel Publishing Co., Dordrecht.

99. R.E.L. Tolbert and J.C. Arnett (1984), 'Design, installation and performance of ARCO solar photovoltaic power plants', Paper presented in the Institute of Electrical and Electronic Engineers' Photovoltaic Specialists Conference, May 1-4, 1984, Orlando, Florida.